# Hauturu

The history,
flora and fauna of
Te Hauturu-o-Toi
Little Barrier Island

# Hauturu

### EDITED BY LYN WADE AND DICK VEITCH

**Manuhiri te tupuna**
**Manuhiri te tangata**
**Manuhiri te tapu**
**Haumi ē, hui ē, taiki ē**

There have been many books, stories, narratives, films and discussions on the history of Te Hauturu-o-Toi. All of them have versions of the early Māori history through to recent times that are a mixture of myths, legends, extracts from Native Land Court records and oral conversations with individuals — versions that have no connection at all to Ngāti Manuhiri or to any true accounts of whakapapa and tikanga. Some of them have good factual content, but none has the traditional kōrero handed down only to those who were chosen by their tūpuna to be the repositories of the history of Ngāti Manuhiri and Te Uri-o-Papa, the last tribal grouping of iwi to live and sustain themselves on Te Hauturu-o-Toi for hundreds of years before they were forcibly removed from the island by government forces.

The island has been inhabited by Māori for more than 400 years. It was settled in the mid-1600s by the ancestors of Ngāti Manuhiri and Ngāti Rehua. These large groupings had already settled in the Mahurangi and Kaipara areas and had then proceeded to raupatu over the offshore islands, including Te Hauturu-o-Toi.

Manuhiri's tūpuna came from Kāwhia and northern Taranaki. They were descendants of the *Tainui* waka. Tribes such as Ngāti Manaia later joined in customary unions with Ngāti Manuhiri people, forming an alliance with the Ngāti Wai tribes from the north.

Te Hauturu-o-Toi gained national recognition in the late 1800s when the government first sought to purchase the island and, when that was rejected, enacted legislation to make it a conservation estate, owned and managed by the Crown. This decision led to many appearances and hearings in the Native Land Court, where Ngāti Manuhiri were put under huge pressure in their efforts to retain their customary ownership. Tenetahi Pohuehue (pictured above right), his wife Rahui Te Kiri (above left), their children Ngapeka and Wi Taiawa and tohunga Hone Paama argued their case against others — European and Māori — who were contesting their ownership entitlement based on customary rights and responsibilities cemented by tikanga.

Te Hauturu-o-Toi will always be culturally and spiritually significant to the identity of Ngāti Manuhiri people. Almost 150 years since the first court action, the Waitangi Tribunal process has returned the island to its rightful owners, with an apology for wrongful actions in the past. Today, Ngāti Manuhiri have customary rights and interests extending from Takapuna in the south to Bream Head in the north and to the offshore islands, including Te Hauturu-o-Toi. We have now gifted this island, as a wonderful nature reserve, to the people of New Zealand.

— Terrence (Mook) Hohneck, Ngāti Manuhiri

**Me piki tāua ki te tihi
ō Hauturu muia ao.
Ka mātakitaki tāua
Kī ngā pōitu ō te kupenga
   ō Toi te Huatahi.
E tama tangi kine ē!
E tama tangi kine ē!
E tama tangi kine ē!
— oroiori o Hone Puumari Paama**

This oriori was sung by my tupuna Hone Puumari Paama (Palmer) (pictured above left), who defended the rights of Ngāti Rehua, Ngāti Manuhiri and Ngāti Wai during the Native Land Court investigations of the 1800s to acquire Te Hauturu-o-Toi. I am his fourth generation direct mokopuna (pictured above right) and have been the Chair of Ngāti Rehua–Ngātiwai ki Aotea since 2012 and their chief Treaty of Waitangi settlement negotiator.

The following is a historical summary that describes the deliberate and consistent pressure applied by the Crown upon my tūpuna to acquire Hauturu and eventually satisfy their land lust. Almost 179 years after the signing of the Treaty, the descendants of those tūpuna have returned to uphold our enduring legacy and heritage.

I dedicate this writing to my tupuna Puumari, an extraordinary man of his time; our Ngāti Wai tūpuna who fought for our mana motuhake and rangatiratanga; and particularly I acknowledge our Rehua and Manuhiri tūpuna Tenetahi Te Riringa Pohuehue and Rahui Te Kiri, who were the last of our people to stay on Hauturu before being forcibly removed by the Crown.

This is our whakapapa; our hītori; our kōrerorero; and our maumahara. He kaitiaki au.

Title to Te Hauturu-o-Toi/Little Barrier Island was investigated in a number of Native Land Court hearings between 1878 and 1886. Arguments over ownership of the land, and the ability to alienate it, dated back to the early 1860s. The first application to the Native Land Court for title to Hauturu for which documentary evidence survives (there may have been earlier applications) was made in May 1878. More applications were subsequently filed in January, April and November 1879, and in January 1880.

The first Native Land Court hearing concerning Hauturu was held at Helensville on 3 July 1878, but was adjourned indefinitely because there was no survey plan. The second was on 16–17 July 1880 at Awaroa. A re-hearing was held on 7–13 May 1881 at Helensville, but no judgment was issued as the decisions of the judge and the native assessor were split. A further re-hearing was held at Auckland on 4–8 June 1881. A fifth hearing was held at Auckland on 1–15 February 1884. After parliamentary intervention to allow a further re-hearing, a final hearing occurred on 5–15 October 1886.

Negotiations for the Crown purchase of the island began around 1881, with the Crown initially considering that the island would be a good location for defence fortifications, and shortly thereafter that it would be an ideal location for a bird sanctuary.

The 1893 Deed of Sale of Hauturu.

On 28 July 1881 the Crown gazetted a formal notice of the Crown negotiating interests in the land, prohibiting private alienation under the Native Land Purchase Act and the 1878 Amendment Act.

The purchase negotiations between the Crown and Māori were delayed for around a decade by a series of disputes in the Native Land Court and between some of the owners and the Crown. The Crown signed a formal agreement with some of the Māori owners of the island in 1891. However, the agreement was not completed because it required the agreement of all owners, which was not secured as disputes between officials and the owners continued over various issues including the manner of distribution of purchase money. In October 1891 the Crown cancelled the agreement on the basis that the terms were not fulfilled and retracted the offer.

Unable to complete a sale, one of the Māori owners sought to arrange kauri felling on the island. The Crown was concerned this activity would ruin the island's value as a nature reserve, and so it published a warning against tree felling and a trespass order based on purported rights arising from the order under sections 16 and 18 of the Native Land Act 1892, securing the Crown exclusive rights to negotiate the purchase of the land and the power to remove trespassers. On 27 October 1892 the Crown replaced the former notification of their intention to purchase with a notice under the Native Land Purchases Act 1892. When a Māori owner began tree-felling on their own the Crown filed an injunction with the Supreme Court against this action on the basis that felling should not be undertaken until the relative shares of the various owners had been determined by the Native Land Court. The Crown also posted a ranger on the island at this time to enforce the injunction and protect the flora and fauna, even though it had not yet acquired ownership of the island.

Eventually, in 1893, the Crown attempted to reinstate the 1891 agreement and pay individual owners for their interests (although this was contrary to the terms of the original agreement) and collect signatures from owners. At least one of the owners who had signed the original agreement opposed the sale at this time, and considered the agreement had been repudiated. Despite this, their original signatures were used as proof that they had agreed to sell their interests.

All but a few owners signed the agreement, though there is evidence some signatories continued to oppose the sale. The Crown, under mounting pressure by conservationists to acquire the land from Māori, promoted the Little Barrier Island Purchase Act which was passed by Parliament in 1894. This Act forced the completion of the sale on the terms originally agreed in 1891 — even though this agreement had been cancelled by the Crown. The owners who had signed this document were forced by the legislation to sell all of their interests in the island to the Crown for a proportionate share of £3000.

Under the Little Barrier Island Purchase Act the two owners who had never signed the agreement were forced to sell their interests and have the value of their interests determined in the same way as the determination of compensation for the taking of lands for public works. The former Māori owners remaining on the island, who continued to protest the sale, were forcibly evicted in 1896.

— Nicola MacDonald, Ngāti Rehua, Patuharakeke, Ngāti Wai

# Contents

**Foreword**
Ruud Kleinpaste — 10

**Preface** — 13

**1. Introduction to Hauturu**
Matt Rayner — 14

**2. Papakāinga**
Paula Morris — 22

**3. Island life**
Lyn Wade and Dick Veitch — 32

**4. Restoration of Hauturu**
Richard Griffiths and Dick Veitch — 68

**5. Geology of Hauturu**
Jan Lindsay — 90

**6. Biota of Hauturu:**
**Flora, fauna and fungi** — 102

**6.1 Aquatic and terrestrial invertebrates**
David S. Seldon — 108

**6.2 Amphibians and reptiles**
Dave Towns, Sue Keall, Richard Walle
and Nicky Nelson — 124

**6.3 Birds**
Tim Lovegrove, Matt Rayner and Kevin Parker — 134

**6.4 Bats**
Stuart Parsons — 160

**6.5 Vegetation and vascular flora**
Ewen Cameron and Maureen Young — 166

Red-crowned kākāriki (*Cyanoramphus novaezelandiae*) (NF)

**6.6 Mosses**
Jessica Beever _____ 214

**6.7 Liverworts and hornworts**
John E. Braggins _____ 222

**6.8 Lichens**
Bruce W. Hayward _____ 228

**6.9 Fungi**
Peter Buchanan _____ 232

**6.10 Stream vertebrates**
Lyn Wade _____ 242

**6.11 Seaweeds**
Mike Wilcox _____ 244

**7. Seas around Hauturu**
Roger Grace _____ 250

**8. The future**
Dave Towns and Matt Rayner _____ 266

**9. Species lists** _____ 280

**Appendices** _____ 345
**Notes** _____ 363
**Bibliography** _____ 373
**About the editors** _____ 382
**About the contributors** _____ 382
**Acknowledgements** _____ 387
**Image credits** _____ 388
**Index** _____ 389

# Foreword
— RUUD KLEINPASTE

Te Hauturu-o-Toi/Little Barrier Island has always been a magical island in my life. I read about it before I migrated from the Netherlands to New Zealand, more than 40 years ago. As a young and mad nature nerd, I quickly realised that the biota and ecology of New Zealand were very special indeed, and that Hauturu was the ultimate ark of endangered species with, arguably, the least modified environment in the north of Aotearoa. It fitted nicely with the concept of conservation; the management of our fragile ecological systems and the safe-keeping of our most endangered species.

Of course, all those years ago we really did not have a workable plan of what to do with these taonga, apart from keeping them safe on offshore islands. There was no bold and audacious predator-free concept, and even the appreciation of nature and ecological values was limited to dedicated scientists, small groups of citizens and the tramping, hunting and fishing communities.

These days there is a fast-growing awareness that we need to act when it comes to our biodiversity, and other environmental issues for that matter. I believe the word 'conservation' has been overtaken by restoration, regeneration and rehabilitation (action, in other words) and by a concept that starts with the fourth 'r': reconciliation (asking nature's forgiveness).

When I was invited to join the Little Barrier Island (Hauturu) Supporters' Trust in 1997, I didn't hesitate to accept. This was the perfect opportunity to visit the island and raise some much-needed funds to assist the Department of Conservation with special projects. We also raised awareness of this unique island, from biosecurity issues to stories about endangered species.

Everybody who visits Hauturu has a different experience. Most reflect on the bigger picture involving the future of Aotearoa, our ecosystems and humanity's chances of survival on the planet. Hauturu has some lessons for us. In a world that has seen life for around 3.8 billion years, the arrival of *Homo sapiens*, a mere 150,000 years ago, has had some catastrophic consequences. The word Anthropocene — the proposed name to describe the current geological time period — is aptly chosen. We all know the main points: carbon (energy), climate, ocean acidification, atmospheric and stratospheric aerosols, water quality, bio-geochemical contamination, land use, habitat loss . . . the list is endless. Economic growth at all cost isn't helping, and few people recognise that loss of biodiversity on a global scale is the largest 'Health and Safety' hazard we are facing!

When I stroll around the island, by day or by night, I see creatures, fungi and plants in a delicate dance of design and collaboration, creating biological relationships that have stood the test of time. I then realise that we know very little about what makes these systems tick or how it all operates. It also becomes abundantly clear that we haven't got a clue about which species are in play and what their role is in the greater scheme of things.

Taxonomy, ecology and ecosystem services are some of the basic concepts of life, and Hauturu has bucketloads of them. As a result, the ark is becoming bigger and more complex with every new discovery. But Hauturu is so much more than just an ark. It shows us an even bigger picture. As a species, *Homo sapiens* is a real Johnny-come-lately. We joined an already existing and sustainable biosphere and managed to dominate the system in a relatively short time. We have grown to huge populations, are making big mistakes and have been likened to toddlers playing with matches. That begs the question of how all the other species in our ecosystem view us: Are we welcome?

I have noticed that our biological control agents (predators, parasites and pathogens) are quite excited about the number of humans on the planet; all our global diseases are becoming very good at playing the genetic arms race. But the species that generate ecosystem services beneficial to humans tend to lose interest, so to speak.

There is no doubt about the fact that we need to rediscover the operations manual of planet Earth. We need to change the way we live here, on Earth. That's pretty hard, seeing we have rapidly lost our connection to nature in the age of technology. Numbers of eyeballs and clicks are the new currency of disconnect. We need nature-literacy, eco-literacy, call it what you like; we need to become bio-lingual.

It is not surprising that education facilities are becoming more and more interested in using the environment as a context for education. It can start as young as early childhood education, following through primary and intermediate school (the years of maximum engagement) to secondary and tertiary levels, where the science and discovery rubber hits the road. Hauturu, literally, is nature's classroom. No, I am not talking about learning *in* nature or learning *about* nature — documentary-style information made famous by Sir David Attenborough and other wonderful media personalities — although that helps. Hauturu is the best place to learn *from* nature. It has an official name, biomimicry, and it steers us in the direction of learning from nature's best ideas and what life already knows. Nature has been writing the book on biodiversity and ecology for billions of years, and I believe it has urgently become required reading for humanity.

Here are some of the best lessons from Hauturu: Nature runs on current sunlight — not fossil sunlight; it only uses the energy it needs; the forests stabilise climate and mitigate weather extremes; everything is recycled (or rather *up*cycled); nature simply doesn't know the concept of waste; it relies on local expertise, and rewards collaboration and beneficial, symbiotic relationships; nature rarely competes, it mostly facilitates; it has limits (growth, resource accumulation, where genes move to . . .); nature does chemistry in water and uses a small and safe subset of the periodic system; life in the ecosystems is generous, full of diversity and productive; and Hauturu creates a wonderful state of mind.

There are many more learnings from Hauturu — you'll just need to go and experience these for yourself, if you get the chance. But the authors of this book have started the journey for you, by describing some of the pertinent human, geological and natural history of this magical island.

Where do you find an effective ark for endangered species, which doubles as a mirror for humanity's actions and has the power to become the most inspirational outdoor classroom in Aotearoa? Somewhere in the Hauraki Gulf, I reckon!

# Preface

Dr William Maxwell Hamilton (1909–92), known as Bill to his colleagues and as Max to everyone else, was raised on the family farm beside the Mahurangi River. He received his schooling at Warkworth and was a founding pupil and, later, a pupil teacher at the district high school.

He worked for some years on the family farm assisting his father and his entry into the scientific world was almost accidental: while he was completing a short dairy farming course at Massey Agricultural College it became obvious that he had potential and he was encouraged to study agriculture. In 1936 he graduated MAgrSci with double first-class honours in field husbandry and agricultural economics; his theses were on the history, botany and geology of Little Barrier Island and on the New Zealand citrus industry.

In 1936 Hamilton joined the Department of Scientific and Industrial Research (DSIR). One of his early tasks there was to carry out a major survey of the New Zealand dairy industry, for which he was later awarded a DSc. In 1953 he became director of the DSIR. Under his leadership it doubled in size, particularly in the biological sciences, and went from being a small and locally focused organisation to one of international standing. His philosophy was to establish the broad lines he wanted research to follow, hire the best people he could get, then leave them to get on with the job, always encouraging the spirit of adventure in science. He fought hard for the development of the department and for proper funding. His strong scientific base, his capability in administration and his personal integrity gained the trust and respect of politicians, and he proved himself to be a great leader of science.

As a young man Hamilton spent much time exploring Te Hauturu-o-Toi/Little Barrier Island. His original thesis on the island was published in 1937 as DSIR Bulletin 54. It was used by many students and researchers interested in this island sanctuary. The bulletin was revised in 1956 with the assistance of a number of scientists, each recording their speciality, and was published in 1961 as DSIR Bulletin 137 — at that time the only comprehensive scientific work on Hauturu. The naming of Hamilton Track on the island acknowledges his contribution.

This book, which was instigated and funded by the Little Barrier Island (Hauturu) Supporters' Trust, updates and replaces DSIR Bulletin 137.

*Kaka*, 1984, by Don Binney. Don Binney was the Little Barrier Island (Hauturu) Supporters' Trust founding patron. Courtesy of the Binney family and the University of Auckland Art Collection.

# Chapter one
## Introduction to Hauturu

— MATT RAYNER

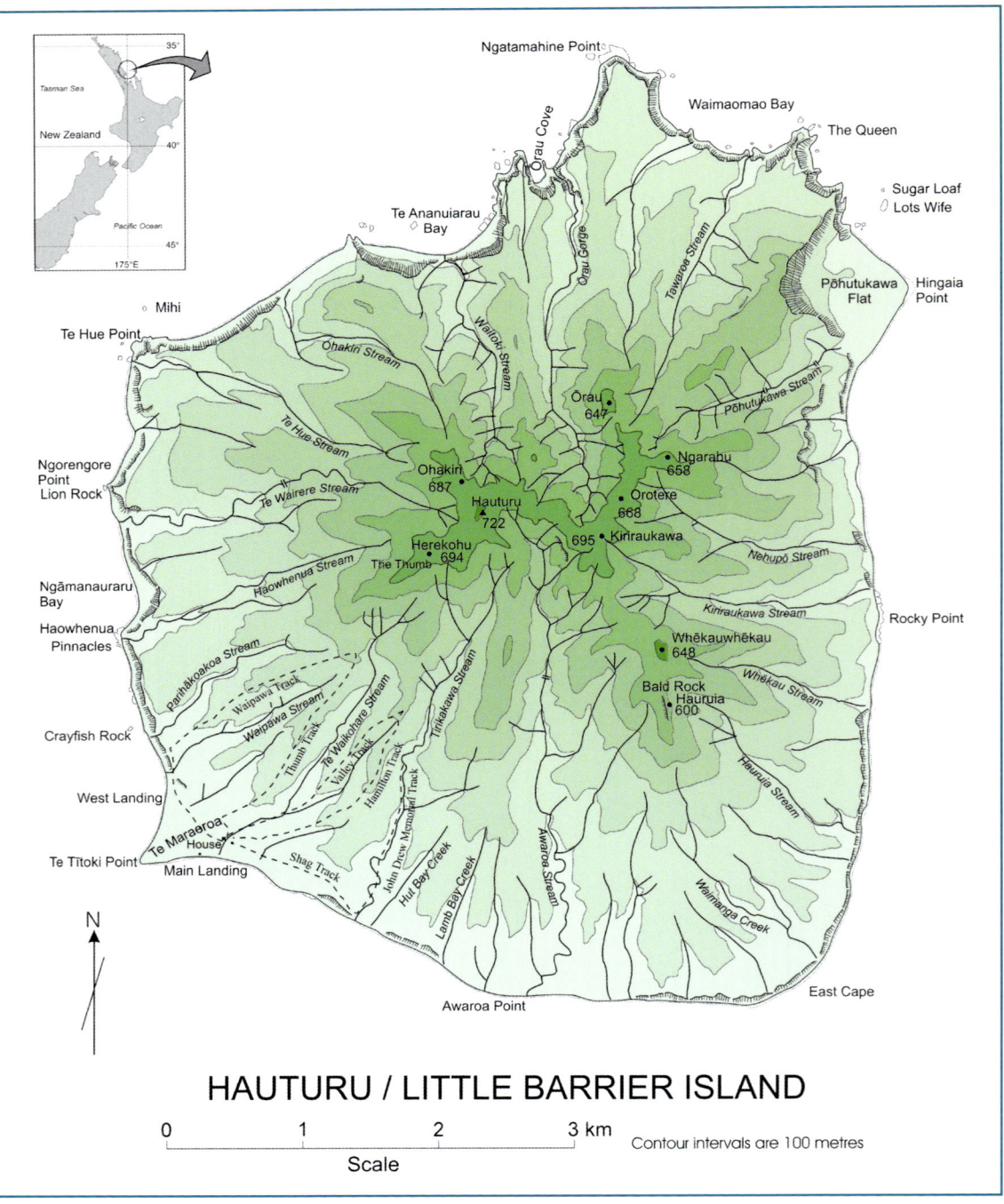

# HAUTURU / LITTLE BARRIER ISLAND

The official area of Hauturu/Little Barrier Island is 2817 hectares. However, there has never been a survey on the ground. At the start of the rat eradication operation in 2004 a helicopter flew the circumference of the island to get a map into the on-board computer systems and that measured the area to be 3083 hectares.

**The remote isolation of the islands of New Zealand in the vast Pacific Ocean influenced the unique evolutionary development of the country's flora and fauna and, as a result, its vulnerability to the impacts of human arrival and settlement from the twelfth century. Over time these human migrant New Zealanders — Māori, Pākehā, Polynesian and other cultures — have developed an intimate awareness and love for Aotearoa's natural island landscapes. This love of islands is central to many New Zealanders becoming global conservation champions and world leaders in the science and management of saving island species. Te Hauturu-o-Toi/Little Barrier Island is a jewel in this conservation crown.**

Positioned at the entrance to Te Moananui-o-Toi/Hauraki Gulf, 70 kilometres from Auckland, Hauturu is named for its highest peak, 'the windblown summit of Toi'. Toi Te Huatahi, a Polynesian navigator, was the first person to discover and name the island, more than 800 years ago. Hauturu was named Little Barrier Island in 1769 by Captain James Cook during his first voyage to New Zealand; and he named the larger island, Aotea, Great Barrier Island.

Hauturu is New Zealand's largest island nature reserve (excluding the subantarctic islands), with a surface area, taking into account the slopes of the land, of more than 4000 hectares. Rising to an altitude of 722 metres above sea level (asl), the island represents the remains of a mid Pliocene to early Pleistocene (3.1–1.2 million years) volcano, with steep valleys radiating out from a series of high peaks. At lower elevations, knife-edge ridges or more gentle slopes fan out, and many of these terminate in sea cliffs up to 250 metres high. Streams and waterfalls dissect the coastline, which is made up of near-continuous boulder beaches in the east and south and rocky points and high cliffs in the north and west. Prevailing winds are from the southwest; however, winds from any direction interact with the topography to create the frequent cap of cloud over the high peaks for which the island is known.

The flora and fauna of Hauturu have a close affinity with those on the New Zealand mainland, particularly the North Island, and much higher biodiversity than on more remote oceanic islands of similar size. To understand this diversity we must travel back in time by

*Previous*: Aerial view of Hauturu from the south, with Te Maraeroa to the left. (WJ)

around 18,000 years. With the last ice age at its coldest, the sea level around New Zealand was approximately 130 metres lower than it is today and the coastline, as a result, was well beyond Hauturu and neighbouring Aotea. Although they are isolated islands now, Hauturu and Aotea were linked by dry land and, along with the Coromandel Peninsula, were simply high points on a vast coastal plain that is today the Hauraki Gulf. The forests of Hauturu may have resembled those of today because the island was still north of the southern boundary of warm temperate forest. With continuous dry land (other than rivers and streams) there were no barriers for flightless species to cross to reach the higher ground of Hauturu; now-extinct giants such as the four North Island moa species would have wandered on Hauturu's wooded slopes. Subsequently, however, the climate began to warm and sea levels gradually rose. By 12,000 years ago links with the Coromandel Peninsula were becoming submerged, and by around 10,000 years ago Hauturu became isolated from all other land masses. The island formed was still much larger than the one we see today, but as sea levels continued to rise and the island area gradually reduced, some of the species trapped there would have lost access to vital resources and died out. Many others survived historically and are present today as reminders of this history of connection with the mainland — including wētāpunga (the Little Barrier giant wētā), tuatara, a host of lizard species and poorly flighted birds such as tīeke and possibly kōkako. In effect, what remained on Hauturu was a fragment of the fauna and flora typical of the adjacent North Island. It is for this reason that most islands in the Hauraki Gulf are referred to as continental islands, distinct from the oceanic islands such as the Kermadecs and Galápagos that have never had these continental links.

From a distance, Hauturu's forested, mist-clad peaks and deeply cut gorges are awe-inspiring. But to visit and step onto its shores is to realise that Hauturu is more than special, it is a spiritual place, a place that Māori and Pākehā alike have viewed with a reverence that stems from the natural wonders of the island, its life force, its mauri. Hauturu represents a bastion for a devastated natural New Zealand where native biodiversity has been silenced by the axe, fire, or the teeth of introduced browsers and predators. Amazingly, the island was not subjected to the effects of most of the invasive mammals that were deliberately or accidentally introduced to the mainland. Furthermore, although some particularly damaging predators such as ship rats (*Rattus rattus*), Norway rats (*R. norvegicus*) and stoats (*Mustela erminea*) reached other large islands in the Hauraki Gulf, they were never introduced to Hauturu. These mammals became so widespread elsewhere in New Zealand that the largest northern offshore island to avoid them completely was Tawhiti Rahi in the Poor Knights, which is only 151 hectares in size. For an island the size of Hauturu to have been exposed to only two resident invasive mammals — kiore (Pacific rats, *Rattus exulans*) and cats (*Felis catus*) — is extremely unusual.

Now that the last of the introduced mammals has been removed from Hauturu, the island has gained even greater significance as a nature reserve. Although larger islands have been cleared of introduced mammals, they are in remote locations with extreme

Mount Ōrau in the mist from the summit ridge. (LW)

climates. They include subantarctic islands such as Campbell (11,200 hectares) and Macquarie (12,800 hectares), but these islands have nowhere near the range of plant and animal species found on Hauturu. From an international perspective, in terms of both conservation values and its rich species diversity, Hauturu is unique — there is nothing quite like this island elsewhere.

Hauturu is frequently described as the most intact native ecosystem in the country. The list of its natural assets is long: a plant, moss, liverwort, lichen and fungal community of more than 1700 species in a landscape where two-thirds of the forest cover has never been logged or touched by introduced browsers such as possums (*Trichosurus vulpecula*) or deer (order Artiodactyla); the most diverse native bird community in the country (over 40 breeding species), including endangered species found nowhere else, and a massive seabird population offering a seamless ecological link between the sea and the land through tonnes of marine-derived nutrients; the most diverse assemblage of native reptiles in the country with 14 known breeding species; and a diverse invertebrate fauna for which many species are yet to be described. From a research perspective this combination of size, elevation, type and extent of forest cover provides the only opportunity to study many of the interactions between plants, invertebrates, reptiles, birds and bats that were once a feature of mainland forests — a vital ecological baseline for New Zealand's natural environment and what it can be, in the face of the continuing decline of our natural world beyond the island's shores.

This natural gem could easily have been lost but for the whims of fate. The story of Hauturu is not just of its impressive wildlife, but of people: men and women who have lived a remote and isolated life protecting the island, researchers working in the challenging terrain to understand its flora and fauna, and conservation innovators bringing new techniques to eradicate the introduced pests in efforts that have inspired the conservation world. Auckland Museum was instrumental in petitioning the government to gazette Hauturu as a nature reserve in 1895 — an action that saved much of its natural biodiversity from human exploitation, but led to the eviction of the mana whenua, Ngāti Manuhiri, in an act that caused lasting pain and deprivation. Today, however, Hauturu is being managed in a spirit of partnership between Māori and Pākehā. After their settlement with the government under the Treaty of Waitangi claims process, Ngāti Manuhiri gifted most of Hauturu to the people of New Zealand, and they are now moving forward in the spirit of co-management with the Department of Conservation, charting new ways for conservation in New Zealand.

It is the purpose of this book to outline the stories of these people and the rich biodiversity of Hauturu. In 1961 Dr W. M. Hamilton and co-authors published the only comprehensive account of the human and natural history of Hauturu. That publication was the benchmark by the standards of the day. But there has since been an explosion of knowledge on the biodiversity of Hauturu and an additional 50 years of human history surrounding the island that is still undocumented in one source.

Hamilton and his colleagues would no doubt be surprised and delighted with progress in the understanding of Hauturu, and our ability to manage and protect it. We live in a world where access to remote Hauturu is so much easier, where damaging pests can be removed on a massive scale, and where native animals can be introduced successfully and monitored remotely using new technologies. These capabilities bring added responsibilities and an array of problems and questions regarding the management of Hauturu. How do we protect this ark from invasion by old enemies such as rats and weeds, or new ones such as kauri dieback? How do we decide the right path for ongoing restoration of the island and navigate the inevitable value judgements that are made when deciding which species to reintroduce and for what reason? How do we study the ecological changes we set in motion through our powerful conservation management actions? And what will be the consequence of these changes to the island that others may document in another 50 years?

This book seeks to address these questions while documenting the accumulated knowledge and history of Hauturu in the almost 60 years since Hamilton's book was published. It is written by experts in a range of fields, from human history to the biology of organisms both large and small.

# Chapter two
# Papakāinga
— PAULA MORRIS

**Te Hauturu-o-Toi: there's a story right there in the island's full name, a legend, a history. The name stretches back in time to Toi Te Huatahi, the great explorer and one of the early ancestors. Toi, in search of his missing grandson, crossed the Pacific following the path of the stars. He didn't find his grandson but he found Aotea and Hauturu, and decided to stay for a while.**

This is the story from the point of view of Paratene Te Manu, one of the last Māori owners and inhabitants of Hauturu, imagined in my novel *Rangatira*:

> Hauturu is not like other islands. It's a place of secrets and resistance. The mountain always wears a cloak of cloud. Some say that an atua lives up there, and when it descends, disguised in swirls of mist, no one is safe. The waves crash and the winds blow, warning us of its descent.
>
> The island is guarded by its sheer ridges and slashed with deep gullies, which is why it was easy to defend. The streams only flow during the rains of winter. There are no beaches, and if the wind blows from the southwest, it's impossible either to land or to leave. Toi himself couldn't find a landing place, so he sent a slave, guarded by Toi's beloved dog, Moipahuroa, to find a suitable spot.
>
> Perhaps the dog was reluctant to swim ashore, or perhaps the slave just wanted to escape his watch. For one reason or another, near the wet boulders near Titoki point, the slave dashed the dog's head with a sharp rock. He'd forgotten that the dog was Toi's property and therefore tapu. Before his eyes, the dog turned into a slab of stone. You can see it there now, lying in the water, as flat as the boulders around it are round. It's still tapu. If anyone strikes that stone, he will die within one month.
>
> The slave was frightened, so he ran away through the trees and disappeared up the mountain. He should have stayed and faced Toi's punishment. Instead he was taken by the patupaiarehe — sprites, they call them in English — and held there forever. You can still hear his voice in the whistling wind when a storm lashes the island, crying out to Moipahuroa for forgiveness.[1]

Ngāti Wai and Ngāti Manuhiri have deep roots on the island. Whakapapa is our unbroken generational connection to the past, the land and our tūpuna. Our history, like all histories, is shaped by battles, alliances and strategic marriages. Manuhiri was the son of

---

*Above*: *Hauturu*, oil on canvas, by Aroha Gossage. Courtesy of ARTIS Gallery. *Below*: Near Te Tītoki Point, looking east, 1893. (ATL, 1/1-020-406-G) *Previous*: Panorama of the summit ridge. (AA)

Maki, seventeenth-century conqueror of Hauturu, and our mana whenua extends from him as well as from Toi. For hundreds of years the island was a place of refuge, of mahinga kai, of wāhi tapu marking battles and burials.

But for over a century we've been estranged from Hauturu. For us it's a psychic landscape as well as a physical one. Across the water it looms, a repository of history and memory, of dispossession and loss. The estrangement is personal. It's a place we can see from every part of the Ngāti Manuhiri domain — Pakiri, Leigh, Omaha, Te Hāwere-a-Maki (Goat Island), Tāwharanui — but needed permission to visit.

**STEP BACK INTO THE NINETEENTH CENTURY.** Te Kiri is a rangatira of Ngāti Wai and Ngāti Whātua, of Te Kawerau and Ngāti Manuhiri. He establishes a stronghold in Pakiri, and becomes a dominant figure in the coastal areas of the region. In 1858 he buys the schooner *Industry* as the centrepiece of his business operations, transporting passengers and goods such as kauri logs and gum to and from Auckland, Tauranga, the Coromandel and ports in Australia and the Pacific. In 1864 he sails to Kawau to rescue 180 Māori prisoners taken during the war in the Waikato — many captured at the battle of Rangiriri the previous year.

When Te Kiri dies in 1872 he passes on mana whenua in Pakiri, Omaha and Hauturu to his daughter Rahui Te Kiri. She and her second husband, Tenetahi, live on Hauturu. Both are skilled sailors and Tenetahi, in his cutter *Rangatira*, enjoys a number of famous wins at the Auckland Regatta. The ornithologist Andreas Reischek hears the story of the 1883 sinking of the *Rangatira* firsthand from Tenetahi during a visit to Hauturu, and publishes it in *Yesterdays in Maoriland*.

> [Tenetahi] had sailed to Catherine Bay to extract blubber from a whale which the natives had caught, and had been overtaken by a storm while at anchor there. In trying to run out, the anchor got wedged, and he and the crew had to jump overboard to save their lives, while their boat was battered to pieces on the rocks.
>
> Tenetahi, his wife Rahui, two men and a boy then set off in a whaling-boat for Tiharea, their Hauturu settlement, but some miles off the Great Barrier another storm came on. The boat turned turtle, and they lost an oar, whereupon Tenetahi ordered the men to keep her steady while he righted her, and baled her out. Rahui, who was a good swimmer, jumped in and swam after the oar, but by the time she brought it back, one of the men had been washed away. They had a frightful job to get the second man and the boy back into the boat, and the latter died soon after from exposure and cold. The man, too, would have died if they had not made him row for all he was worth to get warm.

*Clockwise from top*: Wiremu Tenetahi, Rahui Te Kiri's second husband, outside the huts on the flat at the stream now known as Grave Stream, 1893. (ATL) A portrait of Tenetahi by Robert Fredrick Way. (ATL, A-114-010) Rahui Te Kiri with her daughter Ngapeka Te Roa, seated, on Hauturu. (ATL, 1/1-020598-G)

> Three times after this the boat filled with water, and but for Rahui's skill and courage all would have been up. She was fourteen hours in the water without food or drink, battling with the waves, and when they finally reached the shore, was so exhausted that she could not move a foot. Rahui, I may say, is 5 feet 10 inches tall, and the possessor of a fine if muscular figure.[2]

Rahui is in her fifties when the *Rangatira* sinks. Some of her considerable energy is spent continuing her father's legal battle with the Crown to keep a reserve of Māori land at Omaha, and blocking the sale of land at Pakiri. In 1897 she secures title to the Omaha reserve, almost 40 years after her father began the fight for it.

But the Crown is determined to take Hauturu — initially as a military base, then as a timber resource, and finally as a native bird sanctuary. In October 1886 the Native Land Court names 14 people, including Rahui Te Kiri and Paratene Te Manu, as owners of Hauturu. The next 10 years are spent in fraught negotiations with the government over the sale price and the issue of a small Māori reserve on the island. The Little Barrier Island Purchase Act, which compulsorily acquires all remaining shares in the island, is passed by Parliament in October 1894, despite arguments against it led by Hone Heke Napua, the Northern Māori MP.

Ignoring its centuries of Māori history and areas of wāhi tapu, the Crown will not agree to a Māori reserve on Hauturu, or take into consideration the value of the kauri, livestock and cultivations. An eviction notice is served. All 'natives' are supposed to leave by 10 December 1895.

Earlier that year, Gerhard Mueller, the Commissioner of Crown Lands, visits Hauturu accompanied by police and an interpreter. He tells Tenetahi that the island now belongs to the Crown, and reports back Tenetahi's reply to the Surveyor General:

> First, I refuse to leave the island because I do not consider that the purchase is a proper purchase. Second, neither myself nor my wife have sold our shares — I do not recognise the sale which has been effected. Third, my reply regarding the preservation of the birds on the island is that they are all mine and I have always preserved them to the present time and have therefore never allowed bees to be introduced which would have exterminated them.[3]

The journalist James Cowan visits Hauturu that year, and sees Paratene Te Manu receive his eviction notice. Cowan publishes an account in his 1930 Lindauer monograph, *Pictures of Old New Zealand*.

> The ancient warrior, bent with age, would not touch his summons so it was laid on the ground at his feet. He picked up a mānuka stick and danced feebly around the obnoxious paper, making digs at it as though he were spearing an enemy. The old man said he was not going to court, and was not going to leave the island; it was his, and he was going to die there.[4]

This map is part of the title deed for 'Hauturu or Little Barrier'. Tenetahi made his first recorded title claim to the Native Land Court in May 1878, but the hearing did not take place due to the absence of a survey plan. This plan was then created by enlarging Admiralty charts. The court adjudged that 'the plan produced was not in accordance with the provisions of the act . . . an actual survey on the ground is necessary previous to investigation of title'. However, despite the absence of such a survey, the area of the island was adjudged, and accepted, to be 6960 acres. This was marked on the plan and remains as the official area (2817 hectares) today. There has still been no survey on the ground. (Land Information New Zealand)

The Crown makes its first attempt to enforce the eviction in January 1896, sending the steamer *Nautilus* to carry the Māori residents away, only to find that they return within days to continue farming. More force is needed.[5]

At 5 am on 19 March 1896, the government steamer *Hinemoa* drops anchor at Hauturu and lands two boats carrying Lieutenant Hume and a number of officials, along with an interpreter and '21 men of the torpedo corps' each with '20 rounds of ball cartridge' — according to Charles John Alexander, a bemused passenger on the *Hinemoa*. 'From the ship,' Alexander writes, he can see 'two or three natives about their hut, and it appears that these are the people this small army has been sent out to evict'.[6]

Paratene Te Manu is not there: he died in Ngunguru in January. The soldiers remove five people from Hauturu — two women and three men. The women, Rahui and her daughter, Ngapeka, are dropped off at Little Omaha. There her son, Wi Taiawa, is arrested and brought onto the ship. When the *Hinemoa* arrives back in Auckland late that afternoon, the four male prisoners — named in newspapers as Tenetahi, Wi Taiawa, Kino Tamihana and Kiri Tenetahi (my great-grandfather, then about 24 years old) — are marched away to prison cells, charged with 'wilful trespass' on crown lands. Warrants are out for the arrest of four others, according to breathless reports in local papers across the country, though ultimately all charges are dropped. Hume, Alexander writes, seems 'ashamed' of the part he played in the comedy of 'the taking of the Little Barrier'.[7]

The *New Zealand Herald* describes Tenetahi as a 'haughty, defiant chieftain' but admits there is 'something pathetic in this unfortunate Maori being dragged away from his ancestral rocks by force and arms'. It suggests that the Crown appoint him 'caretaker of the birds and beasts' on Hauturu, and declares that such duties 'can hardly be of that complex and intricate kind which would overtask the capacity of a Maori'.[8]

**ANCESTRAL ROCKS: THAT'S ONE WAY** of seeing Hauturu, a place of cliffs and stones where it's difficult — and often impossible — to land. For Rahui and Tenetahi, it was home, ancestral and current. After the military show of their eviction, they continued to return to the island to graze or move cattle, and to harvest crops. In October of 1896 Tenetahi — it's alleged — demolished the island's only remaining inhabitable dwelling. In the early diaries of Robert Shakespear, the first caretaker of the new sanctuary, he complains about Tenetahi's constant visits to Hauturu. He grew to like and respect Rahui, describing her as 'most interesting' and 'a plucky old thing'. As late as August 1897, she was still sailing to Hauturu to harvest kūmara, and sharing with Shakespear some of the secrets of the island.[9] Shakespear was paid a salary of £150 a year to live on Hauturu. Tenetahi's share of the island's sale price, a one-off sum he refused to accept, was £300 — reduced to £234 after costs for the eviction were deducted. Rahui Te Kiri's share was just £158 and 10 shillings.[10]

Tenetahi continued to petition the government over the forced sale of Hauturu, but by the time of his death in 1923, the exile was complete. Rahui died in 1930, in what was believed to be her hundredth year. They are both buried in the urupā of the Omaha marae, just outside Leigh.

My grandmother, born in 1902, was the oldest child of her family, the mātāmua, and she was sent to live with her grandparents Rahui and Tenetahi. She learned weaving and the piano, and rode around on her horse. All her life, her siblings referred to her as Lady Jane. As far as I know, she never set foot on Hauturu. Neither did my father, born in Pakiri in 1933. He and I talked many times about going, but in old age he didn't feel up to the landing. The island was a lost place, forbidden, reserved now only for wildlife and its keepers and observers, where we would be intruders and outsiders. Two of my cousins — the artists Star Gossage and Aroha Gossage — have spent time there in recent years, working towards paintings. For all of us, Hauturu looms large, an ancestor we see in our dreams and recreate in our imaginations.

In 2011 Ngāti Manuhiri and the Crown signed a Deed of Settlement, returning 1.2 hectares on Hauturu for a reserve. At some point we will have a place to visit and gather there, however small. We'll no longer be in forced exile from home.

All of us, as descendants of other people, have an inheritance: that inheritance is our ancestors themselves, our tūpuna. The lines to them — tātai — are unbroken, and there are many lines, like the lines that make up a net. This is an apt metaphor for Hauturu, as that's part of its creation story as well. When Taramainuku threw his seine across the waters of the Gulf, Hauturu was its centrepost, ensuring the net didn't drift away or sink to the bottom of the sea. Rahui and Tenetahi clung to that post for as long as they were able. It will stand forever, with its crown of clouds and whispering trees, beckoning us back.

---

A version of this essay was first presented as a talk at the Hauraki Gulf Forum's annual Hauraki Gulf Maritime Park Seminar in 2018, under the title 'Imagining Place'.

# Chapter three
# Island life

— LYN WADE AND DICK VEITCH

**Before the government began any conservation action for Te Hauturu-o-Toi/Little Barrier Island, the island was visited by people with the intention of seeing, recording, collecting or killing whatever they could find. Some were successful and put specimens of plants and animals into museums and private collections, or published papers recording their findings. Others failed.**

Thomas Kirk, a botanist, visited in 1867 and recorded his observations in the *Transactions of the New Zealand Institute*. He made subsequent visits in 1871 and 1886, and plants that he collected are in Auckland Museum and the Museum of New Zealand Te Papa Tongarewa. Captain F. W. Hutton spent four days on the eastern side of the island in 1868, and supplied a list of birds he recorded there to the New Zealand Institute. He recorded that tīeke/saddlebacks (*Philesturnus carunculatus*) and kākā (*Nestor meridionalis*) were very common. He noted the presence of New Zealand falcon (*Falco novaeseelandiae*), and a bird he did not know that came out at dusk and had a laughing call, rounded wingtips and was slightly larger than a morepork or ruru; he speculated that this was the whēkau (laughing owl, *Sceloglaux albifacies*). The other 15 forest birds he recorded are all still present on the island.

Austrian naturalist and collector Andreas Reischek may have spent more time than others on the island — he visited five times between October 1880 and December 1883 with the prime aim of collecting hihi or stitchbirds (*Notiomystis cincta*). During some visits he failed to see any hihi but he collected other birds and wrote about them in *Transactions of the New Zealand Institute*. In a letter to the surveyor-general on 17 October 1895, New Zealand naturalist and author Walter Buller recorded that he had purchased a dozen stitchbird skins in Auckland and that these were birds collected by Reischek. He noted that Lord Rothschild, a British zoologist, had 80 to 90 stitchbird skins for his natural history collection. Reischek recommended to the institute that the island be preserved to protect the birdlife. He suggested that one man would be able to oversee the whole island and that the habitat was suitable for many species. His bird list included kiwi, which had been notably absent from previous records: He commented that there were no bees on the island as Tenetahi, one of the Māori owners of the island, would not allow beehives to be brought ashore as he feared they would interfere with the nectar-feeding birds. Reischek said he could not recommend a more favourable place to benefit science and agriculture.

There are a number of records of the condition of the island at this time. Te Maraeroa, the traditional area of occupation and plantation on the southwest portion of the island,

*Above*: Te Maraeroa, 1893 (ATL, 20408). *Below*: Te Maraeroa, 2019. (CRV) *Previous*: House and surrounds, from the hill at the back looking south, 1901. (FS)

**Island life**

was grass and scrub. The hill slopes immediately behind were grass with some kānuka beginning to grow and some patches of pōhutukawa. This area was obviously being grazed by sheep and cattle and those animals would have returned to the flat for water. In summer there would have been little or no water in the streams.

Hugh Boscawen of the Department of Lands and Survey wrote in 1893 of the 'bush cattle' on the island: 'There is hardly a place they will not go, and they break and tear at supplejacks and other undergrowth for feed.' However, Reischek makes no mention of wild cattle. Rahui Te Kiri told the Native Land Court in 1895 that there were 1000 sheep and 30 head of cattle on the island at that time. Hamilton and Atkinson recorded that most of these were grazed on the southern side of the island,[1] but some — as reported by a stock transporter to the caretaker in 1893 — were grazing on the flat-topped planeze above Ngatamahine Point. It is extremely difficult for people to get to that planeze above Ngatamahine, and the report to the caretaker may have been just a story to satisfy his curiosity. Hamilton and Atkinson record that a considerable part of the forest on the southwestern side of the island and on the northern point was felled for firewood — but if that was so, how was the wood transported to the sea and onto a ship to Auckland? They then suggest that these areas were grazed. Undoubtedly the southern, western and northern planezes were burnt at some time, but access to large parts of this area is very difficult for people, and even more so for sheep and cattle. Besides, there is no water there for stock. The firewood must have been taken from the slopes adjacent to Te Maraeroa.

No records have been found of how many kauri were felled and taken off the island. After the government had purchased some shares in the island it began putting the government brand on logs. There are kauri stumps on the ridge between the Tirikakawa and Awaroa streams, which match reports that there were teams of oxen up these streams, although just one team is a more likely scenario. The impact of taking trees out of this forest may have been difficult to discern from the sea. Kauri stumps are also found at the clifftop just south of Pōhutukawa Flat in the northeast of the island where, presumably, the logs were rolled over the edge to crash onto the rocks below.

From this point the information in this chapter relies on diaries kept by the caretakers, their monthly reports and occasional other information now lodged in museums and archives. None of this is a consistent record. Some periods have no records; some are dependent on the interests of the writer and visitors staying on the island.

The first recorded conservation action was an injunction, declared in 1892, to stop kauri logging. Henry Wright was sent to the island as temporary ranger in 1892 to ensure that the injunction was observed. He recorded that there were eight men up the Waipawa Stream felling kauri. About this time Boscawen provided the Department of Lands and Survey with a report on the island supporting its suitability as a bird preserve because, as the terrain was so rugged, it was not suited to much else. Charles Robinson of the Department of Agriculture followed this up in February 1893 with a report to the

---

*Clockwise from top*: Two tents were used as an additional bedroom and the school room. (FS) Tenetahi's boatshed, 1 February 1893. (ATL, 20511) Tenetahi's house. (FS) Launching the Shakespear family's whaleboat, the *Bolivar*, 1900. (FS)

**Island life** 37

Department of Lands and Survey that during his tenure, Tenetahi or his representatives had provided a monthly boat service to Omaha. In December 1895, Kiri Tenetahi, son of Rahui and Tenetahi, made a sworn statement that he had seen a stitchbird skin in Robinson's hut. The Minister of Lands then requested that Robinson be given one month's notice.

Following pressure to set aside the island to preserve some of New Zealand's rarer birds, the Little Barrier Island Purchase Act of 1894 was passed. Under the Act, Hauturu was deemed to be crown land. It was declared a reserve for 'preservation of native fauna' on 26 September 1895. Before this there had been decades of dispute, agreements to sell, refusal to sell, court orders, threats of eviction and, finally, an order for possession in November 1895. Following that, the government arranged on more than one occasion for a Crown Law officer, a bailiff, members of the police and the permanent militia, and an interpreter to visit the island and remove residents. Three of the militia men and Robinson remained on the island.

Tenetahi left behind 15 cattle, 70 sheep, many pigs, fowl and turkeys as well as his crops of maize, kūmara and potatoes. He later removed some of the stock and the remainder was taken off by a contractor to the Department of Lands and Survey and sold in Auckland. Robinson left the island on 25 March 1896 and two of the militia men remained.

It was agreed that management of the island was best overseen by the Auckland Institute and Museum. The government granted £200 to cover the expense of management for one year: the museum was to appoint a resident curator; the government steamer that serviced the lighthouses would be used to land stores; and a sum of £250 was granted to build a house for the caretaker. There were 198 applicants for the caretaker position. Robert H. A. Shakespear was appointed, and £150 of the £200 allocated was used for his salary.

In preparation for his tenure, Shakespear and members of his family visited the island on 30 December 1896 along with the museum president, Mr Donald Petrie, and a number of people from the Auckland Institute. They discovered that the four-room cottage built earlier by Simon Welton Browne, a timber trader, where they had intended to house the Shakespear family, had been removed. Two others were still standing but were not in good condition. The hut occupied by the previous caretaker, Mr Turner (the one remaining militia man), was deserted and the government boat was missing.

Robert Shakespear and his family left their farm at Whangaparāoa and landed on the island from the SS *Kawau* on 19 January 1897 — and there has been a caretaker in almost constant residence on the island ever since. The family would have comprised parents Robert and Blanche, and their children Frances, Robert, Ivy, Ruby, Ethel and Helen, aged between nine and 18. A sixth daughter, Katherine, was born in February 1898. A letter she wrote in 1980 states that two aunts arrived on the island with the family — these would have been Edith Smith, sister of Blanche, and Aunt Alice or Auntie Bolt, who are both referred to in later diaries.

Some members of the family spent their first night on the island in the boatshed built by Tenetahi at the mouth of Te Waikohare Stream; there was rain in the night and the shed 'leaked like a sieve'. They also pitched a tent, which stayed dry. The next day they shifted into Tenetahi's house but continued to live in tents for the next 10 months, using the house as kitchen and living room.

They used a winch they had brought with them to haul their boat onto the beach. They brought a horse and some heifers — and probably some sheep, as other records state that stock owned by Tenetahi had been removed, yet 'the boys' went out to shoot a sheep on 25 January.

The oldest daughter, Frances, created a photographic record of the Shakespears' tenure on the island. She was a self-taught photographer, using a quarter-plate camera. Her entire photo collection has been preserved and many of the albums she created are still with family members.

On 16 February 1897 the family made a circumnavigation of the island in their only boat, the *Bolivar* — a whaleboat about 5 metres long. The diary records that they returned home at 6.10 pm 'after pulling for 6 hours'. There were later occasions when they went out and rowed around the island, but not all were successful — on one occasion a southerly wind rose as they headed south from Pōhutukawa Flat and they hauled the boat ashore at East Cape and walked home. From time to time the family and friends would picnic on a beach well away from home, and some of these outings would have required either two boats or multiple trips to get everyone there.

The house was built between June and August 1897 by F. Horneman and A. J. McCleod, who were paid for 58 days' work. The main part of the building was seven rooms, with kitchen and scullery as a lean-to addition at the back. The overall size of the house was 54' x 28'6" (approximately 16.5 x 9 metres).

An entry in the caretaker's diary of January 1898 talks of the shortage of supplies: it was two months since they had had any meat to eat, and they had trouble catching fish. They must have purchased more sheep but by March of that year they had only five. Apparently the family were busy exploring the island by whaleboat and on foot. The diary also noted the presence of kiwi.

**ISLAND LIFE FOR THE FAMILY** may have been little different to that of a mainland farming family. The diaries show that their lives were ordered, with laundry every Monday and ironing every Tuesday; sewing was on Tuesday and could extend to Wednesday. The children were home-schooled with Auntie (Edith Smith) as teacher. Lessons were every morning, at least. Frances, in her diary, recorded that she joined the classes on Wednesday and Friday mornings to learn specific subjects. Afternoons were free, with frequent walks over the flat, along tracks, or outings in the boat. Frances and Auntie collected and preserved plants that are now in Auckland Museum. Visitors were infrequent for much of the year but some photos show the occasional gathering. For evening and wet-weather entertainment they had board games, a piano and an organ, and a number of family members played and sang. Almost every evening Frances records that she and Auntie read to each other from books and newspapers, or that they were 'toning photographs'.

They had cows to milk and sheep and cattle for meat. There was an extensive vegetable garden. Pat, an employee from Great Barrier Island, and later Blakie, then Jimmie and Mrs

Jimmie, spent time on odd jobs, tarring the paths, boat delivery and weeding. The garden was extensive: 120 cabbages and 60 cauliflowers were planted on one day. They used the horse to pull a harrow to prepare the ground for planting potatoes. There were many endeavours to catch fish for the table. The kitchen stove would have been wood-fired, and wood collection was a regular chore. There was also an open fire in the living room. There is no record of who did the cooking — it may have been Auntie Bolt, who is rarely mentioned in the diaries, and perhaps, later, Mrs Jimmie.

From time to time kitchen leftovers were put at the kitchen door for the birds. When they were bottling peaches, for example, the bellbirds flocked in to feast on the fruit.

In these early years there were no fences to keep stock from wandering. For mutton, one of the family would go out and shoot a sheep — and there were a number of comments in the diaries that the sheep were not fat or the meat was not good. An 1899 photo shows seven cows. In May 1901, when most of the family went to the mainland, 'Papa milked six cows while Auntie and Auntie Bolt fed the calves'. By 1907 they were shipping surplus stock off the island.

The heifers proved to be adequate steeds for the 'Little Barrier Cavalry Corps', and a four-wheeled vehicle was built for the horse to haul goods to the landing and, probably, firewood to the house.

Supplies were shipped from Omaha (either Big Omaha or Omaha Cove/Leigh) every few weeks; the museum paid for this service. In April 1901 Shakespear reported to the museum that he was having difficulty communicating and getting the stores he needed. He asked for £15 — the amount usually paid in steamer services — towards the purchase of a yacht. This money was probably used as part payment for the yacht *Merry Duchess*. This vessel was kept moored offshore from May 1901 to May 1903, when it was replaced by the *Pirate*.

In 1901 the *Akaroa* took 95 kauri logs off the beaches of Hauturu and towed them to Omaha. These were not Hauturu logs but logs that had broken free from rafts of logs being towed in nearby waters.

Robert junior built the *Pirate* on the island, launched it in March 1903 and sailed it to Auckland for finishing. The *Pirate* was sailed mainly by the younger daughters, who went to Omaha for stores and timber and to other parts of the Hauraki Gulf to visit or to take passengers. This vessel was driven ashore in a storm on 24 June 1907. It was repaired, but was driven ashore again on 26 July 1908, this time beyond repair. Robert later built at least three other boats in Auckland, where he undertook a boatbuilding apprenticeship.

In August 1908 Shakespear purchased the *Frances*, built by Logan Bros, a 38-foot (11.5-metre) boat with a 5-horsepower motor. He planned to 'keep her in Oma Cove, getting her out to the island when needed'.

Two tents continued to be used as a bedroom and schoolroom, at least until 1902. Frances Shakespear's photos show two additional buildings to the west of the back door of the house. Close to the house was a storage room and a photographic darkroom

---

*Clockwise from top*: The house under construction, September 1897. (FS) The *Pirate* holed beyond repair, July 1908. (FS) A picnic with family and friends at Pōhutukawa Flat, January 1906. It must have taken two boatloads or two trips from home to bring this many people. (FS) The Little Barrier Cavalry Corps. (FS)

Island life 41

for Frances, built by Robert junior in March 1901. The building further to the west was apparently built by him too, in 1902, as a bedroom for his sisters. Another building to the north of the house, probably built before 1900, was a laundry and rat-proof store measuring 29' x 12'6" (8.8 x 3.7 metres). There was a cowshed with bails for four cows, a neat little aviary (as described by Robert Hunter-Blair, the next caretaker) and, away at West Landing, a fowlhouse with a concrete floor.

The household water supply would have been limited. Tenetahi's house had a small tank to catch rainwater. The new house had three large rainwater tanks near the back door, and each of the additional buildings had a rainwater tank. There was a covered waterhole, 10 feet (3 metres) deep, with a pump, in Te Waikohare Stream near the house. It seems that this water supply was insufficient at times and the diaries record the need to collect water daily from the spring at West Landing.

It appears that a waterhole dug in Turners Creek (now known as Grave Stream) to provide water for the farm stock always held water. Residents bathed there — possibly because it was more accessible than the West Landing spring and the water was warmer.

On 1 April 1905 the Tourist Department took over management of the island. Shakespear was retained, but early in 1910 he resigned and on 30 January he left the island. He died three weeks later and is interred at Leigh. Robert junior took over as official caretaker, but in mid-March he too resigned. He left the island to build a house for his family on the mainland, but he ensured that a family member or other responsible person was on the island until about the end of March. The Shakespear farm stock was removed from the island during this period, apart from two horses and a few cattle.

The replacement caretaker, Robert Hunter-Blair, arrived on the island on 4 May 1910. He had been working for the Tourist Department for some time, including work on the Milford Track. He recorded that the house was in good repair and clean, but that the bath, kitchen range and laundry copper were gone. Hunter-Blair was not in good health when he arrived, and he died on the island in September that year, aged 52. His wife, who had no means of communication with the mainland, buried him on the island. His grave is still there on a knoll near the present pumphouse.

A temporary caretaker, Thomas Pierson Firman, a railway stationmaster from Ōhakune, came to the island on 17 November 1910. His wife Grace and his daughter did not accompany him, but a friend joined him there and they made several trips around the island by boat. Firman then decided that they would go around the island on foot 'by walking the beach to where it was impassable then cutting through the bush . . . we lost our bearings getting lost for four nights and five days'. In his report he describes their condition and notes that they were fortunate to have avoided a serious calamity.

Firman remained on the island until he was replaced by a permanent caretaker, Robert Nelson, on 30 May 1911. Nelson had been a gardener and horticulturalist in the United Kingdom to Sir Thomas MacKenzie, previously a high commissioner in New Zealand. He put some thought into improving the island while he was there. He found and destroyed one gorse plant. He considered blackbirds a pest and 'shot a good many'. On several occasions

---

*Above*: West Flat, 9 February 1893. (ATL, 20411) *Below*: West Landing, 2019. (CRV)

he saw the white kiwi 'which was sent and liberated in April last year [1913]'. He asked for some trees to plant on the now scrubby slopes behind the house; the eventual response from the Department of Lands and Survey informed him that he should 'move some karaka, titoki and kohekohe from the nearby bush'. His May 1916 report records that he had no boat.

Oscar Blundell, the Presbyterian minister at Warkworth, occasionally visited the caretakers on the island. He was a keen explorer, and in January 1916 he traversed the island from Pōhutukawa Flat to the homestead at Te Maraeroa.

Communication consisted of a fortnightly launch service from Leigh provided by Mr D. Matheson. The trips from Leigh, referred to in reports as the 'mail carrier', were not always easy; there were frequent reports that Matheson was late or had had a slow trip in rough seas, and in February 1918 he lost his yacht on the beach — 'a total wreck now broken up'. Matheson continued the mail run until November 1942 when it was taken over by Mr Warren of Great Barrier Island.

Robert Nelson spent 11 years on the island until he retired in May 1922. He would have been happy to stay on, according to a letter he wrote to Thomas Cheeseman at the Auckland Museum, but his wife was keen to leave so their child could go to school.

Nelson's successor was William Cleaver, a Londoner who had spent time in the navy and then joined the New Zealand Expeditionary Force (NZEF). He retired wounded from the NZEF, and later worked for the Department of Internal Affairs. He came to the island on 4 May 1922 with his wife Lilian and their four children.

On 20 July 1923 the *Auckland Star* reported that information had been received by the police that a man named Weidman had died at Little Barrier Island and a party of officers had been dispatched to investigate. The inquest report in the *Northern Advocate* of 25 August 1923 noted that Herbert George Weidman had been on the island since 26 April. Other reports stated that he had landed illegally and that Cleaver had found him in the bush with a sack of kauri gum. Cleaver said Weidman went out fishing on 16 June in an open boat and the sea became rough; 11 days later he found the body on the shore and two days later he buried Weidman. Police collected the body on 19 July.

Cleaver was caretaker for less than 18 months. Robert Nelson returned to the island on 29 August 1923 and continued as caretaker for another nine years, retiring in November 1932.

Permission to visit Hauturu had to be obtained from the Tourist Department, which was then in charge of the island. There were a number of visitors interested in its natural history. In the spring of 1919 Herbert Guthrie-Smith spent 10 weeks there and photographed the stitchbird, robin (*Petroica longipes*) and whitehead (*Mohoua albicilla*).[2] Ornithologist and museum curator W. R. B. Oliver visited briefly in 1921. Robert Falla (who later became director of the Dominion Museum) visited in January 1928 and, in 1934, summarised information on the breeding petrels in his account of the ornithology of the North Auckland area. W. M. Hamilton visited the island several times between 1932 and 1934, researching the history, topography, geology, soils, climate, plant covering and affinities of the flora for his master's thesis at Massey Agricultural College — reproduced in 1937 by the DSIR as Bulletin 54 'The Little Barrier Island, Hauturu', the first comprehensive treatise on the island.

Nelson recorded in his monthly reports how the bellbirds (*Anthornis melanura*) and tūī (*Prosthemadera novaeseelandiae*) gathered around the back door waiting for his wife to come out with their food. Previous caretakers also fed the birds, but it is from Nelson's tenure onwards that there are frequent references to feeding the birds, or the sugar supply not arriving to make syrup for them.

Nelson was replaced by 65-year-old William Hardgrave, a civil servant from Wellington who arrived on the island on 16 November 1932. Hardgrave's wife Martha had died earlier that year, and he was accompanied by his son Len and Len's wife Rene.

Hardgrave recorded that the roof was leaking and the stove was defective, but he added, 'I do not write this to cast a reflection on Mr and Mrs Nelson as they had grown accustomed to it and it was beyond their abilities to make alterations. Two fine old people indeed.' In March 1933 he wrote that he furnished a monthly weather report to the Meteorological Office in Wellington. This must have been very basic rainfall and wind observation data, as temperature and humidity measuring equipment was not installed until February 1947.

In April 1933 he recorded 'cutting out borer infested floorboards' and replacing the piles under part of the house. The Hardgraves made improvements to the house and garden, built a number of fences to confine stock on Te Maraeroa, cleared the summit track and destroyed 360 feral cats and about 6000 kiore. They installed water tanks and connected them to the sink in the scullery, built a cowshed and made repairs to the wire netting around the house 'to keep the penguins out as they are very disturbing at night'.

In September 1933 the monthly report records that they had been given a boat, the *Lady Anne*, about 14 feet (4 metres) long and equipped with an outboard motor; this was a gift from the countess of Orford, Lady Anne Sophia Palmer (née Walpole), daughter of the fifth and last earl of Orford, who was interested in horticulture and must have visited and liked the island. In May 1934 the report notes that they built a boatshed to house the *Lady Anne*, and that the boat was difficult to launch over the boulders. In October 1936 they brought railway tracks ashore to make a ramp, but these were difficult to move. In December 1941 they built a boat ramp and a trolley. They used the boat in January 1942 but then had no more fuel.

Late in 1933 the Hardgraves spent some time digging out convolvulus and columbine, then planted melons, pawpaw and other seeds that Frank, another of William's sons, had brought from Tahiti. They cut out some of the poplars (*Populus nigra*) from the creek near Robert Hunter-Blair's grave but left others standing. In June 1941 they cut down the pines in front of the house.

There are many records of visitors to the island, including William Hardgrave's six other children. In February 1934 the governor-general and his wife, Lord and Lady Bledisloe, visited with 'several Gentlemen and Ladies'. Captain Burgess and some of the officers and crew from their vessel came ashore for an hour; their mission was to see a stitchbird. All the official party had to be carried to and from the lifeboat. The diary does not confirm whether they saw a stitchbird but does list other birds seen.

In June 1935 a wind generator was built to charge the wet battery, used for a broadcast radio. On 19 June 1940 the RMS *Niagara* was sunk by two mines to the north of Hauturu; on 26 June the diary records that oil from the wreckage was killing many penguins.

William retired in 1941 and Len took over the role of caretaker. By December 1941, electricity — presumably a 12-volt system — appears to have become a part of their daily lives and may have been providing some lighting. The generator caught fire one night but was repaired the next day. A note in one of the ranger's diaries in 1942 mentions William having saved the Crown money by making his own solder from metal toothpaste tubes and tin. In July 1942 Len spent time fixing up the Icyball refrigerator that Mr Warren had kindly given them, and Rene made ice cream. It appears that the wind generator was still in use until June 1944, when it fell apart, wrecking the battery too.

CHARLIE PARKIN, WHO HAD A background as a seaman and painter, was appointed caretaker in late 1943, and he and his wife May took up residence on 28 December. Len Hardgrave stayed on until 2 February 1944, and he and Charlie repiled the house. Charlie got the generator going, dismantled Tenetahi's longdrop, made a sledge out of a forked mānuka branch ready for when they got a horse, and erected the front porch. Charlie and May repainted inside and outside and papered all the rooms in the house. Charlie ordered a wheelbarrow with a pneumatic tyre.

In June 1945 the navy arrived, carried ashore a 300-pound (136-kilogram) case and installed two-way radio communication with Musick Point in Auckland. Charlie wrote, 'We radio Auckland every night at 8:00pm.' After the radio link to ZLD Auckland Radio was established, weather data was transmitted daily; this activity continued until the end of 1990 when all the coastal weather stations were automated.

In January 1946 Charlie dug a 12-foot-deep (3.5-metre) well on the flat near Hunter-Blair's grave. The Public Works Department launch and barge brought an engine, pump and water pipes in February 1949 and plumbers stayed to install the pipes. Once the pump system was established it probably pumped water directly to the house, as there was no water reservoir. In August 1953 Charlie hauled precast concrete slabs from the landing to a site on the hill behind the well, built the reservoir and connected it to the water system. This water supply changed many things. Materials for a flush toilet and septic tank arrived in January 1951 and the flush toilet was installed soon after. In February 1954 the washhouse was connected to the pipeline and in August a shower was installed. In May 1955 the water pipe was extended to the 'Camping Paddock', which may have been near the present bunkhouse, and a fireplace was built there.

A kerosene fridge arrived in February 1947 — with no instructions, but they got it going.

Firewood for cooking and heating received little mention in early diaries, but it was a normal part of daily life, probably involving many hours' work with a crosscut saw. The families were fortunate that there was an adequate supply of kānuka, which was good hot-burning wood. Parkin set out to mechanise his wood collecting, and he brought a sawbench and engine ashore in July 1946 for cutting firewood into lengths to fit in the stove.

*Clockwise from top left*: The Nelson family in front of the house. (*New Zealand Herald*, 25 August 1932) Len and Rene Hardgrave with William Hardgrave at right. (WMH) Charlie Parkin using the radio link to ZLD Auckland Radio. (MP) Charlie Parkin reading the thermometers in the Stevenson screen (instrument shelter) supplied by the Meteorological Service. (MP)

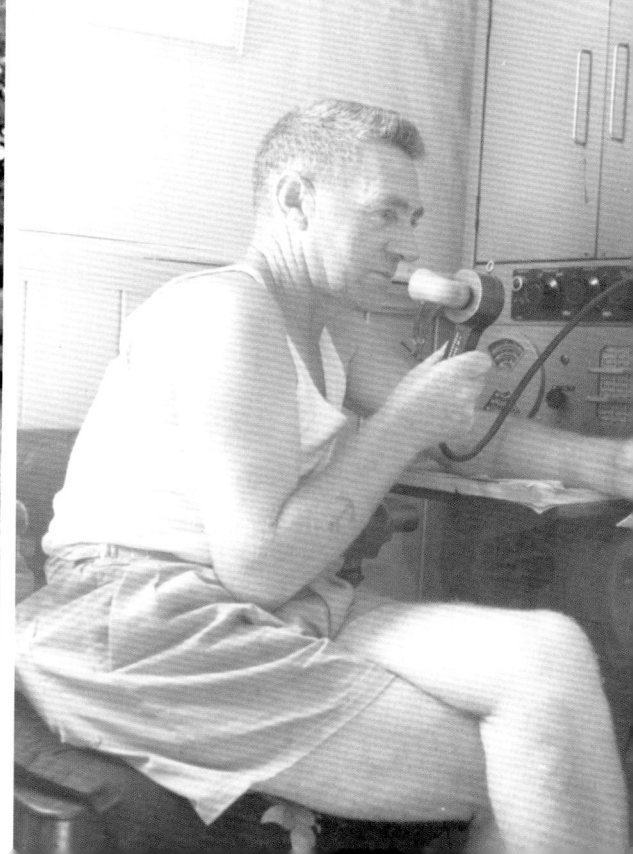

The Gravely tractor may have been on the island as early as 1946 and was certainly there when Charlie wrote in 1952 that he 'got her going and mowed rushes on the flat'. This tractor was fitted with a chainsaw for wood cutting, and he used it to haul a trailer for firewood and many other things.

At the end of January 1947 May Parkin rescued a red-crowned parakeet chick and named it Sammy. It was a very young chick with just the first covering of feathers, and it had a broken leg. May raised the orphan and it would take food from her mouth. In March 1949 Sammy brought one of his offspring down close to the house and they watched it being fed. Sammy lived on for many years and was still there when the Blanshard family arrived in 1958.

The first mention of a kākā feeding with the tūī and bellbirds at the bird feeder was in early 1950. Later that year a kākā came into the kitchen and 'got busy on a tin of dripping', and on 20 July 1953, 'Another kaka has arrived at the trough and the old one who has been here all winter is very indignant.'

Control of the island was transferred from the Tourist Department to the Department of Lands and Survey on 6 December 1951.

It is not known when Parkin started building a large new cowshed, later also used as a shearing shed. The first possible reference to it is that he was assembling the milking machine on 8 August 1953, but there are earlier references to having to hand-milk because the machine had broken. In December he was fitting a shearing machine in the cowshed to work from the milking-machine shaft.

In October 1955 Parkin began building an engine house ready for the 230-volt generator that came on the scow *Jane Gifford* in January 1956. Up to this time the generators mentioned were probably 12-volt, principally to charge radio batteries or to run a single lightbulb. It took a week to wire and install the new system. Shipping fuel for the generator then became a regular task. Floating the 44-gallon drums from the ship to shore was easy; getting a heavy drum up the beach required some effort, and then the sledge was needed to get it to the house.

From about 1947 onwards there was a marked increase in visitors interested in the natural history of Hauturu. Diaries for 1947 to 1958 record visits by groups and parties of scientists: Graham Turbott and Auckland Museum scientists; Ardmore Teachers Training College students; Ken Bigwood, photographer; Dick Sibson and King's College bird club students; Geoff Moon, ornithologist and photographer; the Wildlife Branch of the Department of Internal Affairs; Ross McKenzie and ornithologists; and the Auckland University Field Club. According to notes from the club's journal *Tane* in 1953, 'The ranger and his wife are primarily responsible for policing the sanctuary against unauthorised landing, for extermination of pests (esp. wild cats) and for clearing tracks. They also maintained a weather station and sent daily reports of weather data to Mechanics Bay Meteorological Office, Auckland. This daily link along with fortnightly mail and provisions launch from Leigh make up their main communication with the outside world.'

*Clockwise from top left*: Crosscut saw for sawing firewood. (NM) The Gravely tractor, with chainsaw attached, and Charlie Parkin. (MP) Cutting firewood into stove lengths. (MP)

The boats and boat ramp required constant attention and maintenance; the reports often mentioned parts that had to be ordered, and boats and motors that were obtained or decommissioned. Charlie Parkin's diary makes no mention of a boatshed, but one of May's photos shows a shed approximately where the shed for the *Lady Anne* was built in 1934.

Like young Robert Shakespear, Parkin built several boats during his term on Hauturu. The largest, *Tuhunga*, was about 8 metres long, designed for a motor and ketch-rigged sails. The hull and cabin were built on the island but there was no easy way to install an engine there. Charlie and May loaded their possessions onboard in December 1958, launched the engineless boat and were towed to Auckland by the *Rita W*.

Rodger Blanshard, wife Ani and their four children, David, Gina, Susi and Lisa, aged between four and 10, came from the Lighthouse Service and had previously been on Stephens Island and at Cape Egmont. They arrived on Hauturu on 11 November 1958. Their first task was to set up a school: they turned the storeroom by the back door (previously Frances Shakespear's darkroom) into a schoolroom and built school desks to furnish it.

A storm in February 1960 swept away the Blanshards' dinghy and damaged the boatshed and ramp, but they did manage to rescue the outboard motor and lifejackets. They brought a 14-foot (4-metre) Mosquito Craft boat with them when they returned from leave in April that year. They added a layer of fibreglass and some sacrificial pine runners to the hull so the boat could be easily hauled over the boulders. Their attempts to rebuild the boat ramp were thwarted by storms and the difficulty of getting piles deep into the boulders. They then built a boatshed well back from the water, about where the quarantine shed now stands.

In May 1960 Te Waikohare Stream flooded to level with the house floor and most of the flat area was under a metre of water.

In mid-May work began on building a bunkhouse. Sandy shingle was carried up from the beach for the foundations and deck. The *Rahiri* had brought building materials in March and blocks and cement came in June. Work on the bunkhouse continued through to August 1961, although visitors stayed there during that time. Electricity was connected to the bunkhouse in September. Initially it consisted of three rooms — two bunkrooms with a cold-water shower off each, and the kitchen/living area, which had a sink and a small woodburning stove. The toilet, completed in October 1962, was two cans in an outhouse; visitors emptied the cans into the sea at the end of their stay. There have been many changes to the original building, including a front bunkroom built over the original concrete deck.

In December 1961 a new patrol boat was delivered: it was coated in fibreglass, painted, and bits were added before it was taken for sea trials and declared 'excellent handling at full speed (Seagull est 8½ knots) one person'. The following year the family took charge of a 12-horsepower West Bend outboard.

The diary in March 1962 records destroying a wasp nest in the east paddock. This is the first reference to wasps, presumably German wasps (*Vespula germanica*). A further

---

*Clockwise from top*: Preparing to haul the main part of the generator up the beach. (MP) Hauling a fuel drum to the house by horse and sledge. (MP) The Gravely tractor with trailer, and Charlie Parkin driving. (MP) May and Charlie Parkin sorting the mail. (KB)

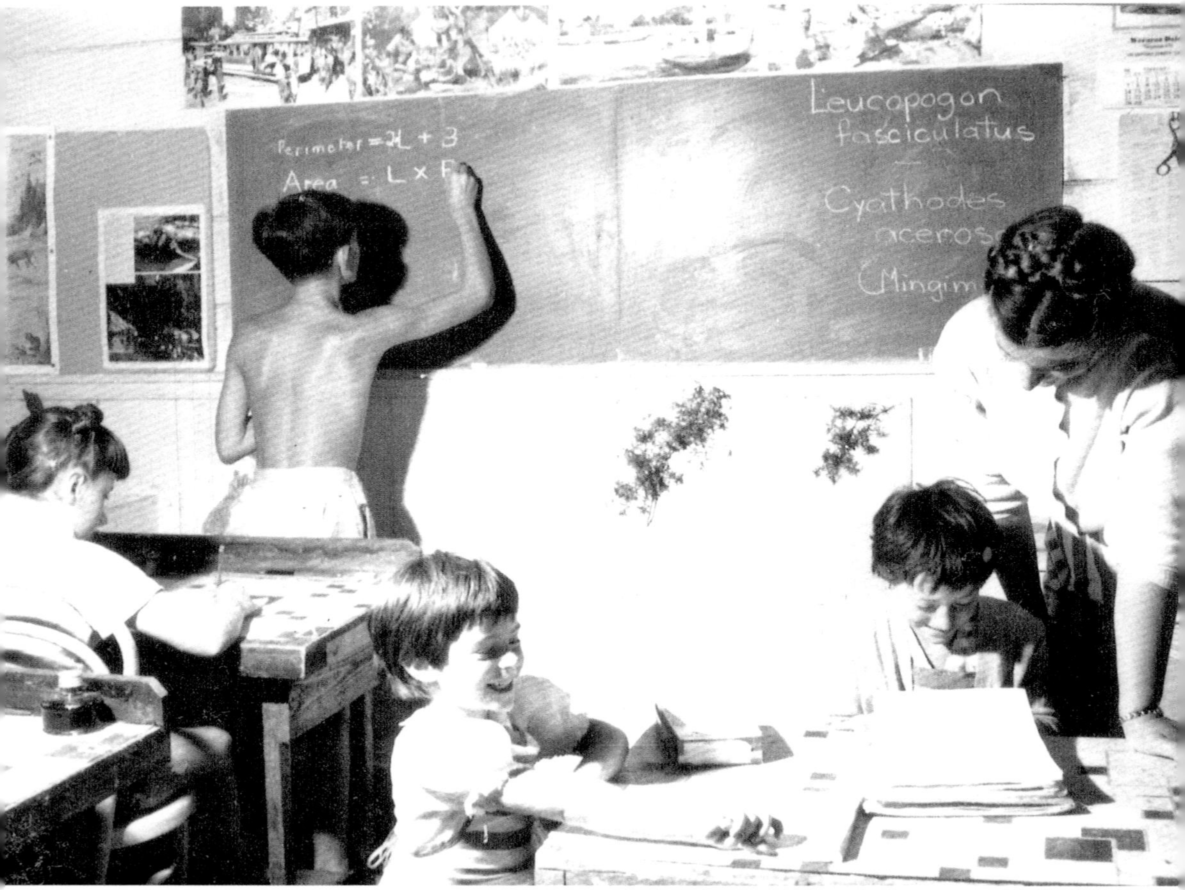
Leucopogon
fasciculatus
Cyathodes

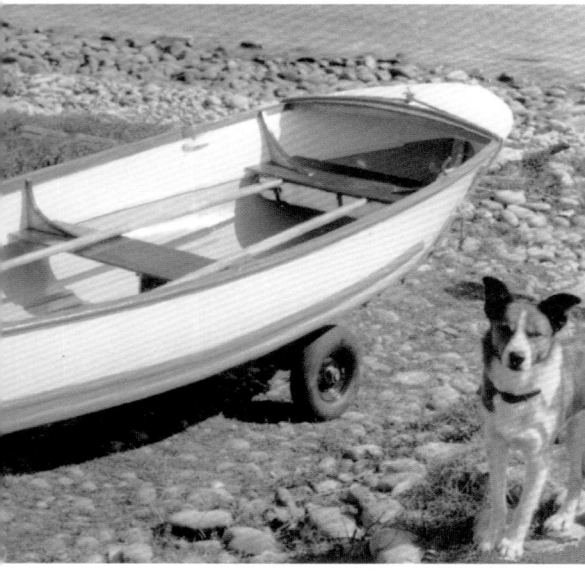

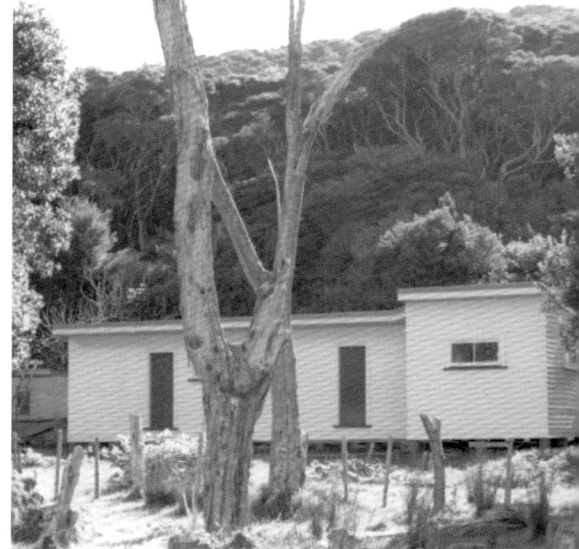

two nests were destroyed on 26 April and they were still searching for more. The timing and location on the island of these wasp nests strongly suggests a queen wasp had come ashore with the bunkhouse materials in March 1960 — despite all the timber being floated ashore from the vessel it arrived on.

In September 1962 the Blanshards demolished the old washhouse and storeroom that had been built by the Shakespears, with the help of Henare Phillips, whom they had got to know through the Lighthouse Service. He spent many holidays with them on the island, helping out. Governor-General Sir Bernard and Lady Fergusson visited the island on 7–9 April 1963.

Scientists and naturalists visited the island regularly in search of a particular bird or through some other area of interest, not always recorded in the diaries. In 1956 Hamilton appears again as secretary (director) of the DSIR, this time with his family and a number of other scientists. From their research Hamilton compiled the revised and updated DSIR Bulletin 137, 'Little Barrier Island (Hauturu)'.

**LIFE ON HAUTURU WAS VERY** different from life on the mainland. Unlike most New Zealand homes, there was no telephone. In the house they had electric light in the evenings only and a kerosene refrigerator; most of New Zealand had full-time electricity and household appliances such as refrigerators and washing machines. Theoretically the electric light was 24-hour but, as David Blanshard related, when there were visitors in the house they had to turn the main switch off at night. The auto-start generator was wonderful, but visitors would turn a light on and, before the generator had properly started and powered up the light, they would think the light wasn't working and switch it off again, and this caused havoc with starter motors and electronics. There was no television.

They had two house cows, hens and an orchard to supplement stores brought to the island at about fortnightly intervals on the Lighthouse Service vessel. Mail and store orders could go out this way too, and additional orders could be sent by telegram via ZLD Auckland Radio.

The active young Blanshard family and visitors recorded a number of unusual bird sightings. On 30 November 1962 the diary records 'found *Apt. Haastii* egg & nest Waipawa Stm'. On 9 January 1964 Gina Blanshard and Dot Madell 'clearly sighted large spotted grey kiwi at side of summit track 600 ft alt 1045am (*Apt Haastii*)'. On 19 August 1964 Rodger stalked and shot one exotic parrot near West Landing; the description matched that of an eastern rosella (*Platycercus eximius*). On 9 January 1967 a pair of falcons (bush hawks) were seen near a slip beside Whēkau. Susi noted that a white kiwi lived on the knoll by Hunter-Blair's grave and could be guaranteed to be in the vicinity at night, where visitors could view it. In 1967 a baby white kiwi was sighted. In January 1965 a morepork fledgling was found along the flat on the edge of the forest. The family named him Wol after Winnie

*Clockwise from top*: Inside the schoolroom. David, at the blackboard, recalled that each day his mother would put a new species name on the board for them to learn. From left: Gina, David, Lisa, Susi and their mother Ani. (RB) The new bunkhouse, early 1961, with the front deck yet to be added. (RB) The new patrol boat. (RB)

the Pooh's friend Owl. The morepork was still too young to fly and spent some weeks roosting behind the bathroom door of the ranger's house.

In May–June 1964 there was the first translocation of birds off Hauturu: six kiwi were taken to Pōnui Island in the Hauraki Gulf where, along with a few birds from Northland, they have created a thriving kiwi population.

The Hauraki Gulf Maritime Park was created in 1967, and Hauturu was included. There were no notable changes in management, but the staff were now part of an expanding group of island management specialists, and they got a radio connection to fishing boats and other shore stations.

Some years, Lisa, Susi and David attended secondary schools in Auckland. Susi commented that it was quite an adjustment adapting to life in the city. She was used to conversing with adults and making her own decisions and her peers, in comparison, seemed shallow and immature. On the other hand, she felt she was very naive about life in the city, had difficulty managing money and was too honest and trusting.

In 1967 a generator shed was built on the flat out towards the main landing to replace the shed closer to the house. The original generator brought ashore by Charlie Parkin was installed there and put to use in October.

The Blanshard family left the island on 2 May 1968 when Rodger was promoted to a more senior position on Kawau Island. They were replaced by the Wisnesky family of John, Betty and their sons Danny and Bunny. John came from a career of boat work and, more recently, as a Department of Lands and Survey ranger in the Marlborough Sounds. He was the first Hauturu appointee to have reserves management experience. His job title was ranger.

The Wildlife Service started a cat eradication project in June 1968. Several cats were cage-trapped, dosed with feline enteritis and released; this reduced the cat population by an estimated 80 per cent. Cat-trapping and track-cutting for access to other parts of the island continued. Tragically, on 5 July 1968, while track-cutting, John Drew, a Wildlife trainee, fell down a cliff in the forest and died. The cat eradication project was halted.

A new boatshed was built, this time out of concrete blocks, and a new boat ramp was installed. The ramp piling was mostly kānuka. Just behind the boatshed a large fuel tank was installed; in future, diesel supplies were pumped ashore to fill the tank from the service vessel.

A two-storey house was designed and built, beginning in 1972 and being completed in March 1974, and the old house was demolished and removed.

Dave Smith came to Hauturu from the Lighthouse Service in June 1975 with his wife Margaret and their three daughters, Pam, Lorell and Lynette. Sadly, in November of that year Dave and family returned to the island from a short break away to find the relieving ranger, John Murison, dead beside the tractor at the landing.

By now, life on the island had become more like that on the mainland. Electricity

---

*Clockwise from top left*: The shed built by Rodger Blanshard in 1967 as a generator shed is now used to store weed sprays and sits between the new biosecurity shed on the left and generator shed on the right. (CRV) The fuel tank that was installed behind the boatshed; John Wisnesky is on the left. (BW) The new house being built. (BW)

Island life 55

was still reliant on the sometimes temperamental generator, but the Smiths had an electric fridge, freezer (a specially made brine-tank model) and washing machine. Communication, still by radio, had expanded to include the Hauraki Gulf Maritime Park stations, many boats and shore stations.

There was further action to reduce cat numbers in April 1976 because the population had reached a similar density to that seen before the introduction of feline enteritis. Wildlife Service scientist Mike Imber had determined that predation levels by cats on black petrel (*Procellaria parkinsoni*) were such that no chicks were fledging. The last cat was trapped on 23 June 1980 — a feat many people, scientists included, had thought impossible (see Chapter 4).

In mid-August 1977 the Smith family departed and the Dobbins family arrived. Alex Dobbins came from a background of park management at Trounson Kauri Park in Northland. He was accompanied by his wife Mike, daughter Toni and son Mark (known as Phred). Early in his time on the island Alex worked to build more fences to keep the farm stock off the boulder bank and away from the west and east ends of the flat. The next task was to build a generator shed to house two generators. The low-profile building was completed late in 1978 but was prone to flooding.

Once it was established that Hauturu was free of cats, 22 kākāpō (*Strigops habroptilus*) were transferred from Rakiura/Stewart Island. The kākāpō were kept on the island for 16 years. Sir David Attenborough visited while they were there and was filmed with Richard Henry, the last kākāpō from Fiordland, for a BBC documentary. The kākāpō were taken off the island in 1998/99 in preparation for the rat eradication programme.

With the bunkhouse now in full-time use by kākāpō workers, as well as frequent visitors, the cables from the generator shed to the bunkhouse were inadequate. There was no money for new cables, so the generator that Charlie Parkin had brought ashore in January 1956 was moved to the bunkhouse in late 1979. Later, a building was erected close behind the bunkhouse to accommodate the additional staff involved in the kākāpō supplementary feeding programme. This is now the second ranger's home.

The absence of cats meant birds could once more be translocated to Hauturu. The first release of kōkako (*Callaeas wilsoni*), from Bay of Plenty forests, was in 1981. Saddlebacks were transferred to Hauturu from Repanga/Cuvier Island, and from Lady Alice and Whatupuke islands in the Hen and Chickens group, between 1984 and 1987. Black petrel fledglings were transferred from Aotea to Hauturu over several years from 1987 (see Appendix 2).

Mike Dobbins rescued a kererū chick and hand-reared it. After it was released into the wide world it would wait on the rail near the back door for food, or for the door to be opened so that it could fly through into the kitchen. 'Pidge' lived to 21 years of age and saw quite a few ranger families come and go.

Feeding the birds in the garden was a daily task for the ranger, and attracted large numbers of bellbirds, tūī and kākā. Island visitors enjoyed watching the birds, and

*Clockwise from top left*: Mike Imber with a black petrel, 1972. (DMc) David Attenborough on Hauturu with Richard Henry, the last kākāpō from Fiordland. (AxD) A kākāpō at a feeding station. (DOC) A black petrel fledgling being transferred in a cat box. (TG)

Island life 57

occasionally television crews came to film the activity. The kākā were fed weevily dates supplied by an Auckland supermarket and became quite fearless. They visited the bunkhouse and begged for food, and soon learned to enter the bunkhouse in search of more food, so all the doors and windows had to be closed when nobody was home.

The first mention of Ratbag the kākā was from a volunteer kākāpō feeder in the 1990s when kākāpō were being managed on the island. Apparently, Ratbag routinely sorted out his favourite foods from the kākāpō feeding mix and would follow along after the fresh food was placed at the feeding stations and have a feast. He became a frequent visitor to the bunkhouse, ready to welcome unsuspecting visitors by stealing food just as they were putting it in their mouth. On one occasion he was seen to extract the batteries from the back of a camera, having undone the case and camera back. He became such a nuisance that the ranger, Will Scarlett, built a special table with a cupboard for the food. On one occasion Ratbag himself was placed in the cupboard and the catch was closed, but it wasn't long before he had figured out the catch and was out again.

Over the following years several milestones were achieved: in 1986 the old boatshed was replaced; the Department of Conservation was formed in 1987 and the Hauraki Gulf Maritime Park became part of that department (although this did not change the way Hauturu was managed); in September 1988, in recognition of Bill Hamilton's many years of work, the Summit Track was renamed the Hamilton Track; the arrival of the first Asian paper wasp (*Polistes chinensis*) was recorded in February 1989; the main water reservoir, built by Charlie Parkin, collapsed and a big new tank was built.

In December 1989 the last weather report was sent by radio from Hauturu to the Meteorological Service, and this effectively ended contact with ZLD Auckland Radio at Musick Point. Auckland Radio itself ceased operation on 30 September 1993. Monthly weather reports were still posted to the Met Service until February 1994.

Alex Dobbins and his family left Hauturu on 26 November 1990 and were replaced by Chris Smuts-Kennedy and his wife Robyn. The first telephone arrived on the island very soon after, on 6 January 1991. Robyn described it as 'a brick which usually did not work at high tide'. A few years later, it was Chris's suggestion to visitors to the island that was the beginning of the Little Barrier Island (Hauturu) Supporters' Trust. Over the years the trust has provided funds for weed eradication and for many other projects, including research, as well as offering the opportunity for people to visit the island through its Working Weekend programme.

In February 1991 the tuatara (*Sphenodon punctatus*) captive breeding programme began. Some people believed there were no tuatara left on Hauturu, but the first field survey located two males and two females, which were housed in the aviary. Four more tuatara were found and added later. The first part of the 'tuatarium' was built in 1991, and later additions almost doubled the area to house the eight adult tuatara and their progeny prior to release. Managing the tuatara became an important part of the rangers' work.

The tuatara search turned up unexpected species, previously unknown on the island.

*Clockwise from top*: Phred Dobbins watching Pidge being fed. (AxD) The bird feeder in the rangers' garden. (CRV) Filming at the bird feeder: Mike Dobbins filling the sugarwater trough, Bill Hamilton watching. (AxD) Alex Dobbins with kākā. (CRV)

A chevron skink (*Oligosoma homalonotum*) was found at Te Hue in February 1991 and a striped skink (*O. striatum*) in the Tirikakawa Stream on 20 December 1994. This brought the total number of known lizard species to 13.

The generator that Charlie Parkin had brought ashore, which had been used in three different locations on the island, finally went back to the mainland in April 1991. By December 1992 large batteries had been set up in the generator shed and electronic equipment installed so that the generators charged the batteries and 240-volt electricity was available at all hours, with less frequent running of the generators.

Bellbirds, tūī and kākā were no longer fed from 1995. Many of the kākā had become a problem as house invaders, and some disease was evident that could be attributed to the feeding. There had been many years of enjoyment sharing these spaces with wild birds, but the need to protect every shoelace at the door and close all doors and windows at all times was offputting.

Chris Smuts-Kennedy planted island-grown seedlings and cuttings around the flats in small fenced areas from east of the bunkhouse to the West Landing, and he planted four kahikatea behind the grave. Most of the farm stock was shipped off before he left the island in October 1995, and the remaining stock was gone and fences removed by February 1997, along with Parkin's cowshed and Blanshard's fowlhouse.

Mark Glithrow filled in for nine months from the end of 1995 until Irene Petrove took over in 1996, the first woman ranger on Hauturu. She was accompanied by her partner Sid Marsh and her school-age daughter Natasha.

In early 1997 there was a concerted effort to remove all exotic species (other than vegetables) from the rangers' garden. The ranger's role at this stage was not very different from that of the early caretakers: managing around 350 visitors per year, issuing permits, giving briefings, maintaining or restoring island conservation values, patrolling the coast, educating the public and home-schooling their children. The Leigh fishermen provided another set of eyes and ears patrolling the island, as well as making occasional deliveries. The visits organised by the Supporters' Trust became regular events, and local iwi became more involved with the island. And internet communication had arrived.

When Irene Petrove left in 2001 she was replaced by Will Scarlett, assisted over his three-year tenure by a second ranger — first Shane McInnes and then Marie Francis — responsible for managing weed control on the island.

In 2004 Peter Barrow arrived with his wife Helen Dodson and their three boys, Hamish, Ben and Tim, aged from six to 10. Pete came from a background of farming and carpentry and Helen had worked with DOC in Te Anau. The ocean was an entirely new experience for the family. They were assisted by Hugh Gardiner, fresh back from Raoul Island. For the first time the rangers and their families were welcomed onto the Leigh marae by elders of Ngāti Manuhiri; Rahui Te Kiri's last surviving grandchild, 'Girlie', was present on that occasion.

In 2004 the rat eradication programme was carried out to remove kiore (*Rattus*

*Clockwise from top left*: The new boatshed and ramp. (CRV) Outside part of the tuatarium: volunteers take in bins of leaf litter that will provide insect life for the tuatara. (CRV) The water tank installed by Charlie Parkin collapsed. (AxD) The new tank being built. (AxD) Bill Hamilton at the new Hamilton Track sign, September 1988. (AxD)

Hamilton Track →

*exulans*) from the island, and in 2006 it was declared rat-free (see Chapter 4).

A replacement island boat arrived in 2005, and the boat ramp received an upgrade. Shane McInnes returned as ranger in 2007 for four years, assisted by Liz Whitwell in charge of the weed work, which was ramped up during this period.

Solar power was installed on the island in 2008, with the two generators still there as back-up. The battery bank is large enough to ensure that 240-volt power is available at all times, and the generators are rarely needed. The water pump, located near Hunter-Blair's grave, has its own solar power supply.

McInnes commented that being ranger on Hauturu had been a dream of his since his teens. It far exceeded his expectations, and was more a lifestyle than a job. Highs for him were seeing the beginning of ecosystem recovery post kiore eradication, and managing the tuatara programme, with the subsequent release of young tuatara back into the wild. There were multiple translocations of species from Hauturu to the burgeoning conservation restoration projects around New Zealand during his tenure (see Appendix 3).

In 2011 McInnes handed over the reins to another family — Richard Walle and his wife Leigh Joyce and their children, Mahina and Liam (eight and six respectively), who came from a DOC role on Te Pākeka/Maud Island in the Marlborough Sounds. Walle had come to New Zealand from Germany as a journeyman carpenter years earlier, and both he and Joyce had many years' experience working for DOC around New Zealand. Joyce had studied kākāpō home range and habitat for her PhD thesis, and her indepth knowledge of kākāpō meant she was ideal to assist when the birds were brought back to Hauturu in 2012.

During Leigh and Richard's time on the island biosecurity and quarantine measures for visitors have become much stricter. This follows the arrival in New Zealand of plant diseases such as kauri dieback (*Phytophthora agathidicida*) and myrtle rust (*Austropuccinia psidii*) and animal pests such as the Argentine ant (*Linepithema humile*) and plague skink (*Lampropholis delicata*, introduced from Australia) which, if they made their way to the island, could have a devastating effect on the ecosystem. Sentry stations around the island monitor for any incursions and require frequent checking. Success with the reduction in pest plant species on the island also required stricter quarantine controls.

From 2011 to 2013 Nichollette Brown worked with Walle as ranger and managed the weed programme. She has a master's in biological sciences and came to the island from a role as environmental manager to a large construction company; a stint on Raoul Island and some volunteer work with kākāpō sparked her interest in conservation and island life.

Brown was followed by Pete Mitchell and his wife Cathy in February 2014. Mitchell came from a varied background including, most recently, work on Matakohe/Limestone Island (Whangārei) and Bream Head. Weed work was an important part of his job and took him to all parts of the island. When he left the island in December 2016 he commented: 'Life on the island is never dull, the ecosystems and wildlife are outstanding and every day

---

*Clockwise from top left*: Irene Petrove and daughter Natasha. (IP) This array of solar panels is sufficient for all the island's needs. (CRV) Digging deep to place new piles for the boat ramp. (RW) Shane McInnes, with young tuatara in hand, briefing a team of Little Barrier Island (Hauturu) Supporters' Trust volunteers at the tuatarium. (LW) Robyn and Chris Smuts-Kennedy planting a kahikatea. (CSK)

out in the bush or on the boat has been about as far from a real job as you can get.'

Andre de Graaf took over from Pete in July 2017. After managing the weed programme in the spring, Andre and his partner Polly were seconded to help on other Hauraki Gulf islands, and they left Hauturu in June 2018. Chippy Wood, a long-time DOC employee from the West Coast of the South Island, joined the island team in August 2018.

The success of the tuatara programme during the 20-plus years it operated resulted in the release of more than 290 young tuatara back into the wild. During 2017–18 the breeding programme was wound down and some of the old tuatara enclosures were removed.

Life on the island today is much like life on a rural property on the mainland, with access to internet, cellphones and television. Schooling is, by choice, home-schooling or Te Kura (formerly the New Zealand Correspondence School). DOC vessels make the service/supply run every two to three weeks, and the vegetable garden supplements what is brought by boat. Gas is used for cooking and water heating, with solar assistance, and solar panels provide most of the power. This has meant items such as microwaves and electric kettles can be in everyday use. A 6-metre vessel lives on its rail and trolley system, with a remote-controlled winch. It is much easier now for the ranger and their family to come and go from the island. Rangers are expected to attend training and meetings on the mainland; in the past this would not have been considered. The weather is still the controlling factor in planning trips to and from the island, but with sophisticated electronic monitoring devices forecasts are more reliable.

Visitor numbers still sit at around 350 a year, though there are now more visits from scientists and volunteers; others without a research or a volunteer purpose on the island may only make day visits. All visitors require a permit to land and must go through a stringent biosecurity check before travelling to the island.

THE COMMITMENT OF THE RANGERS and their families over the past 120-plus years has ensured that Hauturu has been preserved as an exemplar of what northern New Zealand's flora and fauna would once have been. Through them and the scientific and conservation work carried out on the island, it is also an example of what can be achieved when many people work together to preserve a special place for so many of New Zealand's extraordinary species.

*Clockwise from top*: Discussion before kākāpō are released on the island: Leigh Joyce, Richard Walle and Lyn Wade face the camera. (RW) Pete Barrow bait loading for rat eradication, 2004. (RG) Pete Mitchell (right) found another tuatara while out doing weed work. (PB) Nichollette Brown making a reptile artificial cover object. (LW)

| Hauturu caretakers and rangers | |
|---|---|
| Henry Wright, Acting Crown Lands Ranger | 19/12/1892–18/2/1893 |
| Charles Robinson, Agriculture Dept. | 19/2/1893–25/03/1896 |
| Two members of the Permanent Militia | January 1896–February 1897 |
| Robert H. A. Shakespear | 19/1/1897–30/1/1910 |
| Robert H. R. Shakespear | 30/1/1910–30/3/1910 |
| Robert Hunter-Blair | 4/5/1910–died 22/9/1910 |
| Thomas P. Firman | 17/11/1910–30/5/1911 |
| Robert Nelson | 30/5/1911–4/5/1922 |
| William Cleaver | 4/5/1922–29/8/1923 |
| Robert Nelson | 29/8/1923–16/11/1932 |
| William H. Hardgrave | 16/11/1932–30/4/1941 |
| Len Hardgrave | 1/5/1941–2/2/1944 |
| Charlie Parkin | 28/12/1943–3/12/1958 |
| Rodger Blanshard | 11/11/1958–2/5/1968 |
| John Wisnesky | 2/5/1968–31/6/1975 |
| Dave Smith | 31/6/1975–16/8/1977 |
| Alex Dobbins | 16/8/1977–26/11/1990 |
| Chris Smuts-Kennedy | 30/11/1990–23/10/1995 |
| Mark Glithrow | 23/10/1995–31/7/1996 |
| Irene Petrove | 31/7/1996–20/1/2001 |
| **Two-ranger roles commenced** | |
| Will Scarlett | January 2001–January 2004 |
| Shane McInnes | July 2001–July 2002 |
| Marie Francis | July 2002–February 2003 |
| Peter Barrow | January 2004–January 2007 |
| Hugh Gardiner | January 2004–May 2005 |
| Shane McInnes | August 2005–January 2007 |
| Shane McInnes | January 2007–May 2011 |
| Liz Whitwell | January 2007–January 2011 |
| Richard Walle | May 2011– |
| Nichollette Brown | May 2011–September 2013 |
| Peter Mitchell | February 2014–December 2016 |
| Andre de Graaf | July 2017–June 2018 |
| Chippy Wood | August 2018– |

The Hauturu rangers' vessel *Hine Moana* returning to the launching ramp. Richard Walle, ranger, is at the helm. (NF)

# Chapter four
# Restoration of Hauturu

— RICHARD GRIFFITHS AND DICK VEITCH

**The biological riches of Te Hauturu-o-Toi/Little Barrier Island have long been valued by Māori and European alike. Indeed, it was these values that provided the impetus for the island's protection as a nature reserve in 1895.**

At the time, protecting offshore islands as refugia for New Zealand's biodiversity was the logical but also the only viable strategy in the face of rapidly declining native species populations on the mainland. In 1895 existing and future threats to Hauturu and other offshore islands were poorly understood. Individuals who championed nature reserve status for Hauturu undoubtedly recognised the impacts of logging and burning, but with the information available to them they could not have grasped the ongoing influence of kiore (Pacific rats, *Rattus exulans*) or fully appreciated the detrimental effects of cats (*Felis catus*). Nor could they have envisaged future hazards such as a changing climate or kauri dieback (*Phytophthera agathidicida*) that have only recently appeared. Management of Hauturu as a nature reserve has evolved over time in step with our growing awareness.

Here we describe the past and current threats to the biodiversity of the island, the management actions undertaken to address them, and the conservation outcomes those actions have generated. The taonga that is Hauturu today would not exist without the careful management that has taken place. But the biodiversity of Hauturu remains vulnerable and no more resilient to many of the pressures present on the nearby mainland than it was back in 1895. Ongoing care and attention are vital if we are to maintain the island's natural heritage into the future.

## Past and existing threats to Hauturu's biodiversity

The first invasive species to be introduced to Hauturu was likely the kiore, although dogs (kurī, *Canis familiaris*) may also have accompanied the first visitors to the island. Kiore travelled across the Pacific and to New Zealand with early Polynesian voyagers and are thought to have arrived on Hauturu early in the period of New Zealand's settlement, around 1200 CE. This early arrival on Hauturu is suggested by the absence of tree species such as *Streblus banksii* and *Nestegis apetala*, large land snails and slugs on the island — species that are found on New Zealand's northern islands that have never had rats.[1]

Within three years of their introduction to Hauturu, kiore would have overshot the island's carrying capacity and reached population densities potentially greater than 100 rats per hectare. The island would literally have been crawling with rats, with dire

The coastal cliffs, talus forest and boulder beach of Hauturu's southern coastline. (OB) *Previous*: Kauri forest, Hauturu. (AA)

consequences for biodiversity. Larger flightless and ground-dwelling invertebrates would have been the first to disappear, followed soon after by the smaller of the island's ground-nesting seabirds as fledging rates fell to zero. Recruitment of young tuatara (*Sphenodon punctatus*) would have ground to a halt and by the time of their rediscovery in 1991, the island's population was functionally extinct.[2] We know now that Hauturu's 14 surviving reptile species were heavily suppressed and it is possible that three species were lost from the island. Populations of larger ground-nesting seabirds such as Cook's petrel (*Pterodroma cookii*) and grey-faced petrel (*P. gouldi*) were also severely affected by the rat infestation.[3]

The effects of kiore on Hauturu's terrestrial bird populations are harder to discern. Inevitably they did have an impact, but with no subfossil record for the island it is difficult to estimate how extensive this was. We assume that species such as the koreke (New Zealand quail, *Coturnix novaezelandiae*), piopio (*Turnagra tanagra*), huia (*Heteralocha acutirostris*) and others were lost from the island, as occurred on the main islands of New Zealand. Spotless crake (*Porzana tabuensis*) were absent in the earliest surveys undertaken by Europeans, and snipe (*Coenocorypha barrierensis*) were extremely rare, likely due to the presence of kiore.[4]

Studies on Hauturu and other northern offshore islands have shown that through seed and seedling predation, kiore suppressed the recruitment of at least 11 species of coastal tree.[5] These impacts took time to manifest but it is clear that over hundreds of years the species composition of Hauturu's forest changed. We can only surmise the true effect that kiore had on Hauturu's plant life, but some species were almost certainly extirpated, while the distribution and abundance of others, for example parapara (*Pisonia brunoniana*), were greatly reduced.

Many of the island's flightless and ground-dwelling species would also have been vulnerable to dogs. Seasonality of seabird resources may have limited the impacts across the island, but dogs likely ranged over much of Hauturu and, closer to the island's human population at Te Maraeroa, their impacts could have been more persistent. Dogs may have contributed to the disappearance of some of the island's ground-nesting seabirds and flightless birds, and predation by dogs may explain why kiwi were not seen by early Europeans visiting Hauturu. No free-ranging dogs remained on Hauturu after 1896.

Te Maraeroa, because of its gentle topography and fertile soil, naturally became the centre for Māori settlement and gardening. The flats and adjacent slopes of Te Maraeroa were the areas most modified by humans, although uncontrolled fires may have affected vegetation beyond this, particularly on the southern, western and northern planezes of the island, which now appear as kānuka/mānuka forest on vegetation maps (see Chapter 6.5). Modification of the vegetation of Te Maraeroa and the island's southwest corner continued after the arrival of Europeans in the 1860s who visited the island to collect firewood, probably kānuka (*Kunzea robusta*), and kauri gum. Logging of kauri began sometime in 1870. It ceased in 1892 due to government intervention, but not before many trees had been felled

---

*Above*: Kiore (Pacific rat, *Rattus exulans*). (CRV) *Below*: North Island snipe (*Coenocorypha barrierensis*). (Artwork by Paul Martinson, Te Papa Tongarewa, 2006-0010-1/53)

in the Tirikakawa and Awaroa valleys and hauled to the coast by oxen. Trees were also felled at the top of the cliffs south of Hingaia and probably pushed over the cliffs to the sea.

The introduction of cattle and sheep to the southwest corner of the island sometime after 1860 had an impact on the vegetation on Hauturu. Livestock ranged freely on the flats of Te Maraeroa and across adjacent slopes and valleys, removing understorey vegetation and suppressing regeneration. In 1895 there were said to be 1000 sheep and 30 cattle grazing on the island[6] and, not surprisingly, by 1897 when Robert Shakespear arrived the ridges on the southwest side of Hauturu had largely been reduced to low mānuka/kānuka scrub and scattered patches of grass. The cattle were removed in 1896. Caretakers and their families brought smaller numbers of sheep and cattle to the island and these had access to the forest surrounding Te Maraeroa until William Hardgrave built a fence between the flat and the hill in 1933.

Domestic cats were first brought to New Zealand in 1769,[7] but it was not until around 1867 that they arrived on Hauturu. By 1880 they were well established across the island. This led to the disappearance of tīeke/North Island saddleback (*Philesturnus rufusater*) and, perhaps assisted by kiore, the extirpation of grey-faced petrel. Cats also inflicted heavy losses on Cook's petrel and black petrel chicks (*Procellaria parkinsoni*) and, no doubt, on the island's tuatara and other reptile populations. Past reports from a variety of sources blamed cats for the disappearance of diving petrels (*Pelecanoides urinatrix*), little shearwaters (*Puffinus assimilis*) and white-faced storm petrels (*Pelagodroma marina*), but these species were absent by the time of the earliest biological surveys, or were never present on the island.

Pest plants have also made their presence felt on Hauturu in a slower and less obvious manner. Hamilton and Atkinson identified 92 naturalised plant species on Hauturu,[8] and at least another 110 have arrived and been recorded since then. Fifty-seven of the species found by Hamilton and Atkinson were at Te Maraeroa on the grassy flats or in the rangers' garden; the remaining 33 were located in open coastal communities. But whereas Hamilton found no exotic plants within closed forest communities in 1961, invasive species have since increasingly been found elsewhere on the island, including in forested habitats.

Many of the weed species present on the island were deliberately introduced as garden plants and have since spread from their initial location. Others were carried by sea, wind or birds, or made their way across on the footwear or clothing of visitors to the island. Several of the weed species that have arrived, including some that are now well established, are considered to be 'transformer' species: if left unchecked they will significantly alter ecosystems and change the composition of forest and plant communities. Climbing asparagus (*Asparagus scandens*) is one example — it can germinate and establish even in the low light of Hauturu's forests and can strangle and smother native vegetation. Two other transformers, pampas (*Cortaderia selloana* and *C. jubata*) and Mexican devil (*Ageratina adenophora*), have spread along much of the island's coastline. These species are better at dispersing and are faster growing than native species, allowing them to dominate dynamic ecosystems such as the coastal cliffs

*Above*: The south coast of Te Maraeroa, 1893. (ATL, 020519) *Below*: The same view in 2019. (CRV)

of Hauturu. Within the last 20 years Mexican daisy (*Erigeron karvinskianus*) and mothplant (*Araujia sericifera*) have arrived on the island, and although their current distribution remains limited, these species pose a significant threat.

At least four species of wasps have established on the island, and no doubt many other invertebrates (see Chapter 6.1).

## Management of the threats to Hauturu's biodiversity

Gazetting of Hauturu as a nature reserve in 1895 was the first in a series of management actions that continue to this day, turning the tide on centuries of modification on the island. Its status as a nature reserve heralded a new era: it put a stop to further logging of kauri, led to the removal of the bulk of the cattle and sheep, and very likely averted the introduction of other invasive vertebrates such as ship rats (*Rattus rattus*) or mice (*Mus musculus*). Biodiversity became paramount. However, not all impacts were halted or alleviated, and the organisations and individuals tasked with the island's management have had their hands full preserving Hauturu's natural values.

Initial management efforts dealt with most obvious issues and were simple and straightforward. The farm stock run by island caretakers at Te Maraeroa was excluded from accessing surrounding forests in 1933 and, eventually, from the boulder bank in 1978. In 1995 all sheep and most of the cattle were removed; the last cattle left the island in 1997 and all fences were subsequently removed. Some efforts to rid the island of pests were misguided, however. In 1915 Robert Nelson considered blackbirds (*Turdus merula*) to be a pest and 'shot a good many'. In 1922 William Cleaver reported owls (*Ninox novaeseelandiae*) killing birds and shot at least one before he realised that they also killed rats. And in 1923 Robert Nelson was granted a permit to shoot hawks (*Circus approximans*): he got one.

Other interventions required advances in understanding and methods. It took some time to realise that cats were a major threat to the wildlife on Hauturu, although the caretakers regularly killed them. The Wildlife Service began work to remove cats from the island in 1968. In addition to localised trapping, the disease feline enteritis was released, leading to the first major reduction in numbers on Hauturu — approximately 80 per cent of cats on the island were killed. This provided some relief to the island's black petrel and Cook's petrel populations, but within a few years predation had returned to former levels. Poisoning was undertaken annually across the summit area for several years to counter these impacts.

The operation to eradicate cats began in June 1976.[9] Wildlife Service and Hauraki Gulf Maritime Park (HGMP) staff agreed on a joint approach: HGMP would establish the infrastructure to support the project and the Wildlife Service would undertake the eradication itself. Twenty new tracks were cut that traversed every major ridge across the island, creating a 67-kilometre track network that included the pre-existing Hamilton, Thumb and Valley tracks. Huts were built at Te Hue, Pōhutukawa Flat and East Cape as bases for trapping in addition to the bunkhouse, and the island was divided into four zones.

Cat hunting was undertaken by two teams operating on a two-week-on/two-week-off schedule for the first two years and on a monthly schedule for the third year. The work drew on various government work schemes, along with a considerable number of volunteers and wildlife trainees, and a few contractors — 139 people in all. Project members used several methods of catching and killing cats, some proven and some experimental; leg-hold traps with a variety of baits were the most effective. As trappers walked each track checking traps they noted the location of cat sign. Over time a pattern emerged that revealed the location of surviving cats and, as each of these individuals was removed, the cat population steadily declined. The last cat was trapped on 23 June 1980, ending what had been a Herculean effort and, for its time, one of the largest and most challenging cat eradications to be completed globally. It was the first conservation intervention on Hauturu to affect the ecology of the entire island.

As with cats, the ongoing impacts of kiore were long underestimated and, until the development of technology allowing rodent populations to be targeted on an island-wide scale, there was little that could be done. There were localised efforts to address their impacts. Caretakers trapped and poisoned rats around buildings, and extensive trapping was carried out and bait stations were maintained to protect kākāpō nest sites and feeder stations. But these control measures were inadequate and did little to halt the predation of kākāpō eggs and chicks or the decline in species such as tuatara, other reptiles and Cook's petrel.

The removal of kiore from Hauturu was first seriously contemplated in the 1990s as methods of aerial bait application were proven on increasingly large islands. The utility of anticoagulants for rodent eradication had been discovered as early as the 1960s on islands such as Maria in the Hauraki Gulf,[10] but it wasn't until the development of GPS navigation systems and helicopter bait-spreading technology that larger islands such as Hauturu could be considered for pest eradication. The first rat eradication to rely on aerial baiting techniques in New Zealand was on Moutohora in the Bay of Plenty in 1986,[11] and GPS was used for the first time on Kāpiti Island in 1996.[12]

When Department of Conservation (DOC) staff began consultation with stakeholders, not everyone welcomed the idea of removing kiore from Hauturu. The Ngāti Wai Trust Board, which represented the interests of tangata whenua, sought an agreement for co-management of the island with DOC, and a commitment that kiore would not be eliminated from within their rohe. Ngāti Wai asked for these issues to be addressed before discussions could commence on management actions such as the removal of kiore. In negotiation between the two parties it became apparent that they held irreconcilable viewpoints on the long-term vision for Hauturu. Ngāti Wai were adamant that the island be returned to its pre-European state and that kiore remain on the island. This put the department in a quandary, pitting its legal responsibility to uphold the principles of the Treaty of Waitangi against its legislative mandate to remove kiore as an exotic species and to protect Hauturu's native ecosystems under the Reserves Act 1977. No headway was made over a period of four years despite many meetings, much colourful and at times terse dialogue and strong characters on both sides. Conscious of the continuing decline in the island's Cook's petrel population, and that species such as tuatara and chevron skink could be on the brink of disappearance, and with the emergence of support from local hapū Ngāti Manuhiri and Ngāti Rehua, DOC made a decision to proceed with kiore eradication.

Planning for rat eradication finally began in 2002. Securing consents for the project to proceed required a formal hearing in front of four independent commissioners appointed by the Auckland regional and city councils. DOC presented its justification for proceeding with the project, and others involved gave supporting and opposing arguments. Consents allowing the project to proceed were eventually approved, but the department immediately lodged an appeal because a consent condition required it to translocate kiore from Hauturu and establish a population elsewhere, effectively binding a third party. A second appeal was lodged by environmental lobby group Friends of the Earth, who were concerned about risks to the environment and non-target species. Both appeals were resolved through mitigation and an Environment Court hearing was not required.

Operations on other islands and trials conducted on Hauturu suggested that few native species would be at risk. Nevertheless, potential impacts on non-target species considered vulnerable to primary or secondary poisoning were a focus for project planning. Some of the mitigation measures to minimise the impacts included wrapping the tuatara enclosure in frostcloth to shield it from the bait drop, catching the island's brown teal and relocating them to Pūkaha Mount Bruce National Wildlife Centre and placing Ratbag the kākā, who was forever getting into people's food, in an aviary on the island.

Another considerable challenge for the operation was the transportation of 55 tonnes of rodent bait and 10,000 litres of Jet A-1 helicopter fuel to the island, along with a 12-tonne excavator that would be used to efficiently load three helicopters all spreading bait simultaneously. On 2 June 2004 a 42-metre barge with six shipping containers full of bait, Jet A-1 fuel, helicopter spreading buckets and the excavator reached the island. Sea conditions were near perfect and the excavator was manoeuvred off and up the boulder bank while everything else was offloaded by helicopter. A tent was erected over the bait to keep it dry — and then it was an impatient wait for good weather to arrive.[13]

An ideal forecast arrived just a week later and on 8 June a 24-person eradication team headed to the island in preparation for bait application the next day. The operation got under way at first light with the helicopters spreading bait across the island. Timing of the operation for mid- to late winter took advantage of the fact that kiore would be under the greatest stress due to cold weather and lack of food. However, to be successful the operation still needed to get bait into every rat territory on the island. Achieving this across the deeply incised topography of Hauturu was by no means an easy feat and had been a central consideration of project planning. Because of its elevation, flight lines were flown in a parabolic arc over the island by skilled pilots who, in addition to having to stay true to flight lines, had to maintain a constant speed to ensure uniform bait distribution. Short daylight hours meant the finishing touches to the first bait application, including sowing bait around the island's coastline and across areas with a gradient steeper than 50 degrees, were completed the following morning. The second and final bait application was delayed because of bad weather but was completed in one day on 12 July.[14]

*Clockwise from top left*: Helicopter with a load of rodent bait. (RG) Abseilers search for climbing asparagus on the cliffs in Ōrau Gorge. (AbAc) *Asparagus scandens* rampant in the forest understory. (KM) 'Dope on a rope' dangling beneath a helicopter ready to spray weedkiller on pampas. (RG)

Although initial signs were positive, it was two years before the success of the eradication could be officially declared — allowing time for any surviving kiore to breed and be readily detectable. The species' absence was confirmed through extensive monitoring with inked tracking tunnels on and off the track network, spotlight searches and indicator dogs.[15] Team members were hugely relieved and elated when they detected only tracks of wētāpunga (*Deinacrida heteracantha*), geckos and skinks on the inked tracking cards.

**ATTEMPTS AT MANAGING WEEDS** or pest plants on Hauturu began early in the island's history as a nature reserve. Robert Nelson found and destroyed a gorse bush (*Ulex europaeus*) in 1915 and in 1926 he began cutting down some of the poplars (*Populus nigra*). William Hardgrave removed all the wild fennel (*Foeniculum vulgare*) in 1940 and later that year his son Len felled the pine trees in front of the house. However, it wasn't until 1996 that funding was found for a coordinated weed control programme that became a core component of the island's management. The overarching goal of the weed management programme was and still is to protect the natural ecological processes that occur on the island; climbing asparagus, Mexican devil, mist flower (*Ageratina riparia*), pampas, mothplant and Mexican daisy pose a threat to these processes by interfering with patterns of natural regeneration and outcompeting native plant species.[16]

At first the programme focused on climbing asparagus and other weeds close to Te Maraeroa. Climbing asparagus was first recorded on Hauturu in 1978 in a small area behind the bunkhouse. By 1980 it was spreading rapidly; young plants were found at 90 metres on the Thumb Track and, since 1990, it has been found at many sites between Ngāmanauraru Bay and the start of the Shag Track, with outliers found in the Ōrau Gorge. Systematic search and removal efforts began in 1996, and plots were marked to enable repeated searches and treatment. Herbicide sprays were used initially, but later, after most adult plants had been removed, grubbing to remove the entire plant and root system was the preferred method because this reduced the risk of regrowth. With the control efforts, the number of plants found and plants reaching maturity has steadily dwindled over time. In fact, the results were so encouraging that in 2004 DOC revised the programme's objective from zero density to eradication, although it was recognised that to achieve such a goal, a sustained and concerted effort would be required.

The weed eradication programme has grown to include widespread species such as pampas and Mexican devil — both species that likely established on the island via windblown seed. Mexican devil was first recorded on Hauturu in 1940,[17] and pampas in 1974. These species were extremely localised in the beginning but have since spread to the extent that pampas is distributed right across the island in most open sites from ridgetops to the sea, and Mexican devil has a near continuous distribution along the south and west coasts.

Different options for the control of these species were explored. By far the most cost-effective solution for dealing with pampas across the steep and challenging terrain of Hauturu turned out to be the 'dope on a rope' method, involving a helicopter ferrying an underslung crew member on a 15–45-metre rope between infestations across the island to spray herbicide on each plant. This method has recently been replaced by a lance — effectively a very long hose nozzle — used out the side of a helicopter to spray herbicide on individual plants. Specialist abseiling teams are used to work the steep inland gorges that can't be reached by helicopter. Hand pulling was the primary method used to control Mexican devil, and helispraying (2 per cent Roundup) was tried with limited success. Control efforts have turned the tide on the island's pampas infestation but have so far been ineffective at reducing the extent of Mexican devil. As a consequence, control objectives for Mexican devil have been downgraded from 'zero density' to 'containment', while continuing to control infestations that endanger threatened plant populations. Efforts suggest that effective control of Mexican devil requires frequent and consistent management over a number of years before zero density can be attained.

Thanks to the island's weed programme, 12 species of weeds that were of localised distribution have been eradicated — a significant achievement. If sustained and concerted efforts are kept up it is possible to envisage a future in which a number of the 37 other species on the island's weed management target list, including climbing asparagus and pampas, can be considered to have been beaten (see Appendix 1). Constant vigilance will be required to prevent other invasive plant species arriving and establishing. The Little Barrier Island (Hauturu) Supporters' Trust has sourced much-needed funding towards this weed programme.

**HAUTURU HAS ALWAYS BEEN CENTRAL** to New Zealand's bird conservation efforts. There was a bid to obtain huia (*Heteralocha acutirostris*) for the island as early as 1893, but the species may have already been too rare for easy capture. Richard Henry, who had serious and justified concerns about predation by stoats (*Mustela erminea*), sent kākāpō to Hauturu in 1903, thinking it would be a safe place, but he failed to recognise that cats were present and would kill adult birds. In 1982, after cats had been successfully removed, kākāpō were reintroduced to the island and were managed intensively for the next 16 years. Breeding did occur but the programme was stopped and individuals removed in 1998 because of ongoing nest predation by rats. A population was re-established in 2012 and breeding by some individuals has since been documented. However, whether the population will ever become self-supporting and make an important contribution to the recovery of the species is unknown.

The presence of cats also foiled the first tīeke translocations from Taranga/Hen Island to Hauturu in 1925. However, between 1984 and 1988, following cat eradication, tīeke from Repanga/Cuvier Island and Marotere/Chicken Island were introduced to Hauturu. Kōkako (*Callaeas wilsoni*) are not known to have occurred naturally on Hauturu and, if they had been present, we don't know why they disappeared. Concern for the decline of kōkako and continuing modification of mainland habitats in the 1970s and 80s triggered initiatives to translocate birds to Hauturu. The last kōkako from Aotea/Great Barrier Island were rescued and sent to Hauturu in 1994. There were attempts made to rescue the last of the northern Coromandel population and transfer them to Hauturu, but no birds could be caught.[18] Both tīeke and kōkako are now firmly established on Hauturu. (See Appendix 2 for a complete list of translocations to Hauturu.)

Other translocation ideas were less well thought through. The first kiwi translocations, undertaken between 1903 and 1909, were initiated as a means to 'stock the island' and brought species that did not occur naturally. It is fortunate that some translocations did not eventuate, such as an attempt to import black partridge (*Melanoperdix niger*), from Southeast Asia, and a couple of requests for trees; and Robert Hunter-Blair expressed concern that there were no weka (*Gallirallus* spp.) on the island. No reptiles, amphibians, invertebrates or plants have yet been translocated to the island for conservation purposes. After the kiore eradication it was decided to wait at least 10 years before considering any translocations or reintroductions of these more cryptic species in case kiore were still present.

Tuatara were thought to be extinct on Hauturu until the species was rediscovered in 1991–92.[19] After their rediscovery, eight adults were taken into captivity to ensure the relict population did not go extinct before rats were removed. These individuals began breeding in captivity and the eggs produced were transferred to Victoria University of Wellington for incubation.[20] Hatchlings varying in age from one month to 11 months were then returned to an expanded enclosure on Hauturu.

In the early days there was little demand for translocation of birds off the island, as there were few places suitable for their introduction or reintroduction. The movement of kiwi to Pōnui Island was the exception and, despite the presence of cats and rats but with no resident stoat population, they have thrived there. Now, after more than five decades of concerted effort, most of New Zealand's uninhabited offshore islands and numerous mainland locations are free of introduced predators. Hauturu, which has retained much of its biological diversity, has become the go-to location for many bird translocations to other islands and predator-free mainland sites (see Appendix 3). Wētāpunga have been translocated off the island, and this species, once widespread across northern New Zealand, has now been re-established on Tiritiri Matangi and Motuora islands. Hauturu has also been the source of plants for other locations. In the early 1980s *Pittosporum umbellatum* and *Alseuosmia macrophylla* seeds from Hauturu were used to propagate seedlings for Tiritiri Matangi, and in 2004 *Dactylanthus taylorii* seed was collected and used to establish a population there. At the same time seed was spread close to the bunkhouse, establishing an additional site for the species on Hauturu.

Biosecurity would have been a foreign concept to Māori and European residents and visitors to Hauturu. Early caretakers on the island certainly took no special precautions to keep anything off; in fact, opportunities to 'populate the island' were deemed desirable. In 1909 Robert Shakespear reported to the general manager of the Tourist Department, which was then in charge of the island: 'Re the Black Partridge — lost all young birds — could not obtain any old birds — will do so when he [Shakespear's brother Will] returns to India.' This suggests that his brother Colonel Will Shakespear was attempting to bring these birds to the island. In 1910 Robert Hunter-Blair asked the Tourist Department to 'send me plants of koromiko (veronica) the blue one carries most honey . . . or any

---

*Clockwise from top*: North Island saddleback/tīeke, male (*Philesturnus rufusater*). (NF) Wētāpunga (*Deinacrida heteracantha*). (SF) The purpose-built biosecurity shed. (CRV) A chevron skink (*Oligosoma homalonotum*) found by the Abseil Access team. (AbAc)

flowering shrubs . . . and I could plant them here and there'. There were also plants that were brought to the island, lived a while, then died out, such as the kōwhai (*Sophora* sp.); trees that Len Hardgrave refers to as flowering prolifically and attracting tūī in 1941 and 1942, presumably near the house. Kōwhai do not occur naturally on the south side of the island and it appears that no seedlings resulted from this prolific seed bearer — possibly kiore ate all the seeds or seedlings, as they have done with other species. Credit should be given for some of the decisions made, however. In 1915, when Robert Nelson asked for trees to plant on the scrubby slopes behind the house, the prescient response from the Department of Lands and Survey was that he should instead move some karaka, tītoki and kohekohe from the nearby bush.

It took another 100 years for New Zealand's conservation community to realise the detrimental impacts of invasive species and to understand the need for effective biosecurity. The first precautions to prevent new species arriving on the island were put in place in the 1990s when visitors, on arrival, were asked to open their bags for inspection once they were inside the bunkhouse or the rangers' house. However, the decision to proceed with rat eradication in 2002 precipitated a comprehensive analysis of potential incursion pathways and the preparation of a biosecurity plan for Hauturu. Many measures were implemented immediately, but it was another three years before the plan was formally signed off and fully implemented.

In March 2005 the shed the Blanshards had built as a boatshed, with later additions, was given a facelift and secured to meet biosecurity measures. Thorough biosecurity checks were carried out in Warkworth for visitors to the island and in Devonport for stores being transported to the island. In 2006 the entire biosecurity building on the island was replaced with a biosecurity room and workshop. Biosecurity measures for Hauturu can now be considered world class: they include cleaning and packing all supplies and equipment in rodent- and invertebrate-proof containers inside a quarantine store on the mainland, either in Warkworth or in Devonport, and unpacking everything again on arrival at the island in another secure quarantine room. This process, which all visitors to the island must comply with, greatly reduces the risk of unwelcome organisms arriving on the island.

Other pathways have been more difficult to control. Illegal landings, although rare, have occurred, and over the years a number of small boats have wrecked on the island. Boats anchoring or fishing close to the island pose a risk. Weeds have arrived on the wind or have floated across from the mainland or been brought by birds. Surveillance is routinely undertaken to counter these incursions — methods include surveys undertaken by skilled and knowledgeable island staff, aerial surveys for weeds, and the deployment of devices such as tracking tunnels to detect rodents.

Responding to incursions is the last line of defence in preventing new invasive species establishing on Hauturu. The field office on the island is well equipped with traps, rodent bait and detection devices to quickly remove a potential invader. DOC and Auckland Council have a suite of detection dogs and dog handlers that can readily be mobilised to look for new arrivals. Biosecurity measures have been strengthened in response to emerging threats such as kauri dieback and myrtle rust. However, constant vigilance is fundamental to ensuring the island remains secure.

## Conservation outcomes

The fact that Hauturu was given protection as a nature reserve and then managed for conservation purposes has unquestionably saved a number of species from extinction and preserved the integrity of a unique and irreplaceable island ecosystem. It is a miracle that Norway rats (*Rattus norvegicus*), ship rats and mice did not establish on Hauturu, and it is likely the limitations that nature reserve status imposed on the island's exploitation prevented their introduction. If these species had established, it would almost certainly have assured the extinction of species such as hihi (*Notiomystis cincta*), New Zealand storm petrel (*Fregetta maoriana*) and wētāpunga.

Its status as a nature reserve put an end to the felling of kauri on the island, and the reduction, and eventual removal, of livestock resulted in slow but steady regeneration of forest in the southwest corner. Today, aside from parts of Te Maraeroa where native grasses and *Muehlenbeckia complexa* still dominate, native forest covers all previously modified areas. Areas logged for kauri or otherwise modified are transforming from the primary successional forest dominated by kānuka that colonised the logged, burnt and grazed slopes to something akin to the original coastal broadleaf forest that once covered these areas. Kauri continue to expand from those areas where logging did not reach.

Cat eradication was the first conservation intervention to achieve island-wide benefits for biodiversity. Although it was too late to save snipe, it occurred in time — perhaps just in time — to save the New Zealand storm petrel, a species thought to be extinct for 150 years before it was rediscovered in 2003 and was later found breeding on Hauturu.[21] The island's black petrel colony was also saved. Before cat removal, two-thirds of fledglings and up to one-third of adult black petrels were killed by cats every year, with just 5 per cent of chicks thought to survive to fledging.[22] Today the biggest challenge for the black petrel that now number approximately 600 pairs is the young birds that get drawn away to the larger and much noisier colony on nearby Aotea.[23] Other species also benefited, and the island's hard-won cat-free status permitted the introduction of kōkako and tīeke, both of which are now abundant across the island.

The removal of kiore also had far-reaching effects. Most Cook's petrels breeding on Hauturu do so at higher elevation. Breeding success in these higher altitude habitats averaged just 5 per cent prior to rat eradication but increased to over 60 per cent following rat removal as a result of reduced predation pressure.[24] The resultant explosion in the population was witnessed firsthand by people living north of Auckland. Every March chicks fledging from Hauturu leave the island and fly across the adjacent mainland to head out to the Tasman Sea. On nights with poor visibility these birds, confused by street and house lights, crash into people's gardens, and each March since the eradication, bird rescue centres have been inundated with calls regarding Cook's petrels[25] — an unusual problem that we might just have to get used to.

In the absence of cats and rats, those seabirds that had become locally extinct have begun to recolonise. Grey-faced petrels were discovered breeding in 2009 after a supposed 60-year absence, and anecdotal observations of old colony sites suggest a gradual increase in these populations.[26] Calls of other seabird species such as common diving petrels (*Pelecanoides urinatrix*) and fluttering shearwaters (*Puffinus gavia*) are

frequently heard and may reflect recolonisation of the island by these predator-sensitive species.

Banded rail (*Gallirallus philippensis*), last seen on the island in 1946,[27] have returned and reared young; and spotless crake (*Porzana tabuensis*), never previously recorded, are now present and presumed breeding. Another short-term impact worthy of note is the appearance and establishment of bellbirds (*Anthornis melanura*) at Tāwharanui Regional Park on the mainland, 23 kilometres from Hauturu, in the summer immediately following rat eradication.[28] Invasive vertebrates were removed from Tāwharanui at the same time as the rat eradication on Hauturu and this, coupled with an increase in the number of bellbirds on Hauturu (supported by mist-netting data), may have created the conditions conducive to the dispersal and subsequent population establishment of the birds at Tāwharanui.[29]

The benefits of cat and rat removal to other terrestrial birds have been mixed. While forest bird counts undertaken between 2013 and 2017 recorded significantly higher numbers of bellbird, tomtit (*Petroica macrocephala*), parakeets, robin, kōkako and tīeke, the number of whitehead (*Mohoua albicilla*), tūī, hihi, rifleman (*Acanthisitta chloris*), grey warbler (*Gerygone igata*), blackbird and silvereye (*Zosterops lateralis*) has declined when compared with counts done before and during cat eradication and before rat eradication. Additionally, no marked change was detected in the overall number of forest birds. Call-count surveys for kiwi on the island suggested little change subsequent to rat eradication, but calling frequencies have been consistently higher than those recorded at mainland sites, highlighting the advantages of predator-free habitat for this species.[30] The brown teal that were removed prior to rat eradication were returned to the island along with additional individuals but the population has not expanded, most likely because there is only limited wetland habitat available. The brown teal population is considered permanent, the number of birds present being more a reflection of the species' breeding success on nearby Aotea than on local productivity.

Numbers of reptiles caught in pitfall traps (buried 5-litre buckets) have steadily increased since the rat eradication. Populations of Hauraki skink (*Oligosoma townsi*), moko (*O. moko*) and shore skink (*O. smithi*) have flourished, contributing to an eighteenfold increase in the total number of skinks caught per 100 trap nights since the rat eradication.[31] Before rat eradication, 'Chevy' was the one and only chevron skink (*Oligosoma homalonotum*) to be caught on the island. More individuals have been found since, indicating gradual population recovery. There has been limited monitoring of the island's gecko populations, but spotlight surveys completed in 2009 and 2013 suggest populations are recovering from pre-eradication declines. Sighting rates of Pacific gecko (*Dactylocnemis pacificus*), forest gecko (*Mokopirirakau granulatus*) and Raukawa (common) gecko (*Woodworthia maculata*) in 2013 have more than doubled relative to 2009 counts,[32] and Duvaucel's gecko (*Hoplodactylus duvaucelii*) has been sighted more frequently.[33] Since the rat eradication was confirmed as successful in 2006, more than

---

*Clockwise from top*: Tuatara (*Sphenodon punctatus*). (SF) New Zealand storm petrel (*Fregetta maoriana*). (NF) Forest ringlet butterfly (*Dodonidia helmsii*). (EC)

290 young tuatara, raised in captivity, have been released and all captive adult tuatara have been returned to the wild. Additional adult survivors have been detected, breeding has been noted in the wild population and it appears to be expanding.[34]

The rat eradication also had a profound impact on the island's invertebrate populations. Numbers of wētāpunga have more than doubled since rats were removed and research indicates this trend is set to continue.[35] Relict invertebrate populations have been discovered, too. In 2017, surveys for the endemic forest ringlet butterfly (*Dodonidia helmsii*) confirmed its presence on Hauturu. This species was formerly widespread throughout much of New Zealand but is now rare or absent from many areas of previous distribution, and had not been recorded on Hauturu for a long time. Multiple individuals were found on the island in the 2017 survey, indicating a resident population.[36] Gibbs highlights introduced social wasps as a potential cause of the decline in forest ringlet populations.[37] Curiously, the German wasp (*Vespula germanica*) and common wasp (*V. vulgaris*), previously a considerable nuisance on Hauturu, vanished after the rat eradication, and this may explain the reappearance of the forest ringlet. Indirect effects of the rat eradication may also explain the equally intriguing discovery in 2014 of six new species of mayfly (*Mauiulus luma, Isothraulus abditus, Zephlebia spectabilis, Arachnocolus phillipsi, Ichthybotus hudsoni* and *Neozephlebia scita*), and two species of caddisfly (Trichoptera) (*Oxyethira albiceps* and a Chathamiidae sp.).[38] The survey of aquatic invertebrates replicated one that was carried out in 1963.

It is unlikely these will be the last discoveries to be made on the island. Several more seabird species are expected to return, and highly cryptic species are still likely waiting discovery on Hauturu — for example, only one record of striped skink (*Oligosoma striatum*) has been made on the island, but this species is expected to be found again in the future.

The future course of forest regeneration on Hauturu has also changed. Significantly more seedlings were found for 14 species after rat eradication: *Pisonia brunoniana, Coprosma macrocarpa, Ixerba brexioides, Knightia excelsa, Rhopalostylis sapida, Phyllocladus trichomanoides, Nestegis lanceolata, Dacrycarpus dacrydioides, Ripogonum scandens, Hedycarya arborea, Dysoxylum spectabile, Pittosporum umbellatum, Macropiper excelsum* and *Corynocarpus laevigatus*. Before rats were eradicated, seedlings of *N. lanceolata, R. sapida* and *R. scandens* were rare, but in 2008 *N. lanceolata* was found on most plots, and *R. sapida* and *R. scandens* seedlings were common in moister sites. There was a substantial increase in the number of seedlings of other tree species, too: seedlings of *Beilschmiedia tarairi, C. laevigatus* and *P. trichomanoides* were twice as numerous in *Kunzea ericoides* stands after rat eradication, *D. spectabile* was five times more common and *R. scandens* 41 times.[39] These patterns provide a glimpse of what is to come for Hauturu's forest ecosystem, and the return of seabirds is likely to change things even further. Seabirds are known as 'ecosystem engineers' — they promote friable, aerated, fertile soils that are high in phosphorus and nitrogen with a low pH and carbon-to-nitrogen ratio,[40] and over time are expected to have an increasingly major influence over the composition of Hauturu's forests.

Threatened plants also showed a positive response to the removal of rats. Improved seed set by the endangered *Carmichaelia williamsii* was noted and the endangered

*Euphorbia glauca* colonised new areas. Seed production in the endangered *Dactylanthus taylorii* increased and individuals have been discovered at new locations on the island, including at the site close to the bunkhouse where seeds were hand-sown. As a consequence of the focused efforts of island rangers and the many people who have signed up to be part of the island's weed teams, the forest ecosystems on Hauturu are still relatively intact. If invasive plants such as climbing asparagus had been allowed to run amok, the present condition of Hauturu's forests could have been quite different.

The careful management that has been undertaken over the years must be acknowledged, as it is through these efforts that the island is what it is today. Change is, of course, inevitable and natural ecosystems are never static. But the changes that are likely to occur on Hauturu in the coming years will be fascinating, especially as the influence of seabirds increases across the island and the composition of the forest changes. We hope to see the reintroduction of species that are no longer present on Hauturu but for which the island is clearly within their natural range, and the introduction of surrogate species — for example Chatham Island snipe — to replace the ecosystem functions performed by species now extinct, such as the North Island snipe.

# Chapter five
# Geology of Hauturu

— JAN LINDSAY

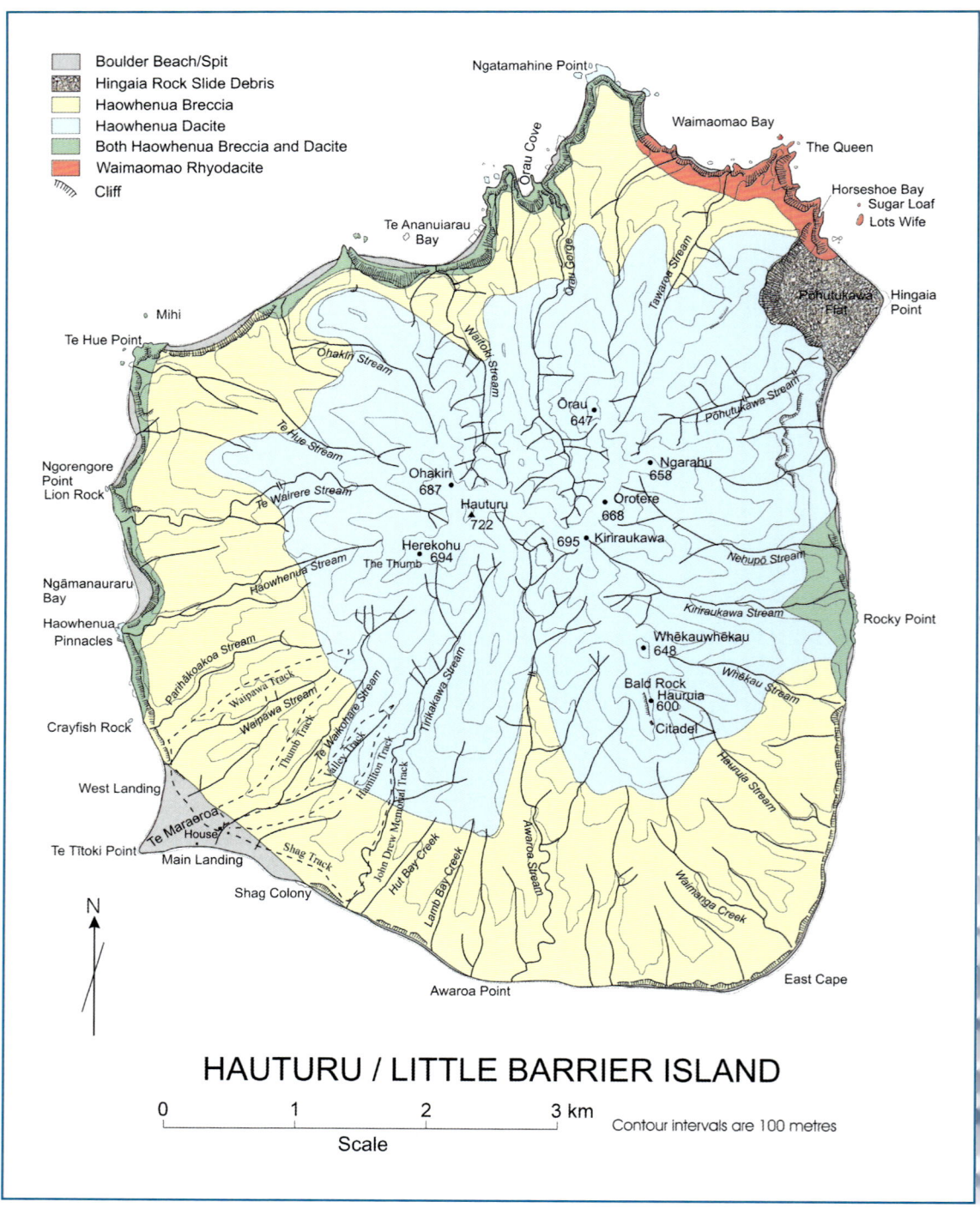

Fig. 1: A geological sketch map of Hauturu.

**Te Hauturu-o-Toi/Little Barrier Island is well known for its status as a spectacular island nature reserve. Less well known is the fact that it represents the emergent part of the largest dacite stratovolcano in New Zealand, whose location in the Hauraki Gulf, miles away from volcanoes of similar age and composition, makes it an ongoing puzzle for geologists.**

The island is roughly circular in shape, with a maximum diameter of 7.5 kilometres from north to south, and 5.8 kilometres from east to west (Fig. 1). Steep ravines radiate from a group of central peaks that rise to a height of 722 metres at the summit (Mount Hauturu). The volcano extends a further 45 metres below sea level, has a basal diameter of 8 kilometres, and a volume of c. 13 km³. On the northern, western and southern sides of the island, ring-plain surfaces form terraces that slope gently down from 400 metres to coastal cliffs generally 20–100 metres high. On the eastern side of the island, coastal erosion has removed these slopes, producing high sea cliffs (up to 200 metres). The island is fringed by a boulder beach, continuous except where rocky headlands meet the sea (for example at the Pinnacles, Ōrau Cove and The Queen). Rapid cliff retreat post 6500 BP (Before Present) has produced spectacular hanging valleys, most at least 10 metres high — good examples of which are visible at the mouth of Hauruia, Te Wairere and Kiriraukawa streams. Hauturu is heavily forested and has very few inland rock outcrops. There are, however, excellent coastal exposures of the main volcanic units that form this stratovolcano.

The geology of Hauturu has received little attention in the literature. Excluding the occasional brief reference to the island, before the 1990s there were only three published accounts and each of these interpreted the island as a late Pleistocene andesite stratovolcano.[1] More recently, Lindsay et al. showed that the island was in fact dacitic in composition (a composition midway between andesite and rhyolite, typically defined by having 63–70 weight per cent $SiO_2$) and mid Pliocene to early Pleistocene (3.1–1.2 million years) in age.[2] This relatively young age makes it an anomaly in its location near the northern end of the Coromandel Volcanic Zone (CVZ) where adjacent volcanic activity of similar composition ceased many millions of years earlier.

The island has a complex stratigraphy — a common feature of andesite and dacite stratovolcanoes. Its lower slopes are made up of lava flows, debris flow deposits and pyroclastic flow deposits. Lindsay et al. divided the lavas of Hauturu into two formations: the Waimaomao Rhyodacite and Haowhenua Dacite.[3] This subdivision combines the Haowhenua and Hauturu andesites of Kear into the Haowhenua Dacite.[4] Samples

*Previous*: Mount Hauruia (Bald Rock), 600 metres above sea level. (NF)

of Waimaomao Rhyodacite have been dated by both the uranium–lead (U–Pb) and potassium–argon (K–Ar) radiometric methods, yielding a consistent age of 2.9–3.1 million years. Potassium–argon radiometric dating of several Haowhenua Dacite samples has shown this unit to be between 1.6 and 1.2 million years old. The reason for this unusually long (c. 1.5-million-year) time gap in the evolution of the volcano is unknown.

The lavas of Hauturu are overlain by a complex series of volcaniclastic breccia deposits, known collectively as the Haowhenua Breccia. This is interpreted as the product of numerous small debris flow events occurring during or after active periods of volcanism, likely resulting from gravitational collapse of steep areas of the volcano and other erosional processes, such as lahars after heavy rainfall. These breccias typically lack the characteristics of deposits produced by column-collapse pyroclastic density currents, such as fall and surge deposits, highly vesiculated clasts, gas escape pipes, vapour-phase crystallisation and post-depositional welding, although some may represent block-and-ash flow deposits from collapse of summit lava domes.

Subdivision of Haowhenua Breccia is difficult. The only common sequence is one that relates the breccia to the underlying Haowhenua Dacite. This sequence is exposed at several places along the western and northern coasts, and consists of grey Haowhenua lava overlain by 15–25 metres of oxidised red brecciated lava, which in turn is overlain by less-weathered Haowhenua Breccia units.

## Main geological units

The Waimaomao Rhyodacite is the oldest rock type on Hauturu. It is a pinkish lava-flow unit that is restricted to the northeastern corner of Hauturu, where it forms a single homogeneous unit up to 150 metres high. In places it exhibits spectacular flow banding (see Fig. 9). These lavas have a higher silica content than the Haowhenua lavas, hence the name rhyodacite — a rock type intermediate between a dacite and rhyolite. The restricted area of the outcrop, thickness of the unit and nature of the flow banding suggests that the rhyodacite represents the remnants of a small volcanic dome.

The Haowhenua Dacite is a medium-dark grey lava, often with pronounced jointing, that forms the central part of the island and is exposed in many places along the coastline around the northern half of the island (see Fig. 6). Tops of the predominantly grey lava flows are characteristically oxidised, resulting in a purple coating. In many localities there is evidence of flow fragmentation in the Haowhenua Dacite, which forms as the viscous upper surface of a lava flow cools and breaks up into blocks while the more fluid lava inside continues to flow. The deposits resulting from this flow fragmentation are known as autobreccia; on Hauturu they are present as deposits up to 25 metres thick on top of some Haowhenua lava flows.

Exposures in the coastal cliffs and valleys along the entire southern coastline from the shag colony in the southwest corner to just north of the Whēkau Stream mouth consist entirely of breccia units. Breccia is also present overlying Haowhenua lavas in cliff exposures along the western and northern coasts, and in the area between the Whēkau and Nehupō streams on the east coast. The Haowhenua Breccia extends some distance inland but, as only a few river valleys are accessible, it is difficult to determine where the breccia outcrops end and the lava begins. A good indication of this transition is the presence of waterfalls in some of the major stream valleys, for example Awaroa and Te Wairere streams.

In many accessible localities the breccia is thoroughly weathered, giving a 'marble cake' appearance (see Fig. 12). Where weathering is not pervasive, the breccia is generally well bedded, with individual depositional units easily distinguishable (see Fig. 5 and Fig. 10). Such beds are typically 0.5–2 metres thick, poorly sorted, and grading is reverse or absent. Clasts are subangular to subrounded and range in size from a few centimetres to over 3 metres in diameter. One very large clast (c. 10 metres) is present in the coastal cliffs north of Nehupō Gorge, but the largest clasts in any one unit are commonly 1–3 metres in diameter. There is no preferred clast orientation. The matrix is bimodal; generally fine-grained but including small clasts up to 2 centimetres in diameter. In fresh outcrops of breccia it is possible to identify individual clast lithologies. The four most common clast-types are grey Haowhenua Dacite, purple (oxidised) Haowhenua Dacite, red Haowhenua autobreccia and a black dense dacite not seen in the form of lava flows on the island.

Notably absent in all breccia units are clasts of the pink Waimaomao Rhyodacite. The fact that clasts in the breccia appear to be derived entirely from the Haowhenua Dacites is reflected in the name 'Haowhenua' Breccia.

## Significant geological features

The following is a brief description of some of the more significant geological features of Hauturu. These descriptions have been reproduced with some modification from Lindsay and Moore.[5]

**Haowhenua Point** (Fig. 2): This is the type locality of the 1.5-million-year-old Haowhenua Dacite. A grey contorted Haowhenua lava flow at the base of the cliff is oxidised to a purple colour at the top of the flow. A layer of autobreccia 15 metres thick overlies this purple oxidation layer, and is itself oxidised to a deep red. This high level of oxidation may have resulted from gases streaming out from the cooling lava. At the top of this autobrecciated unit (which is made up of blocks less than 2 metres in diameter derived from the underlying lava) there is a patchy yellow zone due to alteration of plagioclase to clay (probably montmorillonite). This zone of autobreccia is mantled by a layer of ash, which is in turn overlain by well-bedded Haowhenua Breccia deposits. All these units dip gently to the south.

**Lion Rock and Ngorengore Point** (Fig. 3): Lion Rock (approximately 20 metres across) consists of grey Haowhenua Dacite, oxidised on its upper surface to a purple colour. It is part of the c. 20-metre-thick lava flow forming the adjacent headland (Ngorengore Point), notable for its interesting arcuate jointing/banding pattern. An autobrecciated layer, which overlies the lava flow on the headland, also forms the very top (i.e. the 'ears') of Lion Rock.

**Te Ananuiarau Bay** (Fig. 4): The western end of Te Ananuiarau Bay forms a point that is made up of three separate Haowhenua Dacite lava flows, each one separated by a zone of coarse red autobreccia.

**Ōrau Cove:** The western margin of Ōrau Cove appears to be formed by a dike in coarse red breccia, and aerial photographs reveal a north–south trending lineament reaching the coast at this point. The cliffs in the centre of the cove consist of faulted, well-bedded grey Haowhenua Breccia, which merges on both sides into coarse, red autobreccia associated with the bounding lava flows/dikes. The deposits at the top of the cliff, above the eastern side of the bay (c. 50 metres asl), may be old alluvial gravel.

**Ngatamahine Point** (Fig. 5): Views of Ngatamahine Point from the sea reveal excellent exposures of Haowhenua Dacite lava at the base of the cliff, overlain by a series of well-bedded Haowhenua Breccia units.

**Te Hue Point** (Fig. 6): Te Hue Point forms the northwestern corner of Hauturu, and is a good example of the highly jointed 'platy' nature of many outcrops of Haowhenua Dacite. It forms a very sharp contact with the adjacent red autobreccia, and the jointing in the lava swings sub-parallel to the contact.

**Waimaomao Bay:** This is the type locality of the 3-million-year-old Waimaomao Rhyodacite. The lava is a pinkish colour and locally flow-banded. In the centre of the bay there is an impressive natural arch in the flow-banded lava.

**Hingaia Rockslide** (Fig. 7): The dominant feature of the northeastern corner of the island is Pōhutukawa Flat, the product of a large rockslide. The rockslide deposit consists of large blocks of Haowhenua Dacite up to 17 metres in diameter, and it has a headscarp that rises 420 metres above sea level at its highest point. The slide material was mapped by Kear as the Hingaia Fall Debris,[6] but the name Hingaia Rockslide Debris reflects a more realistic mode of origin as a rock slide rather than rock fall. The thickness of the debris is unknown and it is difficult to assess the volume accurately.

Kear estimated the volume of the debris above sea level as 25,000,000 cubic yards (about 19,000,000 m³) by assuming an average thickness of 200 feet and an area of 120 acres. Despite its name, the surface of the landslide is far from flat; large blocks of

---

*Clockwise from top left: Fig. 2:* A cliff south of Haowhenua Point showing a common sequence exposed around the coast: purple (oxidised) Haowhenua Dacite overlain by autobrecciated lava 15 metres thick, which in turn is overlain by a series of Haowhenua Breccia units, each 1–2 metres thick. See person standing at base of cliff for scale. (JL) *Fig. 3:* Ngorengore Point and Lion Rock (right) from the north, showing an arcuate jointing pattern in a whitish-grey Haowhenua Dacite lava flow at the coast. (JL) *Fig. 4:* Western end of Te Ananuiarau Bay showing three stacked Haowhenua Dacite lava flows. (JL) *Fig. 5:* View of Ngatamahine Point from the northeast. A lava flow of Haowhenua Dacite outcrops at the base of the cliff and is oxidised to purple near the top of the flow; this is overlain by a series of Haowhenua Breccia units. The cliff is approximately 100 metres high. (JL) *Fig. 6:* Jointed Haowhenua Dacite at Te Hue Point. See person at base of cliff for scale. (JL)

lava form ridges and depressions and the surface is generally very irregular.

The age of the rockslide is unknown. Kear pointed out that the slide material seems unmodified by higher sea levels following the Flandrian transgression (when sea levels were approximately 2 metres above the present level), which suggests that the slide occurred less than 2270 years ago. However, it's questionable whether evidence of the Flandrian transgression would be preserved in such an environment. The rockslide does seem to relate to the present-day sea level as it does not appear to be swamped, and its age can therefore be estimated at less than 10,000 years, based on the sea-level rise at the end of the last major glacial period. We do not know what caused the Hingaia Rockslide: it may have resulted from dome collapse or, more likely, cliff instability due to the great height of the cliffs, coupled with the presence of several joint sets weakening the rock mass.

**The Queen** (Fig. 8): There is a good exposure of thick, flow-banded Waimaomao lava at The Queen, a small (20 metres high) island named after Queen Victoria because of its resemblance to a woman wearing a hoop petticoat. The adjacent headland is made up of a thick, strongly jointed and folded lava flow.

**Horseshoe Bay** (Fig. 9): This is a small, horseshoe-shaped cove in the Waimaomao Rhyodacite. The cliff at the back of the cove is approximately 120 metres high and shows good flow banding. Inside the cove there is a cave that has formed by erosion along prominent joint sets, and there is an excellent exposure of very thick, folded, flow-banded lava adjacent to the cave.

**Kiriraukawa Gorge** (Fig. 10): The Kiriraukawa Stream terminates in a hanging valley, carved through semi-weathered, well-bedded Haowhenua Breccia. Streams along the eastern and northern coastline typically terminate as waterfalls and steep-sided gorges such as this one.

**Bald Rock (Mount Hauruia)** (Fig. 11): Bald Rock is the only large inland outcrop on Hauturu. It consists of fine-grained, grey, jointed, relatively resistant Haowhenua Dacite. It forms a north-northwest-trending lineament with the Citadel to the south, and this, along with its resistant nature, suggests it may be a plug or dike.

**South Coast** (Fig. 12): In most coastal exposures, including the area below the shag colony, the Haowhenua Breccia has undergone extensive weathering. In these areas bedding is partly to completely obscured by pervasive chemical weathering, and the breccia has a characteristic marble-cake appearance. Both clasts and matrix appear to

*Clockwise from top*: *Fig. 7*: Pōhutukawa Flat from the northeast, showing bush-clad Hingaia Rockslide Debris. The cliffs forming the headscarp are 420 metres high. This view is approximately 1000 metres across. (JL) *Fig. 8*: The Queen and adjacent headland comprised of Waimaomao Rhyodacite, viewed from the northwest. (JL) *Fig. 9*: Banded Waimaomao Rhyodacite exposed in the cliff opposite Lots Wife rock stack. The view is 10 metres across. (JL)

be equally weathered, suggesting that weathering of the clasts has occurred in situ rather than prior to deposition — for example by geothermal activity.

The weathered breccia gives an impression of being heterolithologic (made up of a heterogeneous mix of different rock types), a result of alteration of the four main clast lithologies — evident in fresher outcrops — to pink, red, yellow and grey weathering products. Because of the intensity of weathering it is difficult to determine the original clast compositions. In a few places, however, weathering of clasts has not been so pervasive, leaving an unweathered core with a highly weathered rim (see top central part of Fig. 12). Good examples of this can be found below the shag colony. Such clasts are almost always of the fine-grained black dacite lava, because of their lower permeability. The unweathered cores of these clasts commonly project out from the cliff.

Weathering en masse of breccia units occurs in stages, adding to the array of colours in these outcrops — initially pinkish hues (due to oxidation), followed by yellow and then orange clays. Most cliffs of this breccia have zones of pronounced honeycomb weathering, particularly around the southwest coast to the north of East Cape, where cliff erosion is not as intensive as along the rest of the coast. In areas of honeycomb weathering bedding is completely obscured. Fretting and cracking of beach boulders through salt weathering is common along the entire coastline.

**Te Maraeroa and Te Tītoki Point:** The only area of flat land on Little Barrier is Te Maraeroa on the southwest corner of the island. This area is bordered by extensive boulder banks on both sides, formed at the meeting point of two prevailing winds, and the area behind the boulder banks is filled with material from several streams that drain into it. Records and photographs taken by past visitors and caretakers show that the shape of the spit has changed considerably during the last 100 years: at times it has had a relatively symmetrical shape, as it does today; at other times it has a pronounced hook at the point. The age of the spit is unknown, but Hamilton suggests it formed during the Flandrian transgression and became exposed during the slight drop in sea level that followed.[7]

*Clockwise from top*: *Fig. 10*: Semi-weathered Haowhenua Breccia exposed at Kiriraukawa Gorge. Streams along the eastern and northern coastline typically terminate as waterfalls and steep-sided gorges such as this. (JL) *Fig. 11*: The resistant Mount Hauruia (Bald Rock) showing the jointed nature of Haowhenua Dacite. This probably represents a volcanic plug. View is from the northwest and the field of view is approximately 250 metres. (JL) *Fig. 12*: Typical 'marble cake' appearance of intensely weathered Haowhenua Breccia at Awaroa Point. Different colours result from multi-stage weathering of at least four different clast types. (JL)

Geology of Hauturu 101

# Chapter six
# Biota of Hauturu: Flora, fauna and fungi

**The following pages provide information on the various groups of biota that occur on Te Hauturu-o-Toi/Little Barrier Island. For a very long time now, people have studied and collected all types of plants, animals and fungi from the island. Some of this material was originally held in private collections, but today museums and other institutions hold most of these specimens. They are carefully stored and curated for their scientific value as records of their presence on the island, as well as being made available for future study.**

In general, fauna collections are referred to as just that, a 'collection'. Each item — be it bug or bird — is carefully labelled and recorded in a database. Plant collections are held in a herbarium, and fungi in a fungarium. A herbarium usually contains vascular plants, that is, plants with a vascular system, i.e., lycopods, ferns, gymnosperms and flowering plants, and non-vascular plants, i.e., bryophytes (mosses, liverworts and hornworts), algae and lichens (a symbiotic association of an alga and a fungus). Most specimens are kept as pressed and dried items. Fungi are usually dried whole, but specimen parts that cannot be pressed or dried, such as fleshy fruit, may be kept in alcohol or other preservative.

The main herbaria known to contain specimens from Hauturu are Auckland War Memorial Museum (AK), Manaaki Whenua — Landcare Research (CHR), Massey University (MU) and the Museum of New Zealand Te Papa Tongarewa (WELT).

The New Zealand Fungarium Collection (PDD) and the New Zealand Arthropod Collection (NZAC) are held by Manaaki Whenua at St Johns, Auckland. The Arthropod Collection contains more than seven million specimens of invertebrates of the large phylum *Arthropoda*, which includes insects, spiders and crustaceans.

'Voucher' specimens are collected and used for a variety of reasons: as proof that a species occurs in a particular place at a certain time; for a DNA sequence to be obtained; when an identification is doubtful; when there's something unusual about the species; or to support a publication reference. The advantages of having a voucher specimen are that it is likely to be included in future taxonomic revisions and that its identification can be checked at any time. These specimens held in public collections provide absolute proof of what species currently occur, or formerly occured, at a particular location.

Today, most of the collection data about these stored specimens is available online

*Lepidium oleraceum* specimen collected on Hauturu by Thomas Cheeseman in September 1898, now in Auckland Museum (AK4470). *Previous*: A forest scene at Tirikakawa Stream. (AA)

via various websites, either directly from the institutions themselves or, in some cases, as pooled data. For example, most of the New Zealand plant and fungal data can be located on the Australasian Virtual Herbarium (AVH), https://avh.chah.org.au/.

There are numerous national and international laws that protect, or allow the management of, species of biota, but some still have no legal protection. Everything that is within the boundary of the Hauturu Nature Reserve is protected. However, plants and animals in the seas around Hauturu have varying levels of protection, ranging from no protection at all to an allowed harvest limit.

Many of the species on Hauturu are also included in lists which attribute a level of threat or endangerment to the species. The IUCN Red List of Threatened Species was founded in 1965. This list of worldwide species continues today. Some countries started their own lists about this time. The New Zealand Threat Classification System was aligned with the IUCN list in 2008,[1] but with some differences to account for local issues, such as naturally restricted distributions of endemic species and status changes for species responding to conservation action. Threat status is reviewed for each taxonomic group at about five-yearly intervals, with the outcomes of these reviews being published in independent peer-reviewed scientific journals. Lists of these publications are to be found on the Department of Conservation website with links to enable the viewer to download pdf files of the published work.[2]

— Dick Veitch

*Above*: Little Barrier snipe (*Coenocorypha barrierensis*) specimen in Auckland Museum, collected in 1870 (LB10066). *Below*: Female giraffe weevil (*Lasiorhynchus barbicornis*) specimen in Auckland Museum (AMNZ96692).

# 6.1
# Aquatic and terrestrial invertebrates

— DAVID S. SELDON

**Without the myriad invertebrate species globally — estimated at 30 million — most ecosystems would not function well. Invertebrates are found in all trophic levels and underpin most functions within an ecosystem. Even though there is high diversity and abundance of invertebrates, they are often the most understudied within an ecosystem because of their small or even microscopic size and the difficulty of identifying them.**

Te Hauturu-o-Toi/Little Barrier Island has terrestrial invertebrates ranging from the intertidal zone to the peaks, located in a wide range of habitats formed by the combination of high peaks, steep slopes and deep stream ravines. The peaks are covered in a cloud forest that is usually damp throughout the year; the broadleaf forest-covered slopes become dry in the summer months; and the stream valleys have relatively high humidity year round. This has led to a high diversity of invertebrates but, as on other Hauraki Gulf islands, there is a surprising lack of true Hauturu endemic species. As the mainland North Island and other islands in the Hauraki Gulf have undergone habitat destruction and have been invaded by introduced mammalian predators, Hauturu has become the last bastion for some relict invertebrates, including some of New Zealand's larger invertebrate species that were previously widespread.

The historical focus of Hauturu has been on becoming a sanctuary for terrestrial and marine native bird species. Conservation management removed the feral cats by 1980 and kiore by 2004 to make the island a safe haven for birds and other fauna and flora, including invertebrates. Kiore and cats would have consumed many large invertebrate species; by removing these introduced predators the island ecosystems will have benefited through an increase in invertebrate abundance.

The omnivorous kiore foraged on a wide variety of plants and seeds, as well as on native vertebrates and invertebrates at all developmental stages and throughout the forest layers from the forest floor to the canopy. Studies of other northern islands have shown that kiore fed on at least 28 different species of insect. Coleoptera (beetles) and Lepidoptera (butterflies and moths) accounted for most of the species eaten by kiore, but the Auckland tree wētā (*Hemideina thoracica*) was the most frequently consumed species.[1] There are also a number of orthopteran species on Hauturu, other than the tree wētā, that kiore probably preyed on, including the nymphs of the wētāpunga (*Deinacrida heteracantha*) and adult ground wētā (*Hemiandrus pallitarsus*).

The problem is that we have no information on what invertebrates were on Hauturu

Coastal copper butterfly (*Lycaena salustius*) on *Muehlenbeckia complexa*. (SF)

prior to Māori and, later, European occupation. These residential occupations could have determined a severe decrease in population size or even extinction of a number of invertebrates, especially the flightless and often long-lived insect species. The German wasp (*Vespula germanica*) was first recorded on the island in 1962 and the common wasp (*V. vulgaris*) in 1995. The Asian paper wasp (*Polistes chinensis*) was first observed on the island in February 1989, and the Australian paper wasp (*Polistes humilis*) arrived at a later, unknown date.

Vespulid wasps are predators of a wide range of invertebrates, yet concentrate mainly on a few orders of Hexapoda, with a very clear division between species of preferred prey.[2] Further division of prey between the *Vespula* and *Polistes* species occurs through preferred habitat (forest or scrub respectively), and where within the vertical stratification of the forest (i.e. canopy to forest floor) a social wasp species hunts. Both *Vespula* species prey on dipteran (fly) larvae and Araneae (spiders), but German wasps prefer orthopterans (katydids), phasmids (stick insects) and large ant (formicid) species. Both of the paper wasp (*Polistes*) species prey almost solely on lepidopteran larvae (up to 39 different taxa) and some dipteran larvae (six different taxa) and, depending on wasp population densities, can take an estimated 15,000 to 478,000 prey loads per hectare per season. This is a significant amount of arthropod biomass that is removed from a habitat, and in direct competition with a number of native invertebrate, reptile, bird and bat species that also prey on these taxa.

One unexpected result of the kiore eradication is the subsequent loss of the *Vespula* wasp species from the island. During late November 2004, after the kiore eradication, there were still a large number of German wasp nests observed in the relatively small area of Te Maraeroa. Some of these nests were quite large, with steady streams of workers exiting and entering. Even though the Hauturu rangers poisoned as many nests as could be found, this would only have made a very small dent in the local population around the southern area of the island. Yet every year since 2004 there has been a decrease in the *Vespula* wasp population, and by 2013 none was observed and the constant hum of flying workers was no longer heard. It is not known what causes the decline of *Vespula* wasps from an island that has had rat removal, but there may be a combination of factors disrupting the nests and/or foraging workers, and the conditions may become unfavourable for wasps even in low numbers. German wasps are still present on at least one island, Tūhua/Mayor Island, where rats have been removed, but there are relevant differences in the history of Tūhua and Hauturu.

What we do know about the invertebrates of Hauturu is mainly from early surveys of the island, which looked at specific fauna — for example earthworms — and a number of short surveys in the 1950s of, for example, land molluscs, earthworms, beetles and aquatic insects. Most of the other invertebrate fauna has been catalogued as part of non-invertebrate studies — for example of the diet of native birds — or from a small study of native pollinators. However, one group of invertebrates — the aquatic fauna — has been relatively well documented with targeted surveys, and these have given the most accurate account of these taxa in the streams of Hauturu.

## Soil invertebrates

### Earthworms

Earthworms are one of the most studied groups of invertebrates on Hauturu, and have been the subject of a range of surveys over the years. Most of us think of earthworms as the common adventive garden variety (*Aporrectodea caliginosa*) or tiger worms (*Eisenia fetida*) that efficiently devour compost, but Hauturu is the type locality for one of New Zealand's less iconic invertebrate giants. *Anisochaeta gigantea* is an earthworm that can grow to a length of 1.4 metres, with a diameter of 11 millimetres, and forms burrows that can be 20 millimetres in diameter and up to 3.5 metres deep. This giant is rarely seen, although it can be relatively common in the forest soils of Hauturu and some northern mainland regions. Most people who visit the island will have heard the slurping sound these worms make when reacting to the vibrations of approaching walkers, moving from the surface to deeper in their burrows. These burrows can be seen in detail after severe rainfall has caused flooding of the streams, which results in erosion of the riverbanks and can expose the wanderings of *A. gigantea* through the clay.

Currently there are more than 200 described species of earthworms in New Zealand. Many are endemic to single regions and offshore islands. Hauturu has 11 described earthworm species, some of which are shared with the mainland, such as *Anisochaeta gigantea*, and a few that are adventive, for example *Aporrectodea caliginosa* (pasture flats). Another is the smallest species of the genus *Anisochaeta* (*A. shakespeari*), length 120 millimetres, which is usually found in rotting logs or under rocks and other wood debris in forest habitats. The wide variety of vegetation and topography has produced a number of different soils ranging from light sandy silts to wet heavy clays, which has resulted in a variety of habitats for soil-dwelling organisms. The most diverse earthworm genus on the island is *Anisochaeta*, a widespread northern North Island genus with three taxa. Based on the diversity of habitats on Hauturu we could assume that other earthworm genera are just as diverse, and there are probably a number of unknown species. In fact a recent study of earthworms by Buckley and Bartlam unearthed two new genera, one of which is endemic to Hauturu, and they commented on the possibility of cryptic species existing on the island.[3]

### Cryptic soil dwellers

There is a huge diversity of other soil-dwelling invertebrates that are even less known, but that often appear in pitfall trap samples. Some, for example springtails (Collembola), are so numerous that their number is estimated rather than counted, whereas others, for example soil or stone centipedes (Geophilomorpha) are less prevalent. Both of these groups are found on Hauturu, along with giant versions of each, which are much more conspicuous yet still rarely observed. Springtails live in damp soil, leaf litter and rotting wood and feed on a variety of foods including fungi, algae, pollen and decaying vegetation.[4] Generally, springtails are 2–3 millimetres long, but in New Zealand there is an endemic genus *Holacanthella* that are the giants of the collembolan world. *Holacanthella duospinosa* can be found in relatively large numbers on Hauturu, under wood debris along many of the stream banks. This species can be up to 15 millimetres long and has a variety of different-sized yellow digitations (spike-like projections), located dorsally and laterally, protruding from its blue-grey body.

The soil or stone centipedes are predators within the soil, burrowing into the substrate to feed on worms, collembolans and other small invertebrates. Apart from many unknown soil centipede species and other smaller centipedes, the largest invertebrate predator that hunts by night is *Cormocephalus rubriceps*. The giant centipede, a predator that can reach lengths of 250 millimetres and with body segments up to 25 millimetres wide, is endemic to Australia as well as to the North Island of New Zealand, and is especially abundant on the rat-free offshore islands. This species can deliver a poisonous and painful bite, yet when disturbed it seeks cover as soon as possible. Individuals are readily found on Hauturu — under logs and stones during the day, and hunting on the forest floor and on tree trunks at night. The females lay eggs in the hollow of rotten logs and maintain parental care over the eggs by coiling around them, as well as cleaning each individual egg.

## Velvet worms

The velvet worm (Onychophora) is another nocturnal invertebrate predator of the forest floor. Its name belies the true form and function of the organism: it looks more like a velvety millipede (with fewer legs) or a centipede (with fewer hard edges). Onychophorans are known as living fossils as their form has not changed since the early Cambrian Period, 540 million years ago (mya), and are thought to be the ancestral connection between Annelida (worms) and the Myriapoda (centipedes, etc.) and Hexapoda (insects). There are two families with distinct geographical distribution: the Peripatidae are distributed through the equatorial and tropical regions, whereas the Peripatopsidae are found in Gondwanan countries.[5]

*Peripatoides sympatrica* is a species found from Hawke's Bay to the Coromandel Peninsula, and is related to the geographically widespread species *P. novaezealandiae*, which is found throughout mainland New Zealand in low population densities. Peripatus are tactile and chemosensory hunters which eject a sticky slime that captures and immobilises the prey ready for consumption. *P. sympatrica* is a viviparous species: the eggs develop within the female, which then gives birth to live young. The Onychophora also appear in folklore and in the art of many different peoples across the world, and are a food source for some Papua New Guinean peoples. Māori also ate peripatus, which they called ngāokeoke (meaning 'to crawl'). Today onychophorans are not so well known, but in the past New Zealand has featured the velvet worm in cartoons, tourism, literature and even as a stuffed toy.[6]

Onychophorans may be the missing link between annelids and hexapods, but Protura (or coneheads) are the most basal of the Hexapoda and live in the soil. The Protura are very small (less than 2 millimetres), have no eyes, wings or antennae, and no pigment; in fact, they are so insignificant that they were not discovered and described until the early twentieth century. Still very little is known about the life of most proturans because of their size, which makes them very difficult to observe in the wild. Hauturu has two species from two different families: *Eosentomon dawsoni* (Eosentomidae) and *Gracilentulus gracilis* (an adventive Acerentomidae). These coneheads are most likely to be found on the underside of stones in moist habitats, where they graze on mycorrhizae — fungi that grow symbiotically along plant roots.

Giant centipede (*Cormocephalus rubriceps*) (SF)

Peripatus/ngāokeoke (*Peripatoides sympatrica*) (CRV)

Giant worm (*Anisochaeta gigantea*) (SB)

## Aquatic insects

A number of abiotic and biotic factors determine the diversity of invertebrates in streams and rivers, especially the material of the streambed and the stream morphology (or fluvial geomorphology). Coarse gravel and boulder streams generally contain more diversity and abundance of invertebrates than streams with silt beds; and streams with more riffles and races (shallow, fast waterflows) than pools tend to also have high numbers of taxa. The riparian vegetation plays an important part in stream morphology by controlling sediment loads, contributing plant debris, reducing sunlight/UV penetration to the stream and providing niches for adult invertebrates. In general, streambeds with coarse material have more grazing leptophlebiid mayflies, stoneflies, caddisflies, dobsonflies, and hydraenid and elmid beetles.

On Hauturu more than 36 streams radiate outward from the peaks, forming steeply eroded and sinuous gorges through volcanic rock, some with high waterfalls, especially along the northern and eastern coasts.[7] Aquatic invertebrates living in these streams have to contend with high water velocities in the catchment areas and an often erratic water supply; most of the streams dry out, except for a few pools, in mid- to late summer. In the southern and western areas there are only three streams — Awaroa, Te Wairere and Hut Bay Creek (which is spring-fed) — that have continuous waterflow throughout the year, and these usually have broader streambeds in the lower reaches near the sea. Stream surveys of four streams in the southwest of the island were carried out by Mike Winterbourn in 1963 and repeated and extended by Lyn Wade in 2014.

### Mayflies and caddisflies

Even though the aquatic environment on Hauturu can be ephemeral, there is a relatively diverse aquatic invertebrate community in the streams of the accessible southern and western areas of the island. More than 50 species of aquatic insects are found in under a third of the streams surveyed so far. Ephemeroptera (mayflies) and Trichoptera (caddisflies) are the most richly diverse groups of insects found. There are 20 species of mayflies, including the yellow dun (*Ameletopsis perscitus*), which is uncommon but widely distributed throughout New Zealand; 17 species of caddisflies, and nine species of stoneflies. The majority of species in these three groups (including mayflies) inhabit riffles and races of streams. Both Wise and Winterbourn found the greatest diversity of aquatic insects in these types of habitats, which are common in the streams of Hauturu.[8]

### Aquatic beetles

With the abundance of habitat in the stream systems of Hauturu we might expect to find a number of species there of the two families of specialist (non-swimming) aquatic beetles (Coleoptera) — the Elmidae (riffle beetles) and the Hydraenidae (minute moss beetles). However, there are no species of elmids recorded from Hauturu, even though both Winterbourn and Wade describe the prime habitat of these riffle beetles.[9] In fact, at the topmost survey station of Awaroa Stream there is a sheet of moss-covered rock that forms a race of very shallow water running over it — the perfect riffle beetle habitat.[10] Not all is lost, though, as there are at least two species of moss beetles (hydraenids) recorded from Hauturu — *Podaena latipalpus* and an island-endemic *P. hauturu*.[11] Visitors and researchers may not have observed these beautiful beetles, as *P. hauturu* is very small (2 millimetres in length) and dark brown in colour, so is easily confused with specks of dirt.

### Water-associated flies

Another mystery aquatic insect that does not seem to have a breeding population or presence on Hauturu is the biting black fly (Simuliidae), with larvae that live in flowing freshwater. These dipteran biters are found on the mainland and on Aotea/Great Barrier Island, but it seems that they are absent from Hauturu, as neither Wise, Winterbourn nor Wade collected larvae while sampling streams.[12] Craig et al. suggest that the island's small area and lack of continuous flowing water year-round are factors.[13] There is some evidence of simuliid presence with two separate reports of visiting dipterists being bitten by a small black fly in the vicinity of Awaroa Stream. Unfortunately, a specimen (*Austrosimulium australense*) was collected only on the first occasion, in 1984, so there is no confirmation that this was the same species present in 2009, or that there is indeed a Hauturu population.

Although there is some debate about the presence of biting flies, there is no doubt that Hauturu is the home of the dance flies (Brachystomatidae) — an ancient group that live near streams, where they prey on other insects. There is a single genus, *Ceratomerus*, with 45 species that are distributed across three land masses that were part of the former Gondwanaland: South America, Australia and New Zealand. Hauturu has three *Ceratomerus* species, two of which are widespread throughout New Zealand, and one species that is endemic to the island, but is as yet unnamed as it is known only from three female specimens.[14] These were collected from moss beside the Awaroa, which seems to be the only stream that consistently has flowing water, and a high degree of invertebrate diversity. Unusually, there are also five species of Dixidae — a dipteran family of aquatic gnats with long legs that require unpolluted standing freshwater to inhabit.

## Forest invertebrates

### Slugs and snails

The phylum Mollusca is the second largest invertebrate group. The Gastropoda — slugs and snails — is the only class to have species that are solely terrestrial. Almost all (approximately 80 per cent) of the 70,000 species of gastropods are either marine or intertidal, and/or aquatic. New Zealand has an estimated 1400–2000 native species of slugs and snails, which is a substantially greater diversity than in other countries such as the United Kingdom, which has 112 species. The New Zealand species belong to 11 families, most of which are Gondwanan in origin and so are found in only a small number of countries including Australia, New Zealand, New Caledonia and those of South America. Representatives of those 11 families, including Athoracophoridae (leaf-veined slugs), are present on Hauturu, which has 61 species of terrestrial gastropods in 27 genera. Most of these were recorded by Milligan and Sumich when they surveyed the Hauraki Gulf islands.[15]

On Hauturu there are representatives of the land snail family Rhytididae, but these are the smaller-sized species (for example *Delos coresia*, 4 millimetres). There seems to be an absence of the medium to larger-sized species (*Amborhytida* sp.), although a number of *Amborhytida* species are found on surrounding islands such as the Hen and Chickens and the Poor Knights and on the nearby mainland. However, Milligan and Sumich commented that their survey and an earlier survey (1952) probably revealed most of the diversity of terrestrial gastropods on Hauturu. Another interesting absentee is a species of predatory pāua slug, *Schizoglossa*. It occurs on almost all the islands in the Hauraki Gulf, so it might still be on Hauturu even though it has not yet been seen there. Overall, Hauturu supports a higher diversity of land snails and slugs than other Gulf islands such as the Hen and Chickens (22 species).

## Spiders (and others)

Arachnida (Acari, Araneae, Opiliones and Pseudoscorpions) are the least known group of invertebrates on Hauturu as there doesn't seem to have been a survey of spiders on the island. Most of the arachnids listed on the species lists (see Chapter 9) have been collected as bycatch during other studies, with about half the known spiders having been identified from an ongoing pitfall trap survey. Only 45 species of Araneae (spiders) in 36 genera have been recorded, along with three species of Opiliones (harvestmen), two species of Pseudoscorpions and a single species of Acari (mites). Almost every pitfall trap sample produces at least 10 different mite morphospecies, which means that these taxa are very understudied. There will be a much higher diversity than indicated above because many of these groups are not identifiable to species or genus level unless an adult male specimen is available.

## Wētā

There is one species of insect on Hauturu that has been well studied: wētāpunga (*Deinacrida heteracantha*), a once widespread giant wētā species that was found on Aotea and in the regions of Auckland and Northland. This charismatic megafauna's last refuge is Hauturu, where it held on despite the presence of kiore that effectively reduced the number of smaller, juvenile stages. Since the removal of kiore, wētāpunga surveys[16] and radio tracking[17] have shown that the population is increasing each year — so much so that a small number of pairs have founded populations through captive breeding programmes at Butterfly Creek, Manukau, and at the Auckland Zoo. Their offspring are now being released on other pest-free islands in the Hauraki Gulf.[18]

There are other, less conspicuous orthopterans on Hauturu. Two medium to large species — the Auckland tree wētā (*Hemideina thoracica*) and Auckland cave wētā (*Gymnoplectron acanthocera*) — are relatively abundant. As with all wētā species, they are best observed at night. Not all cave wētā necessarily hang out in caves — many of them, like *G. acanthocera*, are also found on overhanging stream banks, in hollows in trees and on the ground, foraging for a wide range of food items for their omnivorous diet. Most people assume that because wētā are found in trees and are related to crickets, they are herbivorous, but most, including the wētāpunga, will scavenge on dead birds and other animals when encountered. Two other species of abundant nocturnal orthopterans are found on Hauturu: the ground wētā and the smaller relatives of the Auckland cave wētā (Rhaphidophoridae), most of which are not yet described.

Leaf-veined slug (Athoracophoridae sp.) (PBe)

*Cambridgea* sp. (SF)

*Dolomedes* sp. (SF)

Tunnelweb spider (*Hexathele hochstetteri*) (SF)

Wētāpunga (*Deinacrida heteracantha*) (RG)

Auckland tree wētā (*Hemideina thoracica*) (SF)

There are a number of the smaller orders of Hexapoda found in the forests of Hauturu, including five species of Phasmatodea (stick insects) and five species of Blattodea (cockroaches). As well as the relatively diverse Thysanoptera (thrips), with 10 species recorded, four species of Phthiraptera (lice) have been found, all of which are parasites of birds, and a single species of Siphonaptera (fleas) that is associated with *Rattus* species (i.e., kiore).

## Flies

One of the five major insect orders on Hauturu are the flies (69 spp.), including some common families such as the dance flies and aquatic gnats mentioned earlier. Some Hauturu dipterans can be found on the seashore: shore flies (Ephydridae) and the extremely abundant kelp flies (Coelopidae), with adults that swarm out from between boulders to cover everything on or near the shore in mid- to late summer as they feed on seaweed. The larvae of the grass flies (Chloropidae) are also vegetarian and feed on grass stems, and there are a number of families that feed on fungi, for example the sun flies (Heteromyzidae). However, the fly fauna seems to be predominantly predatory, and a number of different families, including robber flies (Asilidae) and dagger flies (Empididae), have piercing mouthparts and often catch other insects on the wing.

Six species of the New Zealand-endemic genus *Parentia* (Dolichopodidae) are present on Hauturu. These metallic blue-green flies prefer a moist environment in a wide variety of habitats from intertidal to forest, and can often be observed resting on flowers. They are predatory and horticulturalists view them as beneficial species because they feed on many garden pests.

There are two fly families that are associated with other animals: one is an obligate parasite, the louse flies (Hippoboscidae), and the other is the endemic New Zealand family, the bat flies (Mystacinobiidae), described by Holloway.[19] The widespread avian parasitic fly (*Ornithomya variegatus*) has been found on 20 bird species throughout Australasia, including several natives — the grey warbler, saddleback, whitehead, New Zealand robin, North Island tomtit, North Island fantail and silvereye.[20] The unique New Zealand bat fly, *Mystacinobia zelandica*, is unlike other bat flies worldwide in that it is not a blood-sucking parasite but lives in a strange symbiotic relationship with the lesser short-tailed bat (*Mystacina tuberculata*). These blind, wingless flies live on and around the bats, forming substantial colonies within roosts to manage the large amounts of guano produced by the bats. Both the adult and larval bat flies are coprophages — they feed on bat poo — which results in the roosts being available for many generations of both bat and fly.[21]

## Bugs

There are 94 species of bugs (Hemiptera) recorded from Hauturu in 76 genera, contained within 28 families — which is about an eighth of the New Zealand fauna. Many of the New Zealand genera are shared with Australia and the Pacific Islands, yet 82–90 per cent of the species are endemic to New Zealand. Hemipterans are found in a very diverse range of habitats, including on the sea surface. They are parasites of birds and mammals, feed on fungi and on all parts of the seed plants (from roots to seeds) and prey on other arthropods. There are no hemipterans unique to Hauturu, but the large number of known taxa can be attributed to good identification keys, ease of specimen collection, and the numerous studies of bugs due to their ability to spread a variety of plant diseases.

## Bees, wasps, sawflies and ants

Hymenoptera (bees, wasps, sawflies and ants) are a megadiverse order with a worldwide estimate of 100,000–150,000 species, but in New Zealand they are not well known; only 900 species in 47 families have been described. Hymenoptera perform a number of different processes in the ecosystem, including pollination and biological control of many lepidopteran larvae and some wood-boring beetle larvae. Native parasitoid wasps are generally very small (1 millimetre), but the largest, *Certonotus fractinervis* (40 millimetres), parasitises the larvae of the elephant weevil (*Rhyncodes ursus*). This species association has not been recorded from Hauturu, but the glowworm (*Arachnocampa luminosa*) larvae are parasitised by a flightless parasitoid, *Betyla fulva*, that inhabits the leaf litter. A new species of parasitoid, *Casinaria* sp. (Ichneumonidae), was recently discovered during a survey of parasitised native lepidopteran larvae from the Auckland region.[22] This species of parasitoid is probably endemic to Hauturu as it was not collected from any other site during the study, and there are no specimens at Auckland Museum or in the New Zealand Arthropod Collection.

A few species of native stinging wasps have been recorded from Hauturu, along with the two better known introduced social wasps, *Vespula* spp. and the *Polistes* sp. The feeding habits of native stinging wasps differ greatly from the introduced species, as they are solitary hunters of arachnids and other hexapods. They sting the prey to paralyse it then take it back to the nest where the female lays an egg on it, and the emerging wasp larva then devours the spider or insect alive. Both the golden hunter (*Sphictostethus nitidus*) and large black hunter (*Priocnemis monachus*) wasps are found on Hauturu, and can often be observed along tracks and forest edges searching for their spider prey. They both take different large ground spider species with only a slight overlap: the golden hunter prefers large brown vagrant spider (*Uliodon* spp.), tunnelweb spiders (*Porrhothele* and *Hexathele* spp.) and the nursery web spider (*Dolomedes minor*); the black hunter preys on tunnelwebs but most often hunts for trapdoor spiders (e.g. *Stanwellia hapua* and *Migas insularis*).

There are no honeybees (*Apis mellifera*) to compete for nectar and pollen resources on Hauturu, which means there is an abundance of native bees. There are two species of *Leioproctus* present on the island. These smaller (5–12 millimetres) black-grey versions of honeybees are solitary but can often be seen nesting in clay banks with hundreds or even thousands of individual tunnels in close proximity. They feed on nectar and store pollen in the nest for larva to feed on, but do not produce honey. There is at least one species of the solitary *Lasioglossum* bee present on Hauturu. These are small (4–8 millimetres), black or greenish in colour, generally less hairy, and nest in the soil.

The fact that there are only 47 species of hymenopterans recorded from Hauturu, where surely there are many more present, is more a reflection of the overall poor state of this order's taxonomy, as well as the fact that a large proportion of the species are very small (less than 5 millimetres) and cryptic.

Native bee (*Leioproctus* sp.) (PBe)

Moth (*Tatasoma* sp.) (SF)

Stick insect, mating (*Spinotectarchus* sp.) (NF)

Stick insect (*Clitarchus hookeri*) (SF)

Giraffe weevil, male (*Lasiorhynchus barbicornis*) (LW)

New Zealand striped longhorn beetle (*Coptomma lineatum*) (LW)

Darkling beetle (*Tenebrionid* sp.) (SF)

Huhu beetle (*Prionoplus reticularis*) (PBe)

## Butterflies and moths

Butterflies are notably lacking in the New Zealand Lepidoptera, but it does have an extremely diverse moth fauna, both day- and night-active species that are found from the seashore to alpine regions. A very large proportion of these — 1600 of the 1800 estimated species — are endemic to New Zealand, and Lepidoptera is the third largest order in terms of diversity, after Coleoptera and Diptera. Robert Hoare and David Pattemore, who visited Hauturu in November 2008 as part of a wider survey on native pollinators, recorded most of the listed lepidopteran species (43 out of 61 species).[23] No species were found to be locally endemic to Hauturu, but there is a high proportion of endemic New Zealand species present, as well as a small number of adventive species. In terms of diversity, 61 taxa is very low; it is expected that this number will increase by hundreds with further intensive surveys throughout the years.

There is a range of species with larvae (caterpillars) that are specialists on single plant species (or genus, for example *Coprosma*). Others are polyphagous, and there are even species that are known only by the adults, so their biostatis is unknown. A number of lepidopteran species are associated with mosses (seven *Eudonia* spp.), ferns (two) and a single species, *Hygraula nitens*, that is commonly found near freshwater as its larvae feed on aquatic plants. Three species of *Izatha* (lichen tuft moths) are present, with their range of colours and tufts that allow them to blend into bark and lichens, especially the striking *I. peroneanella* (green and black forewings). Species from the family Micropterigidae (micromoths) with wingspans of 4–6 millimetres also occur on Hauturu; most of these genera have links to Gondwana.[24] Ancient in origin, they do not have the coiled proboscis to suck nectar; instead, these moths have mandibles that they use to feed on a variety of spores from the ancient plant groups such as bryophytes, lycophytes and ferns. However, they are survivors — so some species (for example *Sabatina chelcophanes*) will feed on the pollen of angiosperms, especially sedges and grasses.

Hauturu is also home to a few unique species. A single specimen of the rare *Circoxena ditrocha* (larva unknown) has been collected, but several more have been observed. *Pseudocoremia dugdalei* (larva unknown) is a geometrid species with a very disjunct distribution; to date this moth has only been collected in the Waitākere Ranges, Hauturu (Hamilton Track), Golden Bay and Orokonui Sanctuary near Waitati, north of Dunedin. Another lepidopteran that is rarely observed is the forest ringlet butterfly or Helm's butterfly (*Dodonidia helmsii*): individuals have been observed flying above *Gahnia* sp. (sedges) on the Hamilton Track and on Kauri Ridge on a number of occasions.

## Beetles

Beetles (Coleoptera) are the most diverse insect order with over 400,000 described species worldwide, which equates to 40 per cent of all insect species, and 25 per cent of all described animals. In New Zealand the beetle diversity is estimated to be 5000 species; they can be found from the coast to the mountains, but mostly in forests. Some are well known, for example the huhu beetle (*Prionoplus reticularis*) because of its large and tasty larva or grub, and the giraffe weevil (*Lasiorhynchus barbicornis*), the longest New Zealand beetle (up to 100 millimetres), shaped like a waka. Both these are found on Hauturu, along with 119 other species in 114 genera, within 32 families — a small number of species for this island with its great variety of habitats. There are several families that are relatively well known: this reflects the specialist areas of the researchers who have been to Hauturu, but also dependence on the availability of well-written identification keys. For example, there are eight species of anthribids (fungus weevils) identified — a relatively high number — largely due to reliable keys and recent taxonomic revision.

Weevils (Curculionidae) are a hyper-diverse beetle family with more than 70,000 described species worldwide, and in most ecosystems make up about 17 per cent of the total beetle taxa. In New Zealand there are 300 described weevil species, a number that is low by world standards, mainly due to the lack of taxonomic work on this incredibly diverse family. In contrast, Hauturu has only 19 species of weevils (recorded), which reflects the lack of specialists in this field. Weevils are a significant beetle group with a very close association with flowering plants (angiosperms). Many species of weevil, for example *Stephanorhynchus crassus*, live their entire lifecycle on a single plant species. The adult of this weevil feeds on the pollen, then the sticky yellow pith of the seedpods, while the larvae feed on the seeds of *Pittosporum crassifolium*. It could be predicted that the weevil fauna is five to 10 times higher than the recorded species, and that these 19 genera are just a small indication of the diversity in the forest and leaf-litter of Hauturu.

Many weevils are considered pests because of their wood-boring and plant-eating habits. Others are viewed as beneficial and are used as biocontrols for weeds; and some adults (for example *Scolopterus penicillatus*) inadvertently pollinate while they feed on pollen. The huge diversity of weevils has resulted in many species with hidden behaviours, which can answer a wide range of questions, including sexual selection of certain traits. For example, many male giraffe weevils are large with elongated rostrums that they use to fight for and then guard females, yet smaller males are able to sneak below larger males to mate.[25] Hauturu is home to the giant (20–25 millimetres) flax weevil (*Anagotus fairburnii*), a flightless and long-lived beetle (adults can live for longer than 12 months) that is associated with *Phormium* (flax) species. The adults feed on flax leaves and leave a characteristic notch in the leaf edge that is sometimes the only evidence of their presence. The fate of the flax weevil on Hauturu is unknown — specimens were last collected in the 1960s and there have been no confirmed sightings since.

Another beetle super-family that is poorly represented on Hauturu is the Staphylinidae (rove beetles) — the most diverse family in New Zealand. These beetles are characterised by the shortened elytra (wing case), which exposes the abdominal segments, so they are often mistaken for small earwigs. Rove beetles are found in a wide range of habitats, from the seashore to alpine regions, and yet the recorded fauna on Hauturu is nine species. There will be many more staphylinid species on Hauturu, but it will require reliable New Zealand species keys, and a substantial amount of time and effort, to sort out their taxonomy.

A research bias towards Carabidae (ground beetles) is reflected in the number of known species for the island, which is 23 (including two adventive species). This is probably still short of the total number — there may be three to five more taxa. There is one carabid that is endemic to Hauturu, *Mecodema haunoho*, which was named by Ngāti Manuhiri[26] and is closely related to its sister species *M. aoteanoho*, an endemic to Aotea/Great Barrier Island. This medium-bodied, flightless and nocturnal predator of the forest floor is a species within a highly diverse genus found throughout the New Zealand mainland and on many offshore islands.

## Conclusions

The invertebrate diversity of Hauturu as a whole is greatly understudied; however, some groups (aquatic) and individual species (wētāpunga) have been researched over a number of decades from the early twentieth century. Most groups have been studied only in retrospect or as a part of a wider project (for example arachnids), and therefore very little is known about the potential diversity of terrestrial invertebrates on Hauturu. Not all is lost, though, as there are two outstanding data sets that will considerably increase the known taxa of many invertebrate groups on Hauturu: the Department of Conservation has continued collection of altitudinal invertebrate samples through pitfall trapping, and Drummond et al. carried out an eDNA study using pitfall trap and malaise trap sampling.[27]

Even though the entire aquatic and terrestrial invertebrate fauna is not fully known, Hauturu does have 30 orders, 198 families, 433 genera and 551 species recorded. There are some abnormalities, such as a large number of genera with only a single species present. There are absences from the fauna that cannot be explained until a much more detailed invertebrate survey of the island has been undertaken. A relatively large island like Hauturu (2817 hectares) that has such a wide variety of habitats compressed into a steep topography should produce a much greater invertebrate diversity. The cloud forest along the peaks of Hauturu, similar to those of Moehau Mountain in the Coromandel and Mount Te Aroha, probably has a distinct fauna yet to be discovered. We know that there are invertebrate species that may be locally extinct — for example, the flax weevil *Anagotus fairburnii*, which has not been collected since the 1960s. Some insects may have disappeared, but there are new unique species that have been glimpsed and, in one case, photographed that require collection before identification can be confirmed.

The origins of Hauturu's terrestrial invertebrate species are obvious as many genera and species are shared with Aotea, the Coromandel Peninsula, Auckland and Northland, as well as some of the other Hauraki Gulf islands. Those connections to the mainland are evident in the number of ancient species — some Gondwanan in origin — and the many flightless and/or poor dispersers that are shared with the mainland, including the peripatus, giant worms, wētāpunga, carabids and weevils. However, there are still many questions to be answered in studying the current invertebrate fauna, and many more specific habitats to survey.

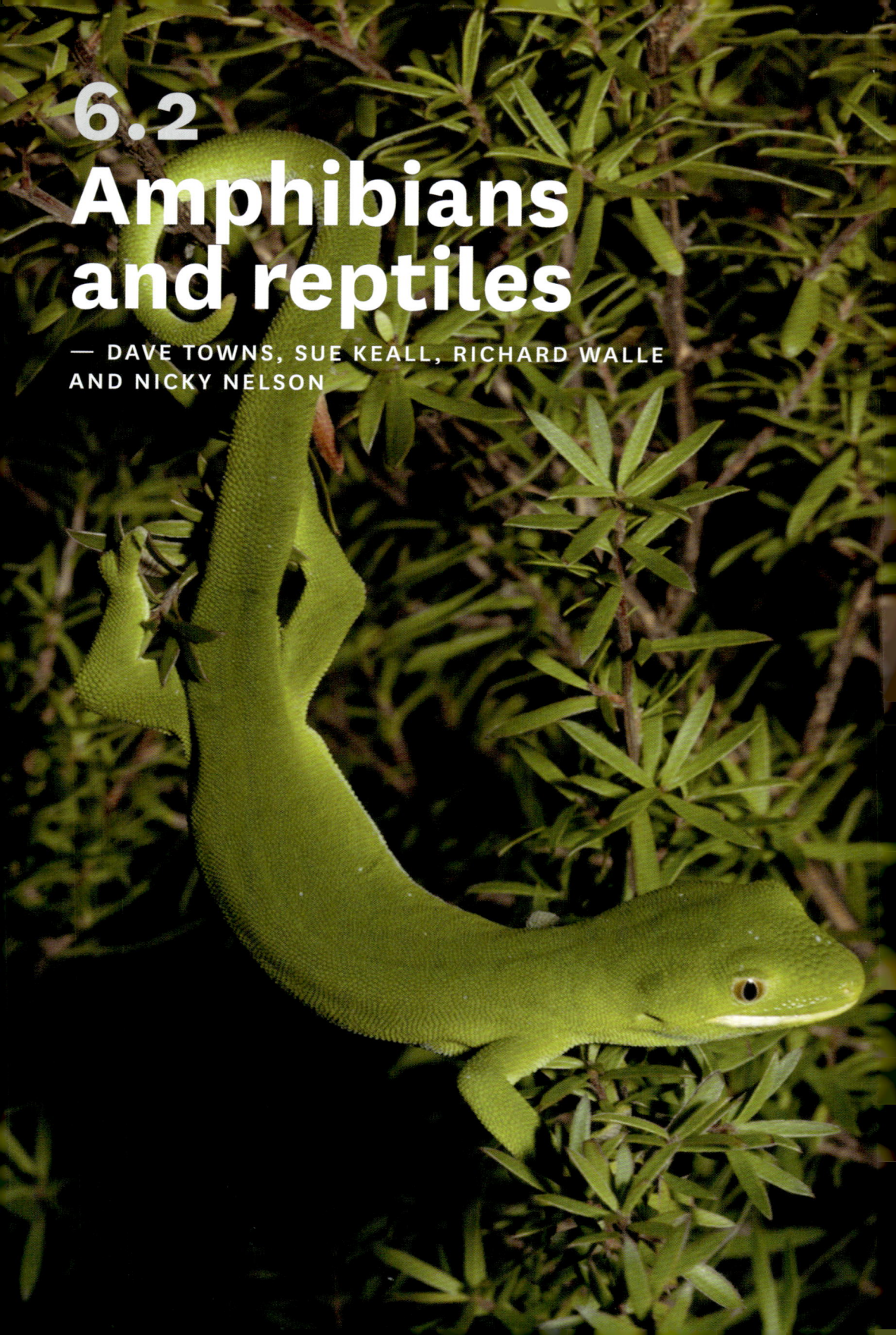

# 6.2
# Amphibians and reptiles

— DAVE TOWNS, SUE KEALL, RICHARD WALLE
AND NICKY NELSON

**Here we provide the first comprehensive account of the herpetofauna (amphibians and reptiles) of Hauturu, using the benefits of modern and intensive systematic studies as well as surveys and long periods of monitoring. Tuatara and lizards have an intriguing conservation history on the island, and they now provide the closest approximation of the composition and habitat use of reptile faunas on the adjacent mainland before human contact.**

When government scientists undertook their studies on Hauturu in the early 1960s the role of non-avian terrestrial vertebrates in the native fauna was poorly understood — so it is not surprising that although the bird fauna was comprehensively described in the resulting volume,[1] herpetofauna received only passing mention. In hindsight there were two reasons for this lack of coverage. First, the uniqueness of native frogs and tuatara had long been recognised, but taxonomy of the lizards was poorly developed and so confused by nomenclature problems and duplicated descriptions that even the more commonly found species were difficult to identify. The bulletin published in 1955 by Charles McCann of the Dominion Museum was the first comprehensive attempt at unravelling this taxonomic mess. But even with McCann's newly published volume in hand, there would have been few people in the country with the required expertise to use it. Perhaps it was a convenient outcome that the expeditions encountered few reptiles.

The second reason for the lack of coverage stemmed from the reasons behind these few reptile sightings. Kiore may have been on the island for a considerable period and feral cats for at least 70 years. More recent studies have shown that both species can have devastating effects on tuatara and lizard populations. Even after cats were removed in 1980, lizards were rarely encountered unless there were specific attempts to trap them. Until 1991, tuatara were presumed to have died out. The studies for the 1961 bulletin were thus undertaken when reptile numbers on Hauturu were heavily suppressed, and even if lizards were found, their identity would have been difficult to determine.

## Amphibians

So far, native frogs have not been found on Hauturu. However, herpetologists such as Tony Whitaker have pointed out that there is a great deal of potential frog habitat. A systematic search for Hochstetter's frog (*Leiopelma hochstetteri*) was carried out in January 2019,

Elegant gecko (*Naultinus elegans*) (DvW)

though not all streams were searched. There was no search for Archey's frog (*L. archeyi*). It would be premature to assume total absence of either of these species but it seems likely that, if they did occur naturally on Hauturu, they were eaten by kiore.

## Reptiles

### Tuatara

Historic records of tuatara on Hauturu are rare. Their significance as a taonga species for Ngāti Manuhiri and Ngāti Wai is evident and the importance of the island is intertwined with its special fauna, but stories of tuatara on Hauturu and the memories of tuatara from kaumātua are scarce.[2] By the time Europeans recorded observations of tuatara on Hauturu they are likely to have already been only small in number and/or difficult to detect because of the predatory and competitive impacts of kiore. The later introduction of cats added another predator, and the negative effects of occupation and logging would have caused disturbance, particularly of lowland habitat. By the early 1900s tuatara sightings were rare, only single large individuals, and by 1977 it was suspected they had been lost from the island.[3]

Reischek, in 1886, described tuatara as different in appearance to those on other islands, and this resulted in their consideration as a subspecies of tuatara: *Sphenodon punctatus reischeki*.[4] Surveys of all tuatara islands in the early 1990s targeted Hauturu in particular, with the aim of establishing whether tuatara with perceived special differences from other populations still existed there. Eight adults were captured during two large surveys. Blood samples from these individuals recognised that Hauturu tuatara are part of a genetic stock of tuatara on islands off the east coast of the North Island that are somewhat different from tuatara on islands in Cook Strait. However, although island-specific differences are apparent — likely due to the extent of time those islands have been separated from the mainland — all tuatara populations, including those on Hauturu, are best described as one species, *Sphenodon punctatus*, with important geographic variation across the distribution of tuatara populations.[5]

The eight individuals captured on Hauturu were four adult males and four adult females. Members of the search teams were convinced there were likely more tuatara on the large island and, indeed, there were a handful of records of further observations in subsequent years. However, it was the legacy of the eight captured adults that formed the basis for the recovery of tuatara on Hauturu. These adults formed the foundation of a captive breeding stock held on Hauturu, secure from kiore predation. Genetic analysis supported the idea of a relatively recently reduced population, in that these tuatara were genetically diverse rather than inbred, and showed similar levels of genetic diversity to tuatara on Takapourewa/Stephens Island in Cook Strait[6] — an excellent start for re-establishing a thriving population of tuatara on Hauturu.

Captive housing on the island allowed researchers to observe and interact with the adults. Egg induction from gravid females was attempted and eggs deposited in nests were excavated to acquire eggs for artificial incubation at Victoria University of Wellington. Artificial incubation of eggs was used as a technique to increase productivity through

Tuatara (*Sphenodon punctatus*) (RG)

Forest gecko (*Mokopirirakau granulatus*) (DvW)

Duvaucel's gecko (*Hoplodactylus duvaucelii*) (DvW)

Pacific gecko (*Dactylocnemis pacificus*) (DvW)

Raukawa gecko (*Woodworthia maculata*) (DvW)

greater hatching success (in 1996 Thompson et al. found that half of eggs can fail in natural nests from environmental causes).[7] By 2018, approximately 290 offspring from incubated eggs had been returned to Hauturu to be raised there.[8]

All tuatara were kept in captivity on Hauturu until kiore were eradicated. This involved enormous dedication from the island team, volunteers and sponsors, as tuatara were given supplementary feed, and the young were raised in small enclosures to reduce the incidence of predation by older, larger tuatara and to minimise the effects of drought and predation. From the first successful hatching of eggs in 1994 and throughout the following 25 years, offspring were housed in ever-expanding captive facilities on the island.

Research was conducted on the Hauturu tuatara to investigate whether the sex-determining pattern for northern tuatara differed from Cook Strait tuatara, where incubation temperatures above 21.6°C produced more male hatchlings and, below that, more females. Laparoscopies were carried out on juveniles from eggs incubated at different temperatures to evaluate the sex ratio of young produced, and future incubations were managed for an even sex ratio.[9] Paternity was evaluated to understand the representation of genetic diversity from founders in the next generation offspring.[10] One male had sired 78 per cent of hatchlings, so he was removed from breeding opportunities and later released at his original capture site to allow better representation of all four males. One male failed to sire any offspring, even with sole access to females, so his genetic legacy was not captured in following generations of Hauturu tuatara. In some cases maternity was also confirmed using genetic tools, especially where nests had been excavated without good knowledge of which female had built the nest. Female and male tuatara were both observed to be polygamous, with multiple mates — sometimes within the same breeding season.[11]

The conservation of tuatara was one reason why Ngāti Manuhiri decided to support eradication of kiore — removing kiore from the island increased the chances of securing a healthy, sustainable population of tuatara and other native species on Hauturu. From 2006, when the island was declared kiore-free, captive-raised juveniles were released in clusters around the island at locations where adult tuatara had been either captured on the original surveys or seen since; for example researchers and island managers had seen two additional adult males after the 1991 and 1992 surveys. Six release areas received juveniles, subadults and ultimately the original adults from the breeding stock.

In 2015 a survey was conducted to evaluate the population of tuatara on Hauturu. It was held 11 years after the kiore eradication had taken place, and included regions of the island where the first eight tuatara were captured during the rediscovery in 1991 and 1992; where tuatara had been subsequently sighted; and release locations for artificially incubated, captive-raised juveniles. The goals were to evaluate how released tuatara were doing, and whether further tuatara could be located that represented stock that had survived through the kiore habitation and eradication, or that provided evidence for recruitment in the wild subsequent to the kiore eradication. Data from tuatara at release sites demonstrated they are healthy and growing. Four unidentified tuatara were located, providing evidence for successful recruitment in the wild, and that there are likely tuatara that exist on the island that were not observed during earlier surveys.

The tuatara housing facilities on the island were opened up in 2017 to allow the remaining captive tuatara to make their own way out of the enclosures and establish territories around the adjacent area — although most of the adults have stayed in their burrows after the fence was removed. Visitors to the island may now be lucky enough to see the fruits of the labour of many dedicated people and organisations; that is, they now have the opportunity to observe tuatara in the wild on Hauturu, as the beginning of a long comeback towards a self-sustaining population.

## Lizards

There was little systematic surveying for lizards until three surveys led by Tony Whitaker in the early to mid-1990s. The first of these aimed at determining the status of tuatara. However, the focus of the surveys changed after the incidental discovery of a chevron skink (*Oligosoma homalonotum*), a species known previously only from Aotea/Great Barrier Island. The island was then extensively searched for chevron skinks. Arrays of pitfall traps were installed in 1993 in habitats similar to those where chevron skinks had been encountered on Aotea, as well as in coastal sites near Te Maraeroa (1990) and in ridge sites and other valley sites (2000–2005). The pitfall trap monitoring has provided a useful basis for determining lizard habitat use and changes in abundance over time. In addition, the traps captured a previously unknown species from the island, the striped skink (*O. striatum*), at a heavily forested site.

The list of 13 confirmed species on Hauturu (see Chapter 9, species list) is the largest for any offshore island and equals the number of species on Aotea. The highest diversity within any of the habitats monitored is in the rocky coastal sites near the settlement, with four species of skinks and one species of gecko (Table 1). Only copper skinks (*Oligosoma aeneum*) and ornate skinks (*O. ornatum*) were found outside these coastal habitats.

## Table 1. Distribution of lizard species

Distribution of five widespread lizard species in a range of habitat types on Hauturu, following monitoring by Keri Neilson and Rosalie Stamp in the period leading up to and immediately after the eradication of kiore. Presence is indicated as a shaded box.

| Species | Coastal bare rock–low scrub | Stream base in coastal forest | Stream terrace in coastal forest | Ridges with kānuka cover |
|---|---|---|---|---|
| Copper skink | ■ | | | ■ |
| Ornate skink | ■ | | | ■ |
| Shore skink | ■ | | | |
| Hauraki skink | ■ | | | |
| Raukawa gecko | ■ | | | |

Four of the listed species of geckos are largely arboreal (see Appendix 4), so have not been captured in ground-based traps. However, even with extensive surveys in the 1990s Whitaker was only able to collate small numbers of chance sightings for each of these gecko species, as well as for the moko skink (*Oligosoma moco*). This low density was reflected in pitfall captures in the forest, where capture rates of more than 1/100 trap nights (TN) were achieved only at one ridge site during surveys between 2000 and 2005. Far greater numbers were caught on the coast, where total capture rates from 2000 to 2005 fluctuated between 15 and 25/100 TN. Since 2005, after rat eradication, Griffiths et al. (in press) reported captures as increasing eighteenfold, with the largest increases recorded for shore skink (*O. smithi*) and Hauraki skink (*O. townsi*). They also reported regular captures of moko skinks and increasingly frequent sightings of Raukawa (common) gecko (*Woodworthia maculata*), Duvaucel's gecko (*Hoplodactylus duvaucelii*), forest gecko (*Mokopirirakau granulatus*) and Pacific gecko (*Dactylocnemis pacificus*) over the same period.

The egg-laying skink (*Oligosoma suteri*) was reported from traps on the coast at Te Maraeroa in the early 1990s, but there have been no reports there since. However, the species was discovered in 2016 at two other coastal sites. These skinks are coastal specialists that forage just above the high-tide mark, often under rotting seaweed, where they can reach very high densities. There are probably numerous additional sites for the species on Hauturu, so they are likely to be far more widespread than they appear to be at present. However, increases in abundance and the spread of the more rarely seen lizards can also reflect slow recovery rates caused by small clutch sizes and slow growth to maturity. For example, all New Zealand geckos have a maximum clutch size of two. Alison Cree found that in addition to low clutch size, species such as Duvaucel's gecko may only produce offspring in alternate years — an effective output of one per annum.[12]

Increases in the capture rates of coastal lizards and sightings of geckos are consistent with reports from other islands after kiore have been removed. For example, in the Mercury Islands, sighting frequency and capture rates of Raukawa and Duvaucel's gecko and shore skinks in coastal areas greatly increased after rat eradication. For species such as Raukawa geckos these increases may have been accentuated by their migration back to habitats previously unavailable in the presence of rats. Increased capture rates in forest areas have tended to be more modest. Reports from islands other than Hauturu suggest that many species of lizards congregate in the most productive sites, including rocky coastal areas exposed to the sun, such as north- or west-facing areas; sites heavily burrowed by seabirds, with the resulting high nutrient levels and abundance of invertebrates; and sites with trees infested with honeydew scale insects, which provide year-round supplies of carbohydrate (sugars). The honeydew-infested sites attract Raukawa and Duvaucel's geckos, which also congregate around coastal flax and pōhutukawa (*Metrosideros excelsa*) for their nectar-rich flowers.

Will we see similar trends on Hauturu? In the monitored coastal site these increases have already started. However, although species such as Duvaucel's gecko have become more abundant in coastal sites in the Mercury Islands and elsewhere (such as the Hen and Chickens), they are still rarely encountered on Hauturu. Other sites where geckos are likely to congregate include tall kānuka (*Kunzea* spp.) forest, since kānuka is host to the

Copper skink (*Oligosoma aeneum*) (DvW)

Moko skink (*Oligosoma moco*) (DvW)

Chevron skink (*Oligosoma homalonotum*) (DvW)

Shore skink (*Oligosoma smithi*) (DvW)

Suter's skink (*Oligosoma suteri*) (NF)

Ornate skink (*Oligosoma ornatum*) (DvW)

Striped skink (*Oligosoma striatum*) (DvW)

Hauraki skink (*Oligosoma townsi*) (DvW)

**Amphibians and reptiles** 131

honeydew scale *Coelostomidia wairoensis*. Coastal sites with ngaio (*Myoporum laetum*) and karo (*Pittosporum crassifolium*) can support a different species of honeydew scale: *C. zealandica*.[13] This species appears to be susceptible to loss of hosts (karo) in the presence of kiore, and it is unclear if it survives on Hauturu. If it is present, honeydew-infested karo and ngaio will provide more highly productive habitats for geckos.

The currently known lizard fauna on Hauturu is the most diverse on any New Zealand island that has been cleared of invasive predators while still coexisting with its apex terrestrial predator (tuatara). To that extent, the reptile fauna of Hauturu is the closest approximation of historic faunas on the adjacent mainland before human contact. However, there are two species that were once present on the mainland that are likely also to have been on Hauturu. The two large nocturnal species, McGregor's skink (*Oligosoma macgregori*) and robust skink (*O. alani*) disappeared from the mainland, where they are represented only by remains in caves and dune deposits. Both species are now confined to a scattering of small offshore islands that rats have failed to reach. Relict populations of both species are found on islands near Hauturu — robust skinks in the Mokohinau Islands and McGregor's skink on an islet near Taranga/Hen Island. Both of these ground-dwelling skink species occupy heavily forested sites, often in or near seabird burrows, although McGregor's skinks also enter coastal habitats.

There are many distinctive aspects of Hauturu that cause us to raise questions about how each species' distribution, habitat use and abundance will change now that mammalian predation has ceased. For example, unlike much of the mainland and nearby Aotea, Hauturu has sufficient elevation to induce areas of beech (*Nothofagus*) and cloud forest. It is unknown how these habitats will be used as the resident lizard fauna recovers. There are microhabitats, too, such as epiphyte gardens on tall forest trees, including the extensive areas of kauri (*Agathis australis*), that are absent from most of the smaller offshore islands. Whether the partially arboreal chevron and striped skinks will appear in such sites is unknown.

It is also unclear how areas with honeydew scale and seabird burrows will be used by lizards in the long term — for example, whether Pacific and forest geckos will be attracted to areas with honeydew supplies. The affinity of forest geckos for kānuka at other locations may provide a clue. It is also unknown whether the indirect relationship between seabird, tuatara and lizard abundance found on the smaller islands will hold for Hauturu, where soils have been modified by kauri.

There are no models for the long-term recovery of reptiles on islands such as Hauturu. So far, all data have come from small, relatively low-lying islands with extensive seabird burrowing, but Hauturu with its old-growth forest and altitudinal zonation provides stratification at micro and macro scales that are unavailable elsewhere. Furthermore, seabird burrowing may remain more localised than on the smaller islands, thereby increasing heterogeneity at ground level. Given the size of the island and the apparent low densities of surviving reptiles, any answers to how they make use of such environmental complexity may be slow to develop. This is a story that is only beginning to unfold.

Elegant gecko (*Naultinus elegans*) (LWh)

# 6.3 Birds

— TIM LOVEGROVE, MATT RAYNER
AND KEVIN PARKER

**In the late nineteenth century, New Zealand naturalist Sir Walter Buller was one of a number of observers acutely aware that many New Zealand forest birds were in decline, including bellbird/korimako (*Anthornis melanura*), robin/toutouwai (*Petroica longipes*) and whitehead/pōpokotea (*Mohoua albicilla*).**[1]

Buller highlighted the value of islands such as Hauturu and Kāpiti where these species were still common. While habitat loss through forest clearing was an important factor, Buller, Reischek and others also blamed the introduction of carnivorous mammals for the decline of many native birds on the mainland.[2] From the 1880s Hauturu was recognised as especially important because it was the only place where the hihi/stitchbird (*Notiomystis cincta*) still survived.[3] Hauturu has some of the best examples of natural ecosystems in New Zealand and, as one of our most important sanctuaries, harbours a high proportion of the North Island's extant indigenous forest birds. The forest avifauna has recently been enhanced through translocations of kākāpō (*Strigops habroptilus*), kōkako (*Callaeas wilsoni*) and tīeke/saddleback (*Philesturnus rufusater*). In recent years Hauturu has also been an important source of birds to found new populations on a number of restored Hauraki Gulf islands and pest-managed sanctuaries elsewhere in the North Island.

What would have been the composition of Hauturu's birdlife when Māori first arrived? Using Medway's example of Stephens Island, Hauturu might have still supported now-extinct small rails (*Capellirallus karamu* and *Gallinula hodgenorum*), flightless wrens (*Pachyplichas jagmi* and *Traversia lyalli*), bush wren (*Xenicus longipes*) and piopio (*Turnagra tanagra*).[4] The survival of North Island snipe (*Coenocorypha barrierensis*) on Hauturu into the mid-nineteenth century provides a tantalising glimpse of what the island's full complement of birds might have been. Some of these species, especially the small rails and flightless wrens, could have been wiped from the island by kiore or Pacific rats (*Rattus exulans*). Snipe were clearly already very scarce by the mid-nineteenth century. Could huia (*Heteralocha acutirostris*) also have been present, but hunted to extinction during the early Māori era? What about piopio? Surely piopio, if it was present, should have survived into the European era on Hauturu as it did on Stephens Island and the mainland.[5] Was piopio naturally absent from Hauturu for some reason? Why, too, did kōkako not occur on Hauturu as it did on nearby Aotea?

Turbott discussed those species such as weka (*Gallirallus australis*) and kōkako that were never historically recorded on Hauturu, and compared the avifauna of Hauturu with Taranga/Hen and the Poor Knights islands.[6] The smaller islands have lower land-bird diversity: Taranga, for example, lacks the rifleman (*Acanthisitta chloris*), robin and

Stitchbird/hihi, female (*Notiomystis cincta*) (NF)

whitehead, while the Poor Knights lack a much wider range of species. The missing species reflect the smaller size, less altitudinal range and more limited diversity of forest habitats on Taranga and the Poor Knights compared with Hauturu. Importantly, during the period when Māori dwelt on Taranga and the Poor Knights a much greater proportion of the forest was burned or cleared.[7] Māori cleared parts of Hauturu, but a very large area of the inland forest was never cleared. On Taranga and the Poor Knights, however, forest clearance would have greatly reduced the available habitat for some species. Those that persisted were species where viable populations could persist in small habitat patches, for example bellbird, riroriro/grey warbler (*Gerygone igata*), pīwakawaka/fantail (*Rhipidura fuliginosa*) and tīeke; those that could survive in open country, such as red-crowned kākāriki (*Cyanoramphus novaezelandiae*); or mobile species that could recolonise, such as kererū (*Hemiphaga novaeseelandiae*), kākā (*Nestor meridionalis*), kākāriki (*Cyanoramphus* spp.), fantail, tomtit/miromiro (*Petroica macrocephala*), bellbird and tūī (*Prosthemadera novaeseelandiae*).

A number of studies have been carried out to determine the abundance of various bird species on Hauturu, and to measure changes in bird populations over the period since feral cats, and later kiore, were eradicated. Some of the results from these are included in the species accounts that follow.

## Seabirds

The Hauraki Gulf is a global centre of seabird biodiversity and seabirds play a huge role in Hauturu's ecology.[8] Although only a few seabird species breed on the island, they are dominant in terms of their biomass and impacts on the ecosystem. Studies over the past 20 years on New Zealand's northern offshore islands, including Hauturu, show that seabirds are important ecosystem engineers. Their burrowing tills the soil and improves its structure; they improve plant growth by depositing marine-derived nitrogen and phosphorus as guano, dead adults, eggs and chicks, thus increasing the abundance of terrestrial invertebrates;[9] and they increase the productivity of nearshore rocky reef habitats through runoff from the land. Hauturu, with its largely intact ecosystems and enormous seabird populations, is a microcosm of mainland New Zealand before humans arrived.

Of the seabird species recorded breeding on Hauturu, it is the tubenoses or Procellariiformes that have the greatest ecosystem impact. Cook's petrel/tītī (*Pterodroma cookii*), black petrel/tāiko (*Procellaria parkinsoni*) and grey-faced petrel/ōi (*Pterodroma gouldi*) are medium to large burrowing petrels that breed in the forest from the lower slopes to the summit ridges. All three species are still recovering after feral cats and kiore were eradicated from Hauturu, and their growing populations are having a profound influence on the ecological trajectory of the island. For example, it is estimated that the rapidly increasing population of Cook's petrels alone contributes approximately 50 tonnes of nutrients annually to the island's forest ecosystems.

The extended footprint of Hauturu's seabirds is important, too. Recent studies using advanced tracking technologies have confirmed migration routes to far-flung regions of the northern and eastern Pacific, up to 10,000 kilometres away.[10] Rebounding populations of migrant Cook's and black petrels from Hauturu are playing an increasing ecological role as top predators in marine ecosystems far from New Zealand.

### Cook's petrel/tītī (*Pterodroma cookii*)

Summer nights on Hauturu are dominated by the din of Cook's petrels returning to their burrows on the higher slopes. This medium-sized (c. 200 grams) endemic petrel is probably the island's most numerous vertebrate. There are two populations of Cook's petrels breeding at either end of New Zealand — on Hauturu and on Whenua Hou/Codfish Island. The populations are genetically and behaviourally distinct and represent two reproductively isolated subspecies.[11] Tracking studies have shown that during breeding, Cook's petrels from Hauturu forage east and west of the North Island. They frequently cross the mainland north of Auckland to reach the Tasman Sea, and on summer evenings the *kek-kek-kek* calls of returning Cook's petrels are a familiar sound over Rodney district in the so-called North Auckland Seabird Flyway, which is probably also used by black and grey-faced petrels.[12] Cook's petrels from Hauturu may travel several thousand kilometres from home but have little overlap at sea with the Codfish Island subspecies.[13]

Beginning in March, Cook's petrels make an enormous trans-Pacific migration from Hauturu totalling over 40,000 kilometres. Over about 35 days they first fly east then north, crossing the equator to overwinter in the North Pacific convergence zone, which encompasses the waters from about 1000 kilometres north of Hawai'i across to the coast of California and Mexico.[14] The return trip occurs over about 20 days in September.

Hauturu has the world's largest population of Cook's petrels. Originally they were estimated to number around 50,000 breeding pairs.[15] Surveys across the island in 2005 showed they were widespread on all forested ridgetops and slopes that had escaped historic burning and deforestation;[16] it was estimated there were about 286,000 breeding pairs, with a total population of over 1.2 million birds.[17] Since 2005 it is estimated that the breeding population has increased to c. 400,000 pairs. We now see many more freshly dug Cook's petrel burrows; the nocturnal arrival of birds over Hauturu is much noisier than before; increased numbers are heard at night over Auckland and the Rodney district; and observers of pelagic birds in the North Pacific Ocean are reporting larger numbers.

The high rate of losses of Cook's and black petrels to feral cats in the 1940s and 1950s was one of the main drivers of the programme to eradicate feral cats from Hauturu in the 1970s.[18] Mike Imber and Matt Rayner studied the birds' breeding biology over a 35-year period between 1972 and 2007; importantly, their studies covered both the feral cat (1980) and kiore (2004) eradications on Hauturu. Counterintuitively, breeding success declined following cat eradication because of the ecological release of kiore, which were also predators of Cook's petrel eggs and chicks. The benefits of removing kiore were clearly shown when breeding success rebounded in 2004 after the rats were gone.[19]

### Grey-faced petrel/ōi (*Pterodroma gouldi*)

Grey-faced petrels are common, large (550 grams), burrowing seabirds on New Zealand's northern offshore islands and on some mainland headlands.[20] Historically they were very abundant on Hauturu, where they burrowed in soft soils along the coastal clifftops.[21] These

Cook's petrel/tītī (*Pterodroma cookii*) (EW)

Black petrel/tāiko (*Procellaria parkinsoni*) (EW)

Grey-faced petrel/ōi (*Pterodroma gouldi*) (NF)

New Zealand storm petrel (*Fregetta maoriana*) (NF)

Fluttering shearwater/pakahā (*Puffinus gavia*) (NF)

Common diving petrel/kuaka (*Pelecanoides urinatrix*) (NF)

winter-breeding petrels were easy prey for feral cats, and Sibson, Turbott and McKenzie all described the severe impacts of cat predation on adult and fledgling birds.[22] Given the level of predation, it is likely that grey-faced petrels were nearly locally extinct on Hauturu by the time the last cat was removed in 1980; intensive searches in the 1970s and 1990s failed to find any breeding birds. Rayner et al. described the return of this species when they found seven active burrows and four chicks above sea cliffs east of Ōrau Cove in November 2009.[23] Since then the birds have recolonised former breeding sites on steep slopes above the mouths of the Waipawa and Haowhenua streams on the west coast of the island, and many have been heard at night over Pōhutukawa Flat on the east coast. At present, 100–300 pairs are believed to be breeding on Hauturu. A comprehensive survey of the island is needed to assess the size of the current population.

### Black petrel/tāiko (*Procellaria parkinsoni*)

Black petrels were formerly widespread and abundant on mainland New Zealand, including the Auckland region, where they bred in the Waitākere Ranges. Today this large 700-gram endemic petrel breeds only on Aotea and Hauturu in the Hauraki Gulf. Mike Imber carried out the first comprehensive study of their breeding biology on Hauturu between 1971 and 1975.[24] Tracking studies show that breeding birds forage along the edge of the continental shelf and beyond in deeper oceanic waters,[25] and observations at sea show that they migrate to the tropical eastern Pacific Ocean during the non-breeding season.[26] As with Cook's petrel, studies of the black petrel were initiated by the New Zealand Wildlife Service in the 1970s following reports of mortality of fledglings as a result of cat predation on Hauturu.[27] After cats were removed in 1980, 249 black petrel chicks were transferred to Hauturu from Aotea between 1986 and 1990 (see Appendix 2). The goal of this project was to investigate the potential of translocating seabird populations and, although most of the birds transferred were subsequently recaptured on Aotea, this method has since been refined and become a mainstay of global efforts in seabird conservation.[28]

Between 2015 and 2017 a comprehensive population survey of black petrels on Hauturu was funded by the Ministry for Primary Industries and the Department of Conservation. This work was prompted by an observed 1.4 per cent annual decline of the Aotea population, and observations of high mortality in longline fisheries.[29] Extensive island-wide surveys using acoustic recorders, transects and dog surveys found that as with Cook's petrels, black petrels on Hauturu prefer to breed on or near the higher ridges, presumably to aid takeoff from the forest canopy. This population survey revised the estimate for black petrels breeding on Hauturu upwards from 100 pairs to 620 pairs.[30]

### New Zealand storm petrel (*Fregetta maoriana*)

This tiny 35-gram storm petrel, known only from a few specimens collected off the New Zealand coast in the nineteenth century, was presumed extinct until it was rediscovered in 2003, when sighted by a bird-watch group off Whitianga.[31] However, it took another 10 years of intensive searching before Hauturu was identified as the only known breeding site.[32] After its rediscovery, initial work focused on developing a method to catch birds at sea, and to confirm its identity by comparing the DNA from blood samples of captured birds with museum specimens.[33]

Between 2006 and 2012 a Hauraki Gulf breeding location was strongly suspected because birds captured in the Gulf were found with bare brood patches, consistent with incubation at some nearby location in the late summer.[34] The timing of this incubation (February–April) differed from all other Hauraki Gulf Procellariiformes, which incubate in late winter or spring. Further evidence of local breeding was found in January 2011 when a bird at sea was photographed with a leaf petiole stuck to its leg, showing the bird had recently been ashore. The petiole was identified as most likely to be a *Pseudopanax* species found naturally only in northern New Zealand, indicating a local breeding site.[35]

Evidence that the breeding site might be Hauturu came on 4 November 2005 when a bird flew aboard Geordie Murman's fishing boat, which was anchored for the night at Waimaomao Bay on the northern side of the island. When the tiny bird fluttered aboard, Geordie, a former Wildlife Service officer who was very familiar with local seabirds and aware of the search for the mystery storm petrel, knew exactly what he was looking at. With the bird safely captured he immediately contacted the Department of Conservation. Richard Griffiths confirmed its identity as a New Zealand storm petrel the next day; feather samples were taken for DNA analysis and the bird was banded and released.[36] The search for the breeding site focused at first on the northern side of Hauturu, near where the bird was caught. Finally, in 2013 New Zealand storm petrels captured at sea were tracked using micro radio tags to breeding sites in the Parihākoakoa Valley on the western side of Hauturu.[37] Subsequently, five active nests were found using a combination of radio telemetry and trained wildlife detector dogs. Two chicks and several adults were banded and nests were monitored using movement-triggered game cameras.

Between 2014 and 2018, spotlights were used to dazzle and gently bring to earth New Zealand storm petrels flying over Te Maraeroa, in the Parihākoakoa Valley and at Pōhutukawa Flat on the northeastern side of the island. Captured birds were colour-banded, and recapture rates and resightings of banded birds at sea were used as a basis for mark–recapture population estimates. Population surveys using acoustic recorders, dogs and radio telemetry, in combination with mark–recapture population modelling, indicates c. 300–400 breeding pairs and a total population of c. 2000 New Zealand storm petrels on Hauturu. The breeding population appears to be widely dispersed at low density across the island. The species is now recovering from fewer than 1000 birds surviving at the time of rat eradication. Small petrels are highly vulnerable to mammalian predators, so how this tiny storm petrel survived for so long on Hauturu in the presence of both feral cats and kiore is a mystery.

In 2015, wooden nest boxes and an acoustic playback system were installed in an attempt to establish a breeding colony to facilitate studies of the species' breeding biology. Juveniles were captured by spotlighting and their age was confirmed through brood patch condition. They were then placed in the nest boxes. Game cameras and field observations since then show the site has been visited frequently by banded and unbanded New Zealand storm petrels. To date six eggs have been laid, but so far none of the nests has been successful.

## Other tubenoses on Hauturu
Based on the findings of Reischek and other observers, Turbott listed several other tubenoses or Procellariiformes as apparently breeding on Hauturu, including little

shearwater (*Puffinus assimilis*), fluttering shearwater/pakahā (*Puffinus gavia*), fairy prion/
tītī wainui (*Pachyptila turtur*) and common diving petrel/kuaka (*Pelecanoides urinatrix*).[38]
Terry Greene and Paul Scofield found several fluttering shearwater chicks, along with
some small burrows that probably belonged to diving petrels, on Lots Wife rock stack
on 12 January 1991. Their observations showed that the small diving petrel colony found
there by Peter Bull in October 1945 probably still existed.[39] Acoustic recorders placed on
the northern coast of Hauturu in October 2009 near Waimaomao Bay confirmed that both
fluttering shearwaters and common diving petrels were coming ashore, based on ground
calls. Today it is likely that these species are recolonising old habitats after the removal
of cats and kiore. We do not yet know the population sizes and distribution of these
species on Hauturu. There is no recent evidence that little shearwaters or fairy prions
breed on Hauturu, although a fairy prion was found injured on the track to the summit on
26 November 1960, another was captured at the summit on 1 December 1962, and others
have recently been attracted to spotlights during research on New Zealand storm petrels.

### Pied shag/kāruhiruhi (*Phalacrocorax varius*)
There is a small population of pied shags on Hauturu, centred on a long-established
colony in pōhutukawa trees overhanging the cliffs between the mouth of the Tirikakawa
Stream and the eastern edge of Te Maraeroa. Turbott reported about 30 nests in the late
1940s, and Dawson counted 35–40 nests in 1949.[40] Recent counts show that the colony
is smaller now, with c. 10–15 nests. Turbott also recorded a colony at Te Ananuiarau Bay
on the northern side of Hauturu which had about 20 nests in 1945. Turbott described
other colonies too, including one on rocks at the northern end of the island that was
photographed by G. A. Buddle in 1904–05, and another with 15 birds and a single nest at
Pōhutukawa Flat.[41] It appears that none of these other colonies exists now. Pied shags may
be seen fishing and roosting at various places around the coastline.

### Other seabirds on Hauturu
Little penguins/kororā (*Eudyptula minor*), southern black-backed gulls/karoro (*Larus dominicanus*) and white-fronted terns/tara (*Sterna striata*) all breed on Hauturu. Little
penguins breed in many places around the coastline, and under the floors of some of the
buildings; the population numbers possibly 100–200 pairs. About 20–30 pairs of black-backed gulls breed on stacks and rocky headlands around the coast, and there is a small
colony of 5–10 pairs on the boulder bank in the southeast of the island near East Cape.
White-fronted terns have been reported breeding on some rock stacks, including Lion
Rock and Lots Wife.[42] About 50–100 pairs may be present, but as elsewhere with this
species, colonies can be itinerant and may not be present at the same location every year.

## Land, shore and freshwater birds

### North Island brown kiwi (*Apteryx mantelli*)
Despite reports to the contrary, kiwi have been present on Hauturu for a very long time, but
there were times when they may have been very scarce. In 1841 local Māori told German

Pied shag/kāruhiruhi (*Phalacrocorax varius*) (SF)

Southern black-backed gull/karoro (*Larus dominicanus*) (OT)

White-fronted tern/tara (*Sterna striata*) (NF)

North Island brown kiwi (*Apteryx mantelli*) (SF)

Brown teal/pāteke (*Anas chlorotis*) (LWh)

Little penguin/kororā (*Eudyptula minor*) (TL)

naturalist Ernst Dieffenbach that kiwi were present,[43] but later visitors, including Captain Wood, Sir George Grey and E. L. Layard, found none.[44] Captain Hutton, who had a dog with him when he visited the island, concluded that kiwi were either very rare or not present. In 1873 Buller wrote that T. Kirk had collected several kiwi on Hauturu, but those specimens have been lost.[45] The paucity of early records is probably not surprising as kiwi can be hard to find, especially when numbers are low, the search is only during the day, and the searcher has not previously observed kiwi sign. It is also possible that free-ranging dogs from the Te Maraeroa settlement had substantially reduced the population.

In his first week on the island in January 1897 Robert Shakespear wrote, 'I think I saw a kiwi.' In later diaries he records some kiwi sightings but the origin of those birds is unclear because, at the time, brown kiwi were being sent from various parts of New Zealand to 'populate the island'. The first recorded releases were two birds from an unknown location, sent by Thomas Cheeseman on 21 March 1898. There were further translocations, with departures from the mainland noted in newspapers but arrivals at the island not recorded, and vice versa. Other releases include two from New Plymouth before 1903; as well as a South Island brown kiwi 'presumably from Resolution Island',[46] an albino kiwi from the Taupō region in 1913 and a young male from Northland in 1931 (see Appendix 2).

There is at least one extant specimen that pre-dates the first release in 1898 — a brown kiwi that Reischek collected on Hauturu in 1882 and which is now in the Vienna State Museum. It is from this specimen that Ricardo Palma obtained a new species of louse, unique to Hauturu, which he named *Rallicola* (*Aptericola*) *rodericki*.[47] This finding has been confirmed by louse samples taken more recently from other Hauturu kiwi. Further evidence of brown kiwi being natural on the island was provided by Herbert and Daugherty, who found a unique haplotype in brown kiwi blood samples from Hauturu.[48] Brown kiwi on Hauturu today are a genetic mix of the original population and kiwi translocated from various parts of the mainland.

Brown kiwi are widespread and common on all parts of Hauturu. A kiwi call count survey in 2002 confirmed their widespread distribution. During the survey a rate of 24 calls per hour was recorded, and the population was estimated at 240–300 pairs, based on an average territory size of 10–12.5 hectares per pair.[49]

## Great spotted kiwi (*Apteryx haastii*)

Nineteen great spotted kiwi were introduced to Hauturu from the Gouland Downs, northwest Nelson, in 1915.[50] The caretaker, Robert Nelson, reported that they had become established. Turbott notes that after a visit by R. A. Falla in 1928 when some feathers were found, there were no further sightings.[51] However, records from the rangers' diary suggest that this species may have persisted much longer: a nest and egg were found at the Waipawa Stream in November 1962, and there is a clear sighting by Gina Blanshard of a 'large spotted grey kiwi at side of Summit Track at 600ft' on 9 January 1964. There have been no further reports of this species on Hauturu.

## Australian brown quail (*Coturnix ypsilophora*)

Reischek found brown quail on Hauturu,[52] and they were common at Te Maraeroa during Robert Shakespear's caretakership between 1897 and 1910.[53] They were also observed during the 1940s, and Dawson thought there were possibly a few pairs at Te Maraeroa.[54] There have been no recent reports, but brown quail could recolonise from Aotea or Tāwharanui. They thrive in places free of introduced predators. In prehistoric times the extinct New Zealand quail (*Coturnix novaezelandiae*) might have been present.

### Paradise shelduck/pūtangitangi (*Tadorna variegata*)
Paradise shelducks were regularly seen with ducklings at Te Maraeroa during the 1980s, and they have been recorded as visitors. There are no recent records of this species breeding on Hauturu. Regeneration on the flat since livestock was removed means there is very little suitable open habitat for them.

### Brown teal/pāteke (*Anas chlorotis*)
Pāteke have disappeared from most of New Zealand, probably mainly as a result of predation by introduced mammals, especially mustelids. A small population exists on Hauturu, based mainly at Te Maraeroa and the stream mouths along the south coast.[55] A rangers' diary record (18/1/90) describes up to 18 pāteke seen at night in the rangers' garden, where they were being fed. It is assumed that pāteke are capable of reaching Hauturu from the much larger population on Aotea. The Hauturu population might also benefit from birds dispersing from nearby sites such as Tāwharanui, where new populations are being established through the release of captive-bred birds. Over the long term, however, the Hauturu population will always be limited by the small amount of available habitat.

### Swamp harrier/kāhu (*Circus approximans*)
Turbott notes that the harrier was regularly seen on the island, and the same applies today.[56] There are no records of harriers breeding on Hauturu, but since livestock was removed, large areas in the damper parts of Te Maraeroa have reverted to rank sedgeland, which would provide good nesting habitat. Their diet probably consists mostly of birds and carrion scavenged from the beach. Terry Greene noted that kererū regularly fell victim to harriers at Te Maraeroa; he also saw a harrier pursuing a kākāriki through the open kānuka forest and another feeding on a recently fledged kākā chick.

### New Zealand falcon/kārearea (*Falco novaeseelandiae*)
Hutton and Reischek recorded New Zealand falcons on Hauturu,[57] and Reischek found a nest and shot the nesting female.[58] The most recent sighting, according to the rangers' diary, was in January 1967 when a pair was seen 'near a slip beside Whekau'. Falcons are highly mobile and are capable of reaching the island unassisted, so they could be expected to recolonise naturally. However, apart from sporadic sightings they are absent from Northland and most of the Auckland region; the nearest potential source populations are some distance away in the South Waikato and Rotorua districts.

### Banded rail/moho pererū (*Gallirallus philippensis*)
Turbott recorded sightings of banded rails near the homestead from about 1940 and noted that there were a few on the island.[59] A long period followed when banded rails

were apparently absent. Recently there have been numerous sightings at Te Maraeroa and they are now breeding on Hauturu. The feral cat, an important predator, has been removed, and the habitat has almost certainly improved since the pasture became rank after the removal of livestock in the mid-1990s. Sources of colonists include Aotea where they are abundant, and Tāwharanui where they have thrived since the open sanctuary was established.

### Spotless crake/pūweto (*Porzana tabuensis*)
The first spotless crakes were seen on Hauturu in 2015 and, like banded rails, they could have colonised from Aotea or the mainland. As with the rails, extensive areas of rank, damp pastureland and sedges on Te Maraeroa provide good cover. In predator-free environments such as the Poor Knights Islands crakes occur in the coastal forest, and at Tāwharanui they occur in forest and on grassy hillsides retired from grazing, as well as in wetlands. In future they could spread into the forest on Hauturu.

### Pūkeko (*Porphyrio melanotus*)
Turbott lists just two records of pūkeko on the island, in 1908–09 and in 1955.[60] Today, however, pūkeko are resident and breeding on Te Maraeroa. In recent years there has been a small group living in the rank pastureland east of the homestead area, and another small group in similar habitat between the West Landing and the Waipawa Stream mouth.

### Variable oystercatcher/tōrea (*Haematopus unicolor*)
Hutton recorded this species during a four-day visit to the eastern side of the island in December 1867,[61] and Reischek listed it as present.[62] It possibly bred on Hauturu, but there have been no recent records. Variable oystercatchers are common on Aotea and on the adjacent mainland.

### North Island snipe (*Coenocorypha barrierensis*)
Turbott and Miskelly describe in some detail the unique snipe specimen,[63] now held at Auckland Museum, that Hutton described as being obtained from Hauturu in 1870. It is assumed that Hauturu was the last place where the now extinct North Island snipe survived. Snipe vanished early from mainland New Zealand as a result of predation by kiore, and Miskelly speculated that its survival at least until 1870 on Hauturu in the presence of kiore suggests that these rats could have been late arrivals on the island. It is assumed snipe on Hauturu became extinct soon after their discovery, and they would certainly not have survived very long after cats became established.

### New Zealand pigeon/kererū (*Hemiphaga novaeseelandiae*)
Kererū are common at Te Maraeroa where flocks feed on *Muehlenbeckia* and *Calystegia* on the boulder bank. The number of ground-feeding kererū on Hauturu leaves a strong impression on visitors because they are rarely seen on the ground in such numbers on the mainland. Kererū make seasonal movements around the island and to the mainland depending on the fruiting cycles of various native trees. During summer and autumn the fruit of kawakawa, pūriri, nīkau, karaka, *Coprosma* spp., tawa, taraire, kohekohe and

miro are important, while in winter and spring, foliage such as *Muehlenbeckia* forms an important part of their diet. Local movements may account for the higher numbers of kererū found on the high-altitude bird counts reported by Girardet et al.[64] Overall, during the transect bird counts between 1975 and 2017, kererū numbers did not change markedly.

## Kākāpō (*Strigops habroptilus*)

Soon after kākāpō were rediscovered on Rakiura/Stewart Island in 1977, the small population of 100–200 birds was found to be declining quickly because of feral cat predation.[65] Using trained dogs, all of the surviving birds that could be located were rescued and transferred to Whenua Hou/Codfish Island, Maud Island and Hauturu. Only 62 kākāpō were found: 38 males and 24 females. At the time, both Hauturu and Whenua Hou had kiore, a potential threat to kākāpō nests. This was not the first time kākāpō had been released on Hauturu: four birds brought from Resolution Island were released on Hauturu in 1903, and others were released on Kāpiti Island.[66] The first kākāpō on Hauturu probably did not survive long in the presence of feral cats.

After the release of 21 Stewart Island and one Fiordland kākāpō on Hauturu, the birds were managed intensively and were provided with supplementary food including nuts and fruit and vegetables such as apple, carrot and kūmara. There was early optimism when males established track-and-bowl systems and began booming along the summit ridges and some females nested.[67] Two young males fledged successfully but some nests were preyed on by kiore, despite protective rings of rat traps placed around them.[68] When the decision was made to remove kiore from Hauturu using an aerial poison drop, all of the kākāpō that could be found were transferred back to the southern kākāpō island, Whenua Hou. A period followed between 1999 and 2012 when there were no kākāpō on Hauturu. In the meantime, there were successful breeding years in the south, coinciding with rimu mast years, and the population was slowly expanding. After the removal of kiore, kākāpō were translocated back to Hauturu from 2012 (see Appendix 2 and Appendix 3). The returning birds included some of those that had been there previously and, remarkably, some of them moved back into their old territories despite an absence of 13 or more years. Visitors to Hauturu are unlikely to see a kākāpō. They are more likely to see their characteristic feeding sign on sedges and other plants along the ridge tracks.

## North Island kākā (*Nestor meridionalis*)

Hauturu, Aotea and the Hen and Chicken Islands are northern strongholds for kākā (photo on page 159), and between them the islands probably support a population of several thousand birds. The kākā is a conspicuous and noisy resident around the homestead and Te Maraeroa, and this was particularly so in the days when kākā were fed at the homestead. A radio-tagging study showed that kākā move quite freely between the islands, and every winter there is an outflow of young to the mainland with frequent sightings in suburban Auckland. Some disperse even more widely: a kākā banded by Terry Greene on Hauturu was recovered in Gisborne. The thriving Hauraki Gulf kākā population is an important source of founders to restored habitats on the mainland.

Banded rail/moho pererū (*Gallirallus philippensis*) (CRV)

Pūkeko (*Porphyrio melanotus*) (LWh)

Spotless crake/pūweto (*Porzana tabuensis*), on Aorangi Island, Poor Knights Islands (CM)

New Zealand pigeon/kererū (*Hemiphaga novaeseelandiae*) (SF)

Kākāpō (*Strigops habroptilis*) (AD)

Red-crowned kākāriki (*Cyanoramphus novaezelandiae*) (NF)

Yellow-crowned kākāriki (*Cyanoramphus auriceps*) (SF)

## Red-crowned kākāriki (*Cyanoramphus novaezelandiae*) and yellow-crowned kākāriki (*Cyanoramphus auriceps*)

There are now very few places with sympatric populations of red- and yellow-crowned kākāriki. The only Hauraki Gulf islands where both are well established are Hauturu and the Hen and Chicken Islands. The red-crowned kākāriki is the more abundant of the two parakeet species on Hauturu and occurs across a wide range of habitats. Small flocks are common at Te Maraeroa, where they often forage on the ground; they are the parakeet visitors are most likely to see soon after stepping ashore. The yellow-crowned kākāriki, a forest specialist, is more cryptic and much harder to find. There are subtle differences in the calls of kākāriki species, and a soft two-note call given by yellow-crowned kākāriki is diagnostic. In a study of the diets of the two species Terry Greene found that yellow-crowned kākāriki ate significantly more invertebrates than the red-crowned, which fed on a wide range of plant foods including flower buds, flowers, fruits and seeds, as well as some invertebrates such as sixpenny scale (*Ctenochiton viridis*).[69]

Kākāriki are vulnerable to mammalian predators because both species nest and often roost in tree cavities.[70] In addition, the red-crowned kākāriki's ground-feeding behaviour makes it even more vulnerable and this probably explains why this species is practically extinct on the mainland. They also disappeared from some offshore islands where both goats and feral cats occurred.[71] On those islands, goats opened up the understorey, allowing cats to hunt more efficiently. On islands such as Hauturu and Kāpiti that had cats but no browsers, red-crowned kākāriki survived. The yellow-crowned kākāriki still occurs in the larger North and South Island forests, on some Cook Strait islands, on Stewart Island and its outliers and in the Auckland Islands.

Reischek found orange-fronted kākāriki (*Cyanoramphus malherbi*) on both Hauturu and Taranga, and there are two specimens collected by Reischek on Taranga in the Vienna State Museum that Rowley Taylor examined and confirmed as orange-fronted kākāriki. There has been a more recent report of orange-fronted kākāriki on Taranga.[72] Both Hauturu and Taranga should be carefully searched in case this endangered species, believed now to be confined to a few alpine valleys in Canterbury, has been overlooked.

## Shining cuckoo/pīpīwharauroa (*Chrysococcyx lucidus*)

Migrant shining cuckoos reach New Zealand from the Solomon Islands during late September, heralded by their distinctive whistling calls. They are brood parasites of grey warblers, laying mainly during November. Fledgling shining cuckoos, attended by busy warblers, can be easily found in late December–early January by their persistent begging calls. Since rats were removed from Hauturu in 2004 the forest-bird transect counts show that grey warblers have declined. The effect on the local shining cuckoo population of a decline of its host species has not been measured.

## Long-tailed cuckoo/koekoeā (*Eudynamys taitensis*)

The harsh screech of long-tailed cuckoos is a feature of Hauturu during spring and summer. They winter on many islands of the tropical Pacific, returning to New Zealand in early October.[73] They breed during November and December. In the North Island they are brood parasites of whiteheads, and the abundant whiteheads on Hauturu support a strong population of long-tailed cuckoos. Ian McLean estimated a minimum population on Hauturu of about 260 adult long-tailed cuckoos. In a recent study Michael Anderson caught long-tailed cuckoos on Hauturu and fitted them with satellite radio tags. An astonishing series of tracks from just a few birds showed repeated visits to Aotea, Kawau and the adjacent mainland around Tāwharanui and Leigh. Later, the cuckoos were tracked on their northward migration to wintering grounds across a wide arc of the Southwest Pacific from Vanuatu to Sāmoa.

### Morepork/ruru (*Ninox novaeseelandiae*)
Moreporks are widespread and common on Hauturu. The coastal and valley-floor pōhutukawa and pūriri forest has abundant tree holes and epiphytes, which provide numerous nest sites and secluded roosts. Their diet includes a wide range of large and small prey, and they are capable of taking prey that weighs almost the same as their own body weight. Items reported in their diet on Hauturu include Cook's petrel, kākāriki fledglings, kingfisher, bellbird, whitehead, kiore and wētāpunga.

### New Zealand kingfisher/kōtare (*Todiramphus sanctus*)
During spring and summer kingfishers occur sparingly on Hauturu around the coast and in forest in the valleys and on the lower slopes. Nest holes are usually in rotted trees, but around the coast kingfishers will dig nest holes in the soft breccia on the cliff faces. They take a wide range of prey including many lizards. In winter many of them leave Hauturu and make a seasonal local migration to harbours and estuaries on the mainland coast where food is more abundant.

### North Island rifleman/titipounamu (*Acanthisitta chloris*)
For many years it was thought that Hauturu held the northernmost population of rifleman in New Zealand, until they were discovered at Warawara in Northland.[74] There have been sporadic recent reports of rifleman on Aotea, and Oliver also listed Taranga for rifleman, although none has been reported there in recent times. In counts on the three forest-bird transects on Hauturu between 1975 and 1989 the rifleman was most often seen in the low-altitude transect, and overall on the three transects during that period, rifleman numbers fluctuated. The results suggested that feral cats may not have had a major effect on their numbers.[75] However, in the most recent series of counts from 2013 to 2017, carried out after kiore were removed, rifleman numbers have declined markedly, although we don't know why.

### North Island kōkako (*Callaeas wilsoni*)
For some reason Hauturu lacked a natural population of kōkako in historic times. Hutton found kōkako on Aotea, where they survived until the early 1990s. Reischek speculated that: 'The high slopes would be a favourable resort for kakapo . . . crow [and] . . . saddleback'.[76] During the 1970s and 1980s, despite wide publicity and treetop protests, there was extensive government-sponsored clearing of mature indigenous forests on the Volcanic Plateau in the central North Island. Among the birds that lost their habitat, only a few kōkako were rescued; the first kōkako translocated to Hauturu in October 1980

Shining cuckoo/pīpīwharauroa (*Chrysococcyx lucidus*) (SF)

Long-tailed cuckoo/koekoeā (*Eudynamys taitensis*) (SF)

New Zealand kingfisher/kōtare (*Todiramphus sanctus*) (NF)

Morepork/ruru (*Ninox novaeseelandiae*) (NF)

North Island rifleman/titipounamu (*Acanthisitta chloris*) (TL)

North Island kōkako (*Callaeas wilsoni*) (NF)

North Island saddleback/tīeke (*Philesturnus rufusater*) (MS)

were plucked literally from the teeth of advancing chainsaws. Between 1980 and 1988, 32 kōkako, captured at a number of sites across the Volcanic Plateau, and the last two kōkako known to be on Aotea/Great Barrier Island, were released on Hauturu.[77]

After their release on Hauturu the first birds established territories on the higher slopes in mixed forest dominated by tawa and rātā. As the population expanded, kōkako spread onto the lower slopes, and they are now resident and easily seen around the edges of Te Maraeroa in seral and coastal forest near the ranger's house and bunkhouse. Kōkako were not recorded in the forest-bird transect counts from 1980 to 1989, but were regularly logged in these counts from 2013 to 2017. Hauturu probably now supports more than 400 pairs of kōkako.[78]

## North Island saddleback/tīeke (*Philesturnus rufusater*)

Hutton described the saddleback as very common on Hauturu and not uncommon on Aotea, but by the time Reischek visited in the early 1880s it was very rare.[79] By then feral cats were well established and cats almost certainly extirpated tīeke from Hauturu.[80] Until then they had coexisted on Hauturu with kiore. Kiore are occasional nest predators but, unlike the larger European rats, rarely prey on juvenile or adult birds.[81] On the mainland, however, tīeke faced a suite of introduced predatory mammals, including cats, Norway rats (*Rattus norvegicus*) and ship rats (*Rattus rattus*) and, later, mustelids. Norway rats and cats were early arrivals and had probably reduced the population by the time ship rats spread through the North Island after 1860. Ship rats were probably mainly responsible for wiping out the remaining tīeke from the North Island after 1860.[82]

There was an early attempt in October 1925 to restore tīeke on Hauturu. Seven birds captured on Taranga were released near the caretaker's house. These birds would have been easy prey for feral cats. However, after the cat eradication in 1980 it was finally possible to restore tīeke to Hauturu. There were four translocations totalling 188 birds.[83] Three batches from Cuvier Island (1984, 1987, 1988) were released at the bush edges at Te Maraeroa, and the fourth, consisting of birds from Lady Alice and Whatupuke islands in the Hen and Chickens, was released in 1986 at Pōhutukawa Flat.

Tīeke have thrived on Hauturu, and in recent years they have become one of the more common forest birds. They occupy a wide range of forest habitats from sea level to near the summit, but they are most numerous in the coastal broadleaf forest where favoured invertebrate foods and fruit are plentiful and where ancient pōhutukawa and pūriri provide abundant roost and nest holes. Based on distance sampling counts, Toy et al. estimated the tīeke population on Hauturu at around 6800 birds in 2013.[84]

## Stitchbird/hihi (*Notiomystis cincta*)

Hihi were formerly common throughout the North Island but declined quickly during the second half of the nineteenth century; the last mainland records of their existence in the southern North Island were in the 1880s. Like the saddleback, the hihi not only nests but also roosts in tree cavities. This behaviour made these two species especially vulnerable to the arboreal ship rat that swept across the North Island after 1860. Both hihi and saddleback were very early mainland extinctions. Hutton drew attention to a thriving population of hihi on Hauturu,[85] and Reischek's visits in the 1880s were specifically to hunt for hihi and to obtain specimens; Reischek may have collected up to 130 hihi on Hauturu.[86] The threat posed to the hihi population on Hauturu by rapacious collectors such as Reischek provided added impetus for those lobbying in the early 1890s to have the island set aside as a nature reserve. Reischek found hihi to be very scarce on his first visits in 1880 and 1882, but more numerous on a visit in 1883. These fluctuations may have been due to changing food supplies, although Reischek blamed their rarity on 'domestic wild cats, which are very numerous and commit great havoc among them'.[87] Herbert Guthrie-Smith visited Hauturu in 1919 and had no trouble finding hihi; he saw them on his first day there. With no current knowledge that hihi was a cavity nester he searched fruitlessly at first for nests and eventually found five. Guthrie-Smith's exquisite plates in his 1925 book *Bird Life on Island and Shore* were the first ever photographs of hihi.

During the early 1980s, with the support of the Wildlife Service, George Angehr carried out research on hihi with the aim of learning as much as possible about the ecology of the species and to gather information to help inform translocations to other island habitats. Angehr found that the diet of tūī and male bellbirds included a high proportion of nectar with some fruit and invertebrates.[88] Female bellbirds ate the most invertebrates, along with fruit and nectar, while hihi consumed fairly even proportions of nectar, fruit and invertebrates. Based on observations of the numbers of available food plants and the forest types favoured by hihi on Hauturu, Angehr assessed the suitability of a number of islands for hihi translocations.

The first translocations were to Taranga (1980, 1981), Cuvier (1982) and Kāpiti (1983) (see Appendix 3).[89] Although there was early optimism for a successful translocation to Taranga, all of these early translocation attempts failed. Subsequent translocations to Tiritiri Matangi, Kāpiti, Zealandia (Karori Sanctuary), Bushy Park, Rotokare and Maungatautari have been successful, but these populations rely on supplementary feeding.[90] Thus Hauturu is still the only site where hihi survive without the need for human intervention.

Hihi numbers increased in some of the higher-altitude transect counts during the early 1980s, soon after cats were eradicated. It was assumed that hihi feeding low on flowering shrubs such as *Alseuosmia* might have been vulnerable to cat predation. However, recent transect counts suggest hihi numbers have stabilised, and over the entire 1975–2017 period of transect counts their numbers did not change noticeably. While recent transect counts showed stable hihi numbers, distance sampling from 2005 to 2013 showed that, like tūī, hihi numbers had fluctuated, possibly in response to changes in availability of preferred foods.[91] Based on distance sampling counts, Toy and others estimated the hihi population on Hauturu at around 3100 birds in 2013.

Hihi have a highly unusual breeding system, which has been closely studied.[92] Guthrie-Smith commented on groups of males seen near nests,[93] and more recent studies show that hihi have a variable mating system in which males compete with other males for copulations, and where there is pair and group nesting and promiscuity.

Hihi were long regarded as a honeyeater, along with tūī and bellbirds. However, recent genetic research shows that they are not honeyeaters at all, despite their fondness for nectar.[94] The hihi is more closely related to the wattlebirds (tīeke and kōkako) and now sits in a new family, the Notiomystidae.[95]

### Grey warbler/riroriro (*Gerygone igata*)

Turbott, Sibson and McKenzie all described lower densities of grey warblers on Hauturu compared with the mainland.[96] Sibson suggested that competition with dense populations of other small passerines may be the reason for their scarcity on the island. Grey warblers are distributed across a wide area on Hauturu and are most common on the lower slopes, especially in the mānuka and kānuka forest and in the kānuka groves at Te Maraeroa. The forest-bird transect counts between 1975 and 2017 showed that after cats were removed the numbers of riroriro changed very little, but they declined significantly after kiore were eradicated. During the same period the numbers of robins, bellbirds, tīeke and kōkako increased. As Sibson noted, the decline of grey warblers may be due to competition with other endemic bush birds. In addition to possible competition between species, other changes were occurring such as increased understorey regeneration and changes to the invertebrate fauna, including the local extinction of vespulid wasps.

### Bellbird/korimako (*Anthornis melanura*)

Hauturu and other northern offshore islands became strongholds for the bellbird in the north when the species became practically extinct in Auckland and Northland in the 1860s.[97] Disease was thought to be a factor, but it was more likely caused by the spread of the ship rat, which reached New Zealand in the 1860s. Hauturu, the Hen and Chickens, Poor Knights and some other northern islands remained free of ship rats, and there is no evidence that bellbirds on any of these islands declined as they did on the mainland.

Over the period of the forest-bird transect counts on Hauturu between 1975 and 2017, bellbirds have increased markedly, especially after the kiore eradication, and they are now the most abundant bush birds on the island. The removal of kiore probably allowed the quantities of invertebrates, nectar, fruits and seeds available for birds to increase. There have been major changes since rats were removed, including understorey regeneration as a result of reduced seed predation, the disappearance of introduced wasps that preyed on invertebrates and competed with birds for nectar and honeydew, and reduced predation on invertebrates, reptiles and nesting birds. Compared with their mainland cousins, island bellbirds often nest in cavities on or near the ground and, when kiore were present, these nests would have been vulnerable to predation. In 2004, the same year that kiore were eradicated from Hauturu, mammalian pests were also removed from nearby Tāwharanui Regional Park. In about February 2005, bellbirds arrived en masse at Tāwharanui and around Pakiri and Leigh. The arrivals included males, females and young in similar proportions. Their song patterns identified them as Hauturu birds. It is possible that Hauturu bellbirds, including the ground nesters, bred very successfully during the first breeding season after kiore were removed and competition forced some of them to disperse.

### Tūī (*Prosthemadera novaeseelandiae*)

The tūī is a vocal and conspicuous part of Hauturu's avifauna. For many years kākā, tūī and bellbirds flocked to the sugarwater in the rangers' garden, and dozens of birds lining the sides of the trough used to be a highlight for visitors. As with bellbirds and hihi, tūī move to take advantage of seasonally available nectar sources. During the early years of the transect bird counts, in the late 1970s, similar numbers of tūī and bellbirds were recorded, but in the 2013–17 counts with kiore no longer present, tūī declined noticeably whereas the bellbird increased. It is not known why the tūī has declined on Hauturu — at nectar sources they usually dominate bellbirds.[98] Possibly bellbirds, which eat more invertebrates than tūī, have gained an advantage because there are now more invertebrates; or there may be some other process occurring that is rebalancing the forest avifauna as the numbers of robins, tīeke and kōkako increase. In another series of counts between 2005 and 2013 using distance sampling, Toy et al. found that tūī numbers fluctuated, possibly in response to availability of preferred foods: they estimated the total tūī population at 4600 birds.[99]

### Whitehead/pōpokotea (*Mohoua albicilla*)
The whitehead is another forest bird that disappeared from most parts of the northern North Island, with the exception of Hauturu, which retained a strong population, and Rakitū, where a few persisted at least until the late 1950s. Turbott described the whitehead as the most common bush bird on the island.[100] The forest-bird transect counts confirm that this was certainly so during the 1970s and 1980s, and during the period following the cat eradication. However, since kiore were removed in 2004 whitehead numbers have declined significantly, and these counts show that bellbirds are now the most common bush birds. Hauturu's whitehead population is important because it also supports a sizeable population of long-tailed cuckoos. A large number of whiteheads have been translocated from Hauturu to establish populations elsewhere (see Appendix 3).[101]

### North Island fantail/pīwakawaka (*Rhipidura fuliginosa*)
Fantails occur sparingly in most forest types on Hauturu from sea level to the summit, but they are not common in dense bush as they prefer clearings and forest margins. Counts on the three forest-bird transects between 1975 and 2017, covering the periods when both feral cats and kiore were removed from Hauturu, show that fantail numbers have not changed much.

### North Island tomtit/miromiro (*Petroica macrocephala*)
Turbott considered the tomtit to be more numerous than the fantail on Hauturu.[102] Tomtits are widespread on the island, and the 1975–2017 transect counts show they are more numerous in the seral kānuka and mānuka forests on the lower slopes. The counts showed that after feral cats were removed, tomtit numbers increased slightly; however, after kiore were removed, tomtit numbers declined significantly.

### North Island robin/toutouwai (*Petroica longipes*)
Hutton found robins on both Aotea and Hauturu in 1867.[103] By about 1900 they had disappeared from Aotea and, with the exception of Hauturu, they were locally extinct north of the Bay of Plenty and South Waikato. Robins are vulnerable to introduced predatory mammals, especially ship rats, feral cats and stoats (*Mustela erminea*), and kiore are also known to prey on robin nests. On Hauturu, Guthrie-Smith described nests being repeatedly destroyed by rats (kiore) and, at one, even the female was killed. By the 1970s robins were quite scarce on Hauturu, and very few were encountered during the first

Stitchbird/hihi, male (*Notiomystis cincta*) (LWh)

Grey warbler/riroriro (*Gerygone igata*) (SF)

Bellbird/korimako, male (*Anthornis melanura*) (CRV)

Tūī (*Prosthemadera novaeseelandiae*) (LWh)

Whitehead/pōpokotea (*Mohoua albicilla*) (OT)

North Island fantail/pīwakawaka (*Rhipidura fuliginosa*) (SF)

North Island tomtit/miromiro (*Petroica macrocephala*) (LWh)

North Island robin/toutouwai (*Petroica longipes*) (SF)

**Birds** 155

few years of the forest-bird transect counts in the mid 1970s. However, they have thrived after the removal of cats and kiore and are now widespread and very common.

### Silvereye/tauhou (*Zosterops lateralis*)

The silvereye colonised New Zealand from Australia in the 1850s.[104] It seems they had not yet reached the Barrier Islands when Hutton visited in 1867,[105] but they had reached Hauturu by the early 1880s when Reischek visited.[106] Silvereyes are usually quite scarce on Hauturu, especially in the less modified forests, although there are seasonal changes in numbers with small flocks present in autumn and winter. Very few were recorded during the 1975–2017 forest-bird transect counts, and their numbers did not change much over that period. It is assumed on pest-free Hauturu that silvereyes are outcompeted by endemic birds, especially bellbirds, which probably have the greatest niche overlap.

### Welcome swallow (*Hirundo neoxena*)

Swallows are recent colonists to New Zealand. They first bred near Kaitaia in 1958 and rapidly colonised Northland and then the rest of the country. The rangers' diary recorded the first welcome swallow on Hauturu on 13 February 1963. They are now well established there, with several pairs resident and nesting in and around buildings at Te Maraeroa. They also occur around the coast of Hauturu, where nests have been found in sea caves and under rock overhangs.

## Introduced passerines on Hauturu

Turbott listed the introduced passerines on Hauturu.[107] Apart from the common myna (*Acridotheres tristis*), which is absent from Turbott's list, most of these species are still present and probably breeding, but the numbers of most of them have declined since livestock was removed in the mid-1990s. With the stock gone, the former pastureland has reverted to rank grass and sedgeland — suboptimal habitat for many introduced birds. Introduced passerines recorded in recent years include the Eurasian skylark (*Alauda arvensis*), European blackbird (*Turdus merula*), song thrush (*T. philomelos*), common starling (*Sturnus vulgaris*), common myna, house sparrow (*Passer domesticus*), dunnock (*Prunella modularis*), chaffinch (*Fringilla coelebs*), European greenfinch (*Carduelis chloris*), European goldfinch (*C. carduelis*), common redpoll (*C. flammea*) and yellowhammer (*Emberiza citrinella*).

The first myna was recorded in the rangers' diary in November 1962, and small flocks of mynas and starlings were present when sheep and cattle were grazed on Te Maraeroa. However, their populations collapsed after the cattle were removed: mynas and starlings prefer open, grazed pastureland and only small numbers of each species remain. When both species were more common, they often competed with kākāriki for nest holes in the forest at Te Maraeroa. A small flock of about 25 house sparrows survives on Te Maraeroa in the homestead garden and around the buildings. Along with a few greenfinches, goldfinches and yellowhammers, they are often attracted to the seeding giant umbrella sedges. Dunnocks and redpolls are quite scarce in the Auckland region. Their preferred habitat is cool, often exposed scrublands and, nationally, these two species are much more numerous further south. On Hauturu the low gumland scrub on the southwest side of the island and adjacent mānuka and kānuka scrublands provide some habitat. A few pairs of chaffinches survive around the edges of Te Maraeroa where they forage in open places at the bush edge and on the mown lawns and tracks. Skylarks were formerly resident and breeding at Te Maraeroa but there have been few records since livestock was removed. New Zealand pipits (*Anthus novaeseelandiae*) were listed by Hutton, Reischek and Turbott.[108] There are no recent reports.

John McCallum studied the distances that exotic passerines penetrated the forest on Hauturu to test the hypothesis proposed by Diamond and Veitch that exotic passerines would be largely absent from unmodified native forest. McCallum's findings were consistent with Diamond and Veitch.[109] The dunnock was found farthest into the native forest, reaching the kauri/kānuka forest, while blackbirds were found as far as the kānuka forest. The other introduced passerines were mainly confined to the forest margins around the edges of Te Maraeroa. The forest-bird transect counts between 1975 and 2017 show that introduced passerines were mostly in the modified forest on the lower slopes, with very few in the unmodified forest. Counts up to 2017 showed that only a very small proportion (less than 1 per cent) of total birds counted were introduced species and, of these, 80 per cent were blackbirds.

## Visiting and vagrant species

A number of species have been recorded as visitors or rare vagrants to Hauturu. These are listed in the checklist (Appendix 5), along with resident breeding species and those that have become locally or globally extinct during the historic period.

## The future

Assuming there are no negative effects from fisheries bycatch or food limitation through climate change, Hauturu's colonies of Cook's, grey-faced and black petrels and New Zealand storm petrels will continue to expand. Fluttering shearwaters and diving petrels, formerly limited by kiore predation, are likely to establish new or larger colonies, probably initially around the northern coastline near existing remnant colonies. Little shearwaters, fairy prions and possibly even white-faced storm petrels (*Pelagodroma marina*) could also colonise, taking advantage of breeding habitat free of predatory mammals. With warming seas, subtropical species such as black-winged petrel (*Pterodroma nigripennis*) and grey ternlet (*Procelsterna cerulea*), which already have small colonies on some northern islands, could also establish.

Recent forest-bird counts on Hauturu show that some of the older New Zealand endemics such as kākāriki, robin, bellbird, tīeke and kōkako are now forming a greater proportion of total bush birds at the expense of other natives such as whitehead, grey warbler, tomtit, fantail, silvereye and tūī. This pattern of increasing dominance of some endemics is also being seen in other sanctuaries where predatory mammals have been removed.[110] Many of the remaining introduced passerines will probably decline, especially as scrubland and forest regenerates at and near Te Maraeroa, where most of them still have a foothold. In future, banded rails and spotless crakes may become more abundant, and they may expand into and occupy a forest-floor niche, as seen on some other northern islands.

There are a few indigenous land birds currently absent that could recolonise naturally, assuming Hauturu has sufficient habitat. Past records show that falcon and pipit were formerly present and, given their strong dispersal ability, these species could return. Another candidate is the North Island fernbird/mātātā (*Bowdleria punctata*), which occurs on Aotea and, formerly, the Aldermen Islands off the east coast of the Coromandel Peninsula. Although the fernbird is considered a weak flier, we probably underestimate its dispersal ability. It is believed that fernbirds colonised the Aldermen Islands after burning by muttonbirders allowed rank *Poa* and scrubland to establish. When coastal forest regenerated on the Aldermen Islands, fernbirds disappeared. They have recently colonised Shakespear Regional Park, possibly by crossing the 3-kilometre sea gap from Tiritiri Matangi.

What potential is there for further bird translocations? It is assumed there will be more translocations of kākāpō as part of the recovery plan for that species. An obvious candidate for restoration would be one of the snipe species to replace the lost North Island snipe. While open, rank grassland and sedgeland are still present at Te Maraeroa there may be sufficient habitat for a few pairs of takahē (*Porphyrio hochstetteri*) if additional release sites are required for the national recovery programme. These birds could be managed as part of the wider Hauraki Gulf metapopulation. Surrogate species are a possible option to reactivate the niches of extinct species. We now have very few choices of surrogate species to replace extinct ones. Ian Atkinson boldly suggested using the rock wren (*Xenicus gilviventris*) as a substitute for the extinct bush wren (*X. longipes*) and, by re-establishing a forest-inhabiting *Xenicus* species, restoring a missing part of the trophic structure of New Zealand lowland forest.[111] Atkinson originally suggested Matiu/Somes Island as a candidate site for such an experiment, but there are now many other pest-free islands where this could be tried.

North Island kākā (*Nestor meridionalis*) (NF)

# 6.4 Bats

— STUART PARSONS

**Historically, New Zealand has been home to up to seven species of bat. The extant species are the greater short-tailed bat (*Mystacina robusta*), lesser short-tailed bat (*M. tuberculata*) and the long-tailed bat (*Chalinolobus tuberculatus*); both *Mystacina* species are endemic. The greater short-tailed bat's last known population on Taukihepa/Big South Cape Island off the southwest coast of Rakiura/Stewart Island is thought to have been exterminated by ship rats in the mid-1960s, but its extinction has yet to be proven. Four other species of bat have been recorded from Miocene fossil assemblages near St Bathans in Central Otago, including *M. miocenalis*, a recently described large omnivorous species from the same family, and perhaps two additional small mystaciniid species.**

While there is little genetic variation between long-tailed bat populations across the country, its conservation status has recently been elevated to nationally critical. Lesser short-tailed bats consist of three subspecies: *Mystacina tuberculata aupourica* (northern lesser short-tailed bat), *M. t. rhyacobia* (central lesser short-tailed bat) and *M. t. tuberculata* (southern short-tailed bat); their conservation status is nationally vulnerable, declining and recovering, respectively.[1]

Both lesser short-tailed bats and long-tailed bats are found from the far north of New Zealand to the deep south, where their range has been severely restricted because of habitat loss, predation and competition from introduced pests. While the lesser short-tailed bat relies on old-growth forest for roosting, the long-tailed bat is more flexible.[2] The majority of its roosts are in trees — both native and introduced — and it has been recorded as roosting in buildings and caves.

The lesser short-tailed bat has one of the widest diets of any bat in the world: it feeds on terrestrial and aerial insects as well as fruit, pollen and nectar,[3] whereas the long-tailed bat feeds solely on insects caught in flight.[4] The lesser short-tailed bat's ability to access this wide variety of food sources is enabled by its highly terrestrial nature: it has several anatomical adaptations that allow it to move with agility over the ground while not compromising its ability to fly.[5] Lesser short-tailed bats are important pollinators

A lesser short-tailed bat (*Mystacina tuberculata*) flying in to feed on the nectar in *Dactylanthus taylorii* flowers. (DM)

of native plants, including the endangered holoparasitic wood rose or pua o Te Rēinga (*Dactylanthus taylorii*).[6]

Lesser short-tailed bats practise a rare mating system known as lekking, where males occupy 'arenas' from which they display to passing females. Males roost in small cracks or holes in trees and sing to attract females with whom they mate. The males who sing the most appear to have more offspring, and some even share roosts to improve their chances of attracting a mate.[7] These arenas are aggregated around communal roost trees so females leaving or returning to communal roosts have to fly through the lek, increasing the chance of their hearing the males' songs.

Both the northern lesser short-tailed bat and the long-tailed bat are found on Hauturu, although most of the research has focused on the former. Bats are widespread across the island and are regularly sighted around the ranger's house and bunkhouse, and lesser short-tailed bats are commonly seen visiting the pōhutukawa near the bunkhouse when they are in flower. No long-tailed bats have been captured on Hauturu, but lesser short-tailed bats have been mist-netted and harp-trapped at a variety of locations including Pōhutukawa Flat, the Tirikakawa Valley and in the forest immediately behind the bunkhouse. Large roosts of lesser short-tailed bats have been located in large old-growth trees close to the bunkhouse and at Pōhutukawa Flat; each tree contained many hundreds of individuals. The northern lesser short-tailed bat is the lightest of the three subspecies and has the shortest forearm. There is no data available on the morphology of long-tailed bats on the island. Although the echolocation calls of both species of bat vary across the country, including on Hauturu, this is unlikely to have any biological meaning or relevance to monitoring.

The insect diet of lesser short-tailed bats on Hauturu is reasonably well known. A study of insect remains in faecal samples collected from captured bats indicated at least 50 per cent of their diet was made up of Coleoptera, Lepidoptera, Diptera and Orthoptera;[8] smaller quantities of Araneae and Myriapoda were also detected. The bats appeared to eat a wider variety of insects in summer than winter. A later analysis of winter diet using molecular methods showed that Lepidoptera and then Diptera were the most frequently consumed insects,[9] followed by Aranaea and Coleoptera. Interestingly, faecal samples from bats on Hauturu had the second highest diversity of prey (compared with two other populations, regardless of season) as measured by a Simpson Diversity Index, and the highest richness of molecular operational taxonomic units compared with bats in the central North Island and South Island, regardless of season. This is likely due to the relatively mild winters on the island and a lack of introduced competitors and non-native flora. No study of the insect diet of long-tailed bats has been conducted on the island.

A study demonstrating the importance of bats as pollinators on Hauturu showed that inflorescences of pōhutukawa (*Metrosideros excelsa*), and the hebes *Veronica macrocarpa* var. *latisepala* and *V. macrocarpa* var. *macrocarpa* that were caged to keep bats out were up to 2.5 times as pollen limited as open, uncaged inflorescences. The mean

*Clockwise from top*: A lesser short-tailed bat at his lek, singing to attract females. (SP) A lesser short-tailed bat with wings folded away hunting on the ground. (SP) A long-tailed bat (*Chalinolobus tuberculatus*) in hand. (CO) Long-tailed bats are rarely seen on the ground as they do all their hunting while flying. (CO)

seed-set rate of uncaged rewarewa (*Knightia excelsa*) inflorescences was higher than from caged inflorescences.[10] Video footage showed that geckos and insects could access the caged plants but birds and lesser short-tailed bats could not. The bats were the most common visitor to the uncaged pōhutukawa and rewarewa (97.4 per cent and 61.1 per cent, respectively), and they also visited *V. macrocarpa*. Although lesser short-tailed bats have been shown to be important pollinators of *Dactylanthus taylorii*, no studies have been conducted on the island. Analysis of the faecal samples of lesser short-tailed bats on Hauturu showed traces of pollen from rewarewa in early summer and pōhutukawa and kahakaha (*Astelia* sp.) from December to February.[11] Fragments of flowers from the rewarewa and pōhutukawa were also found.

Lesser short-tailed bats on Hauturu have recently been the subject of torpor studies (a sleep-like condition).[12] Bats were captured and fitted with small radio transmitters whose beep-rate indicated the animal's skin temperature. Compared with *Mystacina tuberculata rhyacobia* from Pureora Forest in the central North Island, male and female bats on Hauturu used shorter torpor bouts more frequently and were more likely to wake up and forage on warmer nights. Bats on the island did not choose roosts that buffered them from fluctuations in ambient temperature: instead they chose to roost in ponga (*Cyathea dealbata*). This is probably a reflection of the relatively warm winter temperatures on the island, which allows the bats to use passive warming to rouse in the evening. The availability of insects during winter is likely to have influenced this strategy, too.

Lesser short-tailed bats have a unique phoretic relationship with the endemic bat fly *Mystacinobia zelandica*. The bat fly is a blind, wingless dipteran that lives on the bats or within their roosts, where it feeds on guano. Bat flies appear to show a rare level of sociality: they possess a number of behavioural traits, including males using ultrasound to guard other flies. The flies are dependent on the bats for dispersal. Not all populations or subspecies of lesser short-tailed bats have been found to host bat flies, but they are present on Hauturu and are regularly found on captured bats. No systematic study of bat flies has been conducted on the island.

Pōhutukawa in flower is a favourite food source for bats. (LW)

# 6.5
# Vegetation and vascular flora

— EWEN CAMERON AND MAUREEN YOUNG

**For such a large and rugged island, the vascular flora and vegetation of Te Hauturu-o-Toi/Little Barrier Island are relatively well known. The first botanical visitor was Thomas Kirk in 1867, who collected specimens and published a bare list of around 189 coastal species.[1] This was followed by others nearly three decades later, collecting specimens (listed in Table 1 on page 186). From 1932 to 1981 W. M. Hamilton explored Hauturu extensively, probably covering more territory than anyone else. He published the first account of the vegetation and recorded 434 vascular species.[2] Twenty-four years later he revised and expanded the account with Ian Atkinson.[3] Just over five decades after that, in 2012, the flora was updated by four authors based on their observations and collecting on the island over a period of 45 years.[4]**

This current species list (see Chapter 9) adds 20 species and deletes five from the 2012 list. The additions include: recent discoveries (12), previously overlooked or previously inaccessible herbarium records (4), Department of Conservation archives via Dick Veitch (2), taxonomic revisions (1) and a reinstatement of a historical record (1). The 2950 Hauturu herbarium specimens located in the three major New Zealand herbaria have been invaluable to clarify earlier literature records, verify records and 'discover' new records. An impressive 94 per cent of all records are vouchered.

## The vegetation

The vegetation of about one-third of the island has had a long history of modification by human-related activities up until it was made a reserve in 1895. Two of the vegetation classes — pasture and kānuka forest — are in this highly modified category. The island's vegetation was affected by the seed predator kiore (*Rattus exulans*) until it was eradicated in 2004, and John Campbell went so far as to state that the vegetation composition of the modified part of Hauturu has been determined by rats.[5]

Atkinson and Stanley described eight simplified vegetation zones;[6] two of these have

---

Kirk's daisy/kohurangi (*Brachyglottis kirkii*) (NF)

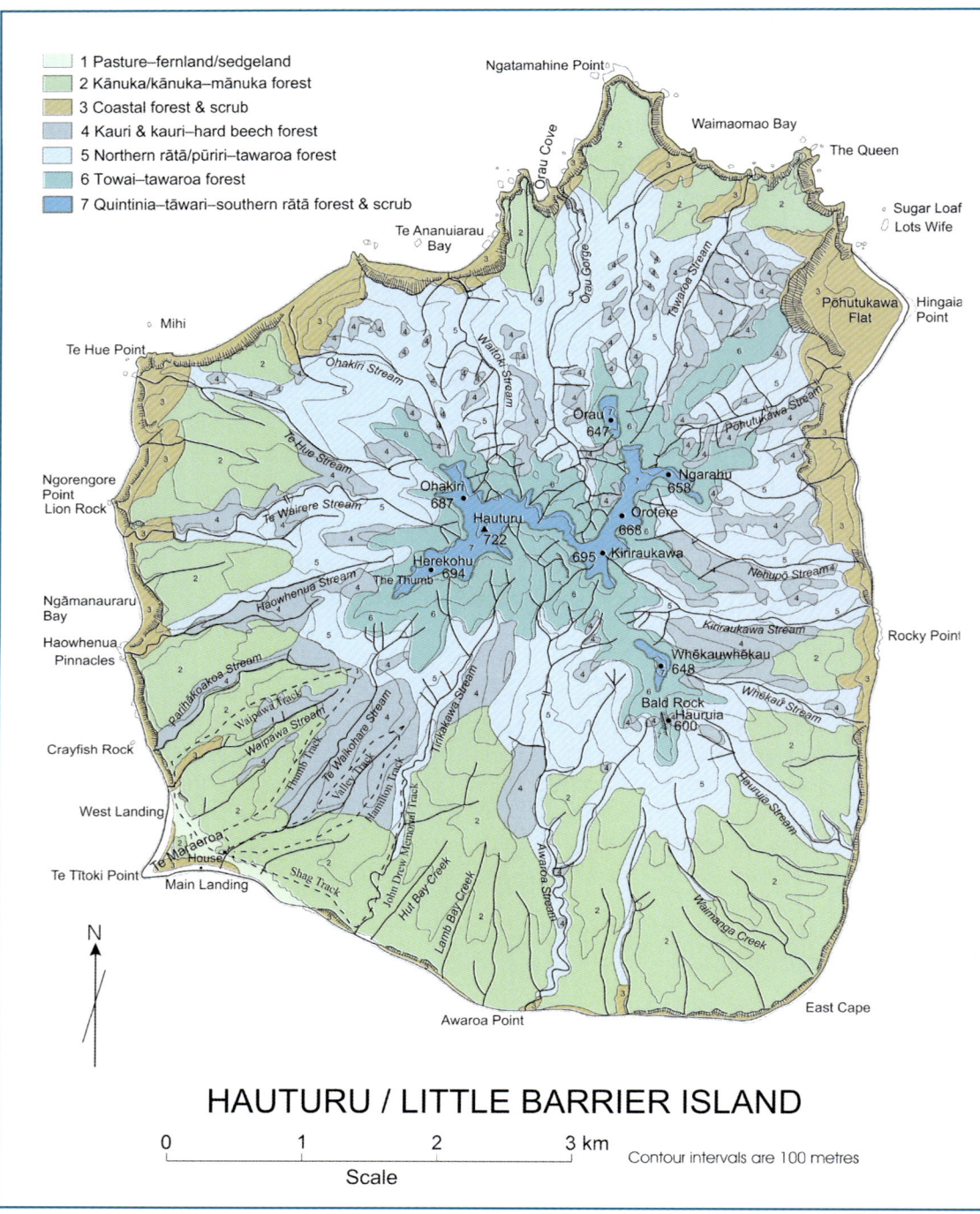

*Fig. 1:* Vegetation map of Hauturu showing the seven zones, based on the map of Hamilton & Atkinson (see Table 2, page 187) but using the simpler vegetation zones of Atkinson & Stanley (2018). Adapted by Dick Veitch.

been combined to give the seven vegetation zones described below (see Fig. 1). This is reduced from 14 zones of Hamilton and Atkinson (see Table 2, page 187). The vegetation descriptions are taken mainly from these two publications except for Te Maraeroa, which we surveyed in early 2018, because this area has changed the most in recent times. The other zones are dominated by long-lived tree species.[7]

Two illustrations of Hamilton and Atkinson's are reproduced here to show the generalised relationship between vegetation types and altitude with respect to ridges or cliffs and valleys (Fig. 2), and change in relative frequency of dominant canopy trees with altitude (Fig. 3). Although these figures and the areas used below for each forest type are based on data and photographs that are nearly 60 years old, we do not expect major changes apart from in the induced pasture and kānuka areas, where the boundaries will be more blurred as they are gradually overtaken by the forest type that they replaced.

## 1. Pasture-fernland/sedgeland (16 ha)
## (zones 1 and 2 of Hamilton and Atkinson's 1961 map; see Table 2)

Since sheep and cattle were removed from Hauturu in the mid-1990s, rank pasture/fernland/sedgeland has developed and dominates about half the vegetation on Te Maraeroa alluvial flat. The flat is bounded by a coastal boulder bank and foothills to the northeast. One stream — Grave Stream, between the Waipawa and Te Waikohare streams — disgorges onto Te Maraeroa. As a result of long human habitation, fire, stock grazing and kiore, the vegetation of Te Maraeroa is the most modified on the island.

### Rank pasture–fernland

The rank pasture–fernland is dominated by pasture grasses of microlaena (*Microlaena stipoides*), cocksfoot (*Dactylis glomerata*) and paspalum (*Paspalum dilatatum*); these are frequently associated with Indian doab (*Cynodon dactylon*), *Geranium homeanum*, pōhuehue (*Muehlenbeckia complexa*), bindweed hybrid (*Calystegia sepium* subsp. *roseata* × *C. tuguriorum*) and prairie grass (*Bromus catharticus*), along with scattered Yorkshire fog (*Holcus lanatus*), *Juncus australis*, *Lotus pedunculatus*, Californian thistle (*Cirsium arvense*), *Vicia sativa*, *Solanum americanum*, tall fescue (*Lolium arundinaceum*) and patches of three fern species to 40 metres across — *Hypolepis ambigua*, bracken (*Pteridium esculentum*) and *Doodia australis*. All of Te Maraeroa would have once been forested and, as was found on Tiritiri Matangi, the rank pasture is generally too thick to allow woody species to establish in the open; instead most of them are creeping in from the forested margins. Partly shaded by a canopy of kānuka (*Kunzea robusta*), pōhutukawa (*Metrosideros excelsa*) or pūriri (*Vitex lucens*) is good regeneration of coastal karamū (*Coprosma macrocarpa*), māpou (*Myrsine australis*), taupata (*Coprosma repens*), māhoe (*Melicytus ramiflorus*), kawakawa (*Piper excelsum*), karaka (*Corynocarpus laevigatus*), taraire (*Beilschmiedia tarairi*), kohekohe (*Dysoxylum spectabile*) and nīkau (*Rhopalostylis sapida*). However, the mown tracks, where they occur along the forest margins, present a hindrance to this type of creeping regeneration.

### Sedgeland

In the lower or more poorly drained areas the pasture–fernland gives way to a sedgeland,

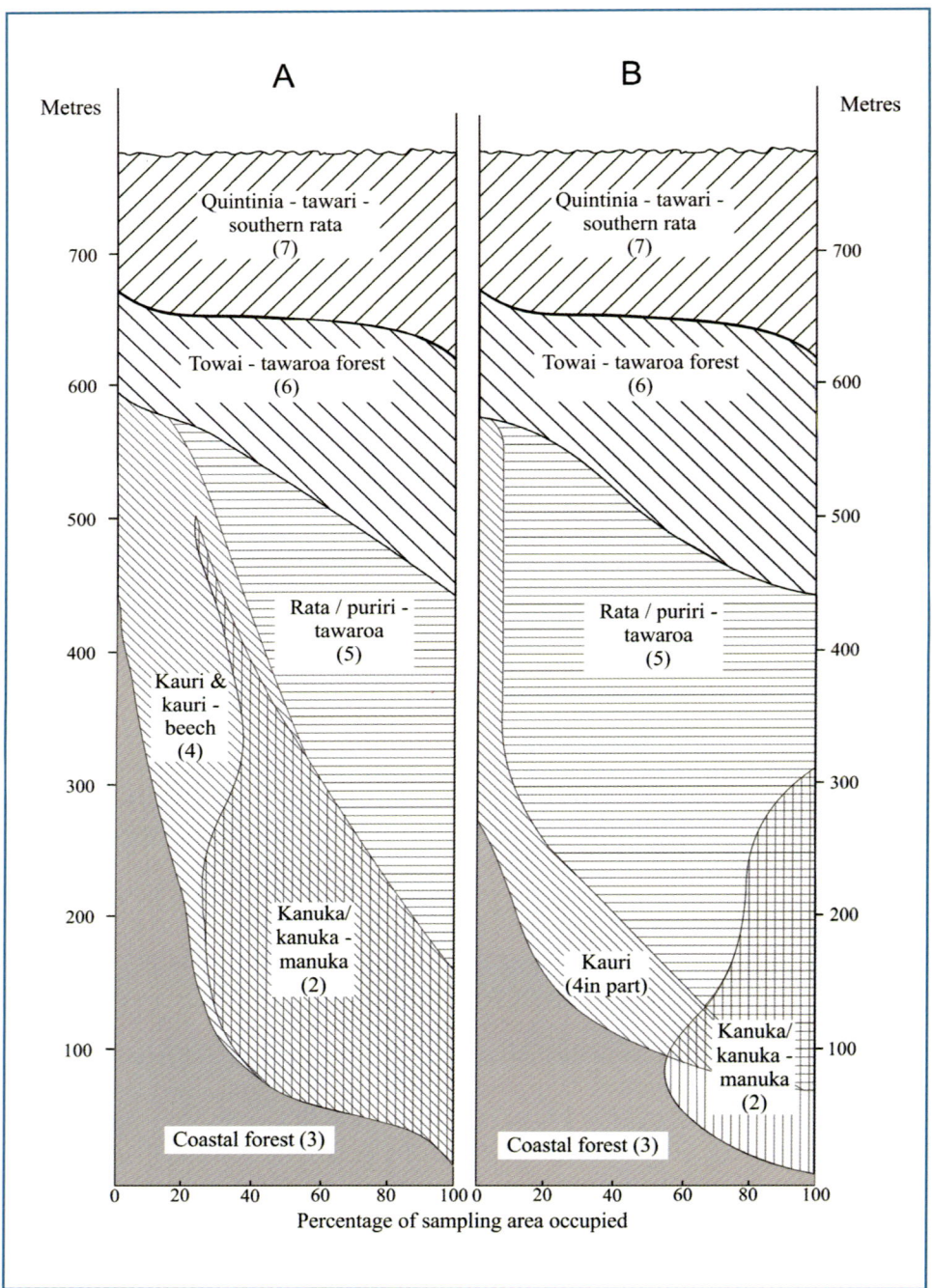

*Fig. 2:* Generalised altitudinal sequence of vegetation types on Hauturu: **A** On cliffs and ridges. **B** In the valleys. Numbers in brackets refer to vegetation zones. (After Hamilton & Atkinson 1961)

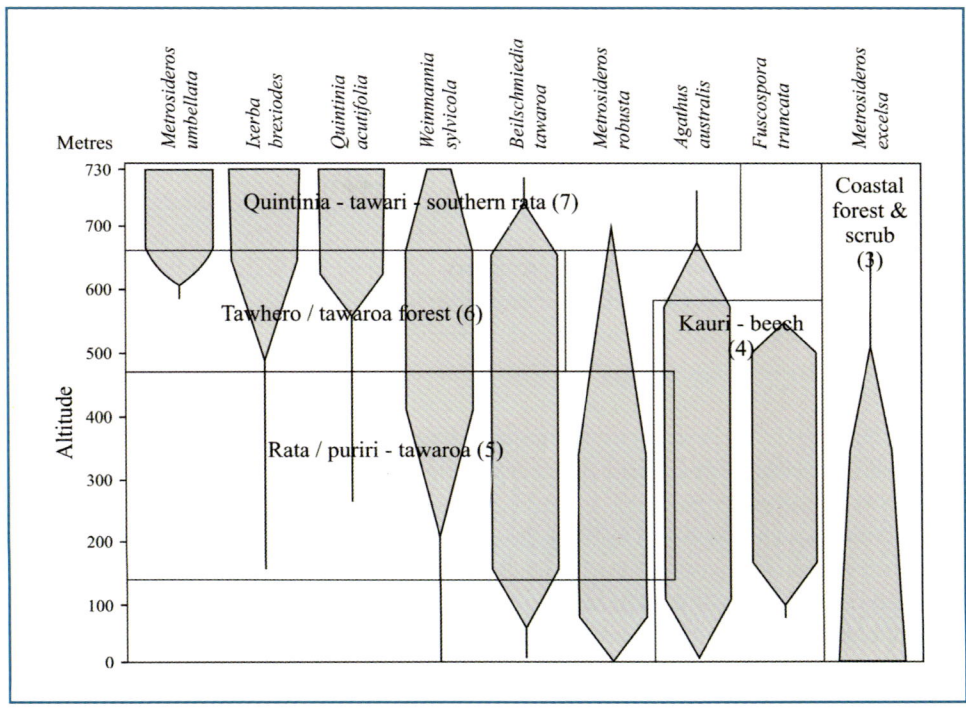

*Fig. 3:* Change in relative frequency of the dominant canopy trees with change in altitude. (After Hamilton & Atkinson 1961)

covering about one hectare. It is dominated by *Cyperus ustulatus* and *Carex virgata*, the former with culms to 1.8 metres tall. Other species are minor components of this vegetation class: *Juncus usitatus*, bindweed hybrid and paspalum. Pōhuehue stops abruptly at the pasture/sedge boundary. Elsewhere on the flat there are localised wet pockets of *Cyperus ustulatus*, with other species such as *Ranunculus urvilleanus*, *Lotus pedunculatus* and bindweed (*Calystegia sepium* subsp. *roseata*). When it is raining, these areas quickly pond.

## Mown lawn

This class of vegetation includes the regularly mown tracks through the rank pasture, the helicopter emergency landing area and the lawn by the ranger's house and bunkhouse. The two dominant species are paspalum and carpet grass (*Axonopus fissifolius*). Scattered through it with varying abundance are white clover (*Trifolium repens*), *Geranium homeanum*, *Oxalis exilis* and *Plantago lanceolata*; and more localised occurrences of *Dichondra repens*, *Cyperus brevifolia*, rat's tail (*Sporobolus africanus*), *Carex inversa*, lawn daisy (*Bellis perennis*), twin cress (*Lepidium didymium*), selfheal (*Prunella vulgaris*), creeping mallow (*Modiola caroliniana*), Bermuda buttercup (*Oxalis pes-caprae*), *Geranium dissectum*, *G. molle* and dandelion (*Taraxacum officinale*). Over 50 species of the smaller exotic herbaceous species, which have been previously recorded for the island but have not been seen for at least 40 years, have probably been crowded out by aggressive grasses, especially carpet grass such as *Veronica persicaria*, *V. serpyllifolia* and *Trifolium pratense*. Bare wheel ruts caused by frequent tractor traffic provide an open habitat for some of the small annual species, especially exotics (*Matricaria discoidea*, *Polygonum arenastrum*), and in a shaded area that is prone to flooding the threatened native annual sneezeweed (*Centipeda minima*) occurs along the tractor tracks.

## 2. Kānuka/kānuka-mānuka forest (910 ha)
### (zone 3 of Hamilton and Atkinson's 1961 map)

This is the largest forest type, covering nearly a third of the island — and much of it is on the old planezes of the volcano below c. 300 metres. Above this altitude the streams have eroded the planezes down to narrow ridges. The vegetation has mainly been burnt off long ago by Māori and today is dominated by kānuka and mānuka. More recent fires may have been related to kauri felling and kauri gum gathering during the 25 years before the island became a reserve, and most of this kānuka forest is unlikely to be older than 110–160 years.

The kānuka is steadily being replaced by other species as old trees collapse, particularly during storms. Haekaro (*Pittosporum umbellatum*) and kauri (*Agathis australis*) are the most common trees replacing kānuka on ridges. On moderately steep slopes, pūriri, kohekohe, māpou, *Coprosma arborea* and haekaro are the most widespread replacement trees. On the accumulated debris slides and slumps in valleys, taraire, rewarewa (*Knightia excelsa*), pūriri and kohekohe are replacing kānuka. Generally, replacement of kānuka is most advanced on ridges and least advanced on midslopes. The collapse of large kānuka is particularly noticeable in the bottom of the southern valleys. A feature of these forests near the streams is the luxuriant nīkau palms. Toropapa (*Alseuosmia macrophylla*) is one of the most abundant understorey shrubs here, as in most of the other forest types. Toropapa is eagerly sought-after by nectar-feeding birds in late winter (August to September), when it produces highly scented long, tubular flowers. It frequently occurs together with the small tree haekaro, and its flowering (August to October) usually overlaps with that of toropapa.

On some planezes the soil is so infertile that mānuka gumland scrub has developed. By the summit of the Shag Track the clifftop vegetation is mostly shrubby vegetation consisting of scattered emergent trees of kānuka to around 6 metres tall; shrubs 2–4 metres tall commonly include mānuka, akepiro (*Olearia furfuracea*), haekaro, mingimingi (*Leucopogon fasculatus*), prickly mingimingi (*Leptecophylla juniperina*), ponga (*Cyathea dealbata*), hangehange (*Geniostoma ligustrifolium*) and *Coprosma rhamnoides*; less common are whauwhaupaku (*Pseudopanax arboreus*) and toru (*Toronia toru*); groundcover includes *Pomaderris amoena*, *Morelotia affinis*, *Schoenus tendo*, *Rytidosperma* spp., *Gonocarpus incanus*, *Drosera auriculata* and abundant bryophytes. A mānuka 'gumland' that scarcely showed any measurable change between 1954 and 2001 is described on the Shag Track by Atkinson.[8] More extreme examples of 'gumland communities', containing mānuka, the sedges *Schoenus tendo* and

---

*Clockwise from top*: Fallen kānuka trunks straddle the Tirikakawa Stream. (EC) Maureen Young beside a kānuka to show the size of the trees by the Lower Valley Track. (EC) The tropical luxuriance of the very large-leaved Hauturu nīkau. (EC)

*Lepidosperma laterale* and *Lycopodium deuterodensum*, are present on some of the western planezes.

## 3. Coastal forest and scrub (262 ha)
## (zones 4 to 8 of Hamilton and Atkinson's 1961 map)

Pōhutukawa is the dominant species in this coastal zone, associated with salt-laden air on cliffs and in valley mouths. The hard rocks of the northern coast largely lack plant cover. However, the cliffs of the southwestern and southern coast are more weathered and often produce deposits of talus behind the shoreline. A rather open forest of pōhutukawa, kawakawa, coastal karamū, karaka and, more rarely, kohekohe and pūriri can develop on these sites. A stand of parapara (*Pisonia brunoniana*) with some tawāpou (*Planchonella costata*), wharangi (*Melicope ternata*) and kohekohe is present on talus behind the shoreline between the mouths of the Tirikakawa Stream and Lamb Bay Creek.

This usually narrow vegetation zone encircling the island at its rocky seaward edge is where the herbaceous plants dominate. Typical plants of this exposed zone include *Asplenium haurakiense*, *Calystegia soldanella*, *Disphyma australe*, pōhuehue, native spinach (*Tetragonia implexicoma*), peperomia (*Peperomia urvilleana*), glasswort (*Salicornia quinqueflora*), *Chenopodium triandra*, *C. trigonum*, *Solanum opacum*, rengarenga lily (*Arthropodium cirratum*), *Scandia rosifolia*, *Wahlenbergia vernicosa* and grasses (*Poa anceps*, *Dichelachne crinita*). Coastal spurge (*Euphorbia glauca*) is a nationally threatened species that occurs in this zone. The best populations are two large patches in the open on either side of Te Tītoki Point; the larger one is over 40 metres long and about 5 metres across, intermixed with pōhuehue. Smaller, scattered colonies exist among the boulders north of the point towards Waipawa Stream.

### Pua Mataahu pōhutukawa forest

West of the boatshed along the boulder bank to Te Tītoki Point there is a continuous canopy of well-spaced, spreading pōhutukawa c. 10 metres tall. This forest extends for about 250 metres along the coast and can be up to 50 metres wide in places. The ground is boulders with virtually no soil. Near the sea edge there are only the canopy trees and a low groundcover; the shrub layer here is absent, giving it a special appearance. The groundcover is locally dominated by leather fern (*Pyrrosia eleagnifolia*) and peperomia, with scattered plants of prostrate *Coprosma rhamnoides*, *Carex flagellifera*, *Oplismenus hirtellus*, *Stellaria parviflora*, *Asplenium oblongifolium*, *A. haurakiense*, and hound's tongue fern (*Microsorum pustulatum*). Prostrate *Comprosma rhamnoides* is unknown to occur anywhere else in its wide New Zealand distribution. Some 20 metres (or more) from the forest sea edge, especially in hollows, there is a continuous low shrubland 1–2 metres tall, dominated by coastal karamū, with the occasional māpou and kawakawa. Karo (*Pittosporum crassifolium*) is occasionally present, but mainly as scattered young individuals. Scattered clumps of *Astelia banksii* are also present, especially towards the

*Clockwise from top left*: The nationally threatened coastal spurge (*Euphorbia glauca*) at Te Tītoki Point — a regional stronghold for the species. (EC) A closer view of coastal spurge. (EC) The open pōhutukawa forest of Pua Mataahu. (EC)

western end of the forest. Locally a few pūriri make it into the canopy and at the western end there is a patch of coastal māhoe.

At the eastern end this forest stops abruptly and the boulder bank here is clothed in low, open, tangled mats of pōhuehue, large clumps of *Asplenium oblongifolium*, vines of the hybrid bindweed and microlaena in the hollows — perhaps the result of pre-European fires. Along the northern edge of the pōhutukawa is a broken canopy of tall kānuka with an understorey of tall māpou, hangehange, māhoe, coastal karamū and, occasionally, ngaio (*Myoporum laetum*). The openness of this pōhutukawa forest is perhaps a combination of past fire, earlier stock grazing and exposure to strong salt-laden storms. Presumably the absence of houpara (*Pseudopanax lessonii*) and the paucity of karo are the result of the past presence of the seed predator kiore. However, the absence of coastal sedge *Ficinia nodosa* from this association is hard to explain. Over time this understorey should thicken up with species tolerant of salt spray.

### Pōhutukawa Flat
Coastal forest occurs on the Hingaia Rockslide at the northeastern corner of the island. The collapse of the cliffs above, over 400 metres high, has resulted in house-sized blocks of rock across Pōhutukawa Flat. Pōhutukawa trees have established on the rockslide itself; some are relatively tall and straight but others, growing on large blocks of rock, are stunted. Tall forest, including some kauri, has established on the fine deposits at the foot of the cliff. The pōhutukawa over this flat are some of the first to flower on the island, attracting dark 'clouds' of tūī that are seen feeding here around the last week of October.

## 4. Kauri and kauri–hard beech forests (373 ha)
## (zones 9 and 10 of Hamilton and Atkinson's 1961 map)
Kauri occurs as scattered patches across the whole island and not as extensive stands — probably because of the steep topography. Kauri is mainly on ridges, spurs and upper valley slopes from 100 to 500 metres asl, and is present in all size classes, although trees with a diameter of over a metre are uncommon. Most of the kauri is regenerating on the previously burnt areas and landslip slopes. Kānuka is usually the primary coloniser, which is then overtopped by kauri or kauri and hard beech (*Fuscospora truncata*) (see Fig. 4). Miro can be an important canopy constituent with kauri, especially at the higher levels. The understorey becomes denser with altitude: kiekie (*Freycinetia banksii*) becomes abundant as this zone passes into the next forest type (rātā/pūriri–tawaroa). Single-dominant stands of hard beech are rare and are confined to small patches on the eastern and southern ridges of the island, although young kauri and sometimes tawaroa (*Beilschmiedia tawaroa*) and tōwai (*Weinmannia sylvicola*) may be present in the understorey. Along a single northern ridge is a unique stand of black beech (*Fuscospora solandri*) at c. 220 metres asl, with hard beech and kauri. In January 2019 a hybrid black beech tree was found further to the east in the upper Tawaroa Stream catchment.

## 5. Northern rātā/pūriri–tawaroa forest (832 ha)
## (zone 11 of Hamilton and Atkinson's 1961 map)
In terms of its canopy composition and the associated species, this is the most varied

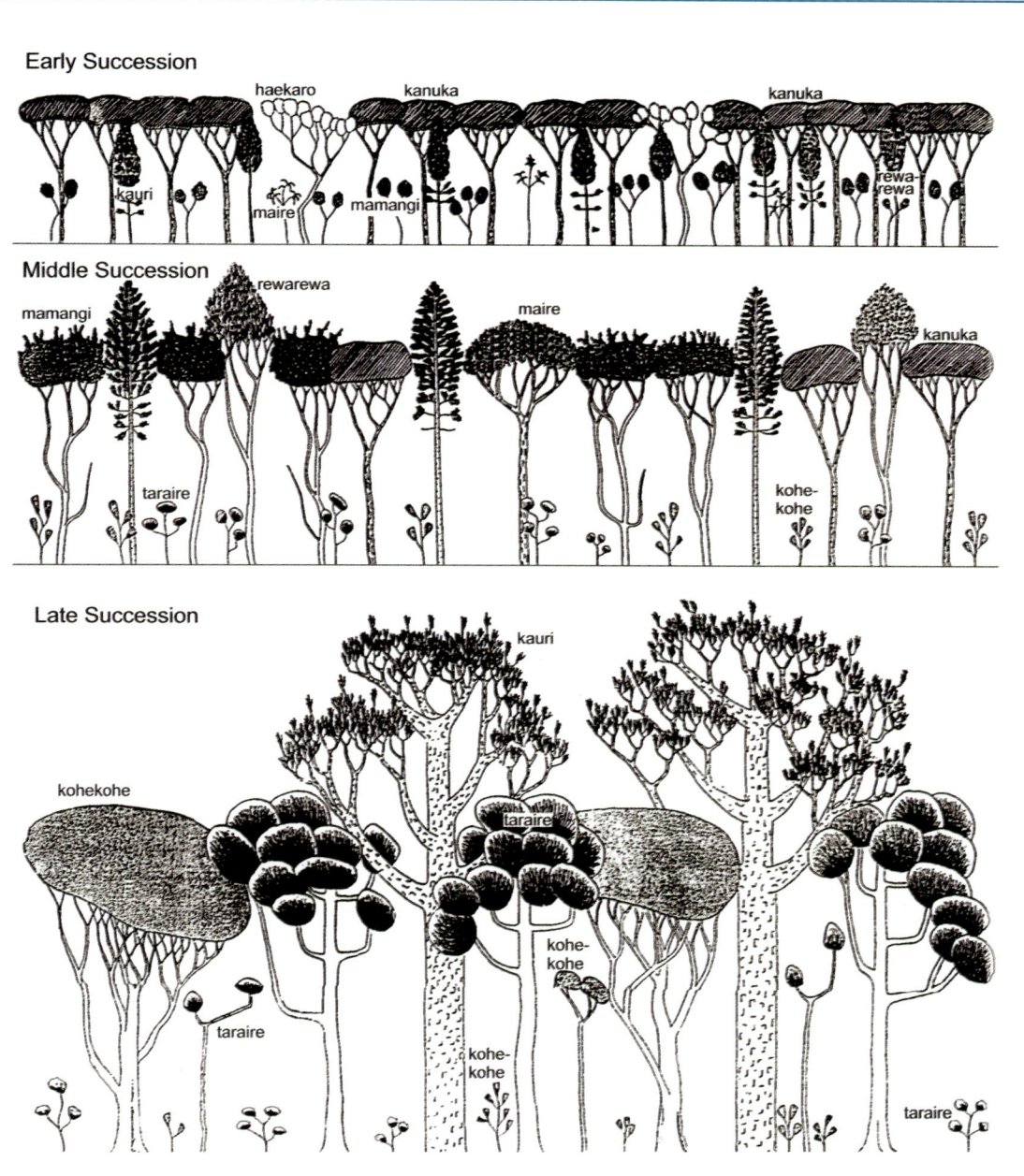

*Fig. 4:* Three stylised stages of a successional sequence from kānuka to kauri–taraire–kohekohe forest, based on valleys and slopes on Hauturu (from Wright 1993).

forest type on the island. It includes most of the forests on the sides of valleys, growing either on relatively stable slopes (including spurs) or on unstable, often concave slopes with frequent landslides. Huge northern rātā (*Metrosideros robusta*) with their fan-like branching and abundant cargo of epiphytes are characteristic of this zone. The larger epiphytes in the light include puka, *Pittosporum cornifolium*, *Astelia hastata* and *A. solandri*, usually accompanied by orchids and ferns. Smaller epiphytes are more abundant on the lower (shaded) branches and trunks, and include ferns, orchids, bryophytes and lichens. These shade epiphytes develop best along the stream margins where the humidity is highest. Most of the emergent rātā started as epiphytes themselves; and Hamilton and Atkinson suggested that pūriri and pukatea (*Laurelia novae-zelandiae*) may be seral to rātā on Hauturu. Pukatea appears to be rare on the island, even where Hamilton and Atkinson said it was abundant (for example Tirikakawa Valley). Rātā was observed establishing mainly on pūriri.

An important feature of this zone is the low number of podocarp trees. Miro (*Prumnopitys ferruginea*) is the main one present; all others occur infrequently. Climbers are common, especially kiekie, mangemange (*Lygodium articulatum*), supplejack (*Ripogonum scandens*) and several climbing rātā species (*Metrosideros* spp.).

Tawaroa, taraire and sometimes kauri have established on the most stable valley sides. Some kauri stands extend downslope nearly to the valley bottom. Where debris slides and slumps have accumulated in a valley bottom, fast-growing species such as māhoe, mamaku (*Cyathea medullaris*) and nīkau and, where there is sufficient light, kōtukutuku or tree fuchsia (*Fuchsia excorticata*), makomako (*Aristotelia serrata*), rangiora (*Brachyglottis repanda*), heketara (*Olearia rani*) and patē (*Schefflera digitata*) have established. When it rains on Hauturu the streams rise quickly, frequently causing flash-flooding and eroding such ephemeral sites.

## 6. Tōwai–tawaroa forest (359 ha)
## (zone 12 of Hamilton and Atkinson's 1961 map)

This kind of forest is dominant in forests above 450 metres asl, with tōwai forming almost pure stands on steep-sided ridges. Nevertheless, tōwai–tawaroa forest is rich in plant species including an undergrowth of *Archeria racemosa*, horopito (*Pseudowintera axillaris*), tāwari (*Ixerba brexioides*), mountain five finger (*Pseudopanax colensoi* var. *colensoi*), whauwhaupaku, heketara (*Olearia furfuracea*), patē, toro (*Myrsine salicina*), Kirk's daisy (*Brachyglottis kirkii*) and the root parasite *Mida salicifolia*. Tōwai–tawaroa forest grows together with quintinia (*Quintinia serrata*)–tāwari–southern rātā (*Metrosideros umbellata*) forest and scrub in the wettest and most frequently cloud-covered zone of the island. Landslides are frequent in the steep terrain here, and the debris descends into the valleys below.

On some of the drier southern ridges, where there is greater exposure to the prevailing wind, tawaroa and emergent tree rātā are more important in the canopy, and the epiphytes lack the abundance and luxuriant growth of the vegetation of the northern slopes — as seen on the drier leading ridge to Mount Hauturu.

## 7. Quintinia–tāwari–southern rātā forest and scrub (65 ha)
## (zones 13 and 14 of Hamilton and Atkinson's 1961 map)

This is the humid bryophyte forest area, often veiled in a cloud cap, that occurs on all ridges above 600 metres. The mineral soils contain high peat content and a low canopy 1–6 metres tall. The outstanding feature of this forest is the density of epiphytic mosses, liverworts and ferns (especially filmy ferns) that clothe the branches, and even some usually non-epiphytes that are acting as epiphytes (*Phormium cookianum*, *Gahnia pauciflora*). The windswept canopy is locally furrowed (parallel to the prevailing wind) — the result of salt-laden southerly storms killing leafbuds. The trees have wide-spreading, often twisted branches, and the more simple structure of the vegetation lacks the successive layers of the taller rātā–tawaroa type of vegetation. In addition to the three main trees — quintinia, tāwari and southern rātā — Hall's tōtara (*Podocarpus laetus*), *Archeria racemosa*, mountain neinei (*Dracophyllum traversii*) and kāpuka (*Griselinia littoralis*) are moderately common, and toatoa (*Phyllocladus toatoa*) is present. Miro and stunted kauri occur up to 700 metres. Where there are openings, thickets of *Phormium cookianum*, *Gahnia pauciflora* and the large white rātā (*Metrosideros albiflora*) grow abundantly, usually as sprawling shrubs, and two local sites support the spindly shrub *Metrosideros parkinsonii*. The shrubby summit vegetation is less than 3 metres tall, and, unlike on the higher Mount Moehau in the Coromandel Range, lacks montane species.

The soils of this uppermost community receive substantial annual inputs of nutrients as a direct consequence of the area being the breeding ground of a large colony of Cook's petrel (*Pterodroma cookii*) and other species of burrowing seabirds. This fertilising will increase as the seabird colony increases now that cats and rats have been eradicated.

# The flora

Hauturu has no recognised endemic vascular plants. The closest to an endemic is the shrub koromiko, *Hebe pubescens* subsp. *sejuncta*, whose type locality is Hauturu. However, it also occurs on the Mokohinau Islands and at a single location on Aotea/Great Barrier Island. The Hauturu flora totals some 639 vascular species and hybrids, made up of 437 indigenous and 202 naturalised taxa — i.e. 68 per cent are indigenous (see Table 3, page 187). These totals include the 37 exotic species that had naturalised sparingly from plantings in or near the rangers' garden; 78 per cent of these are now thought to have been eradicated (see Chapter 4).

## Lost species and a lost habitat?

Fifty of the lost species recorded are treated as historical — i.e. they have not been seen for 40 years — and 36 per cent of these are naturalised species (excluding the targeted eradications). The majority of the lost historical records are herbaceous; only four are woody — two *Kunzea* hybrids (all *Kunzea* taxa are only recently recognised), *Ozothamnus leptophyllus*, and the most extraordinary local extinction of all, the cabbage tree, tī kōuka (*Cordyline australis*). The damp hollows at Te Maraeroa would have been a prime habitat for cabbage trees before they were highly modified by drainage and grazing. Only a single plant was reported over a century ago. Raupō or bulrush (*Typha orientalis*), last observed on the island in 1933, would also have occurred formerly in these damper areas.

Three of the historical records are usually associated with sand dunes (*Ozothamnus leptophyllus*) or damp sandflats (*Carex pumila*, *Cotula coronopifolia*). *Carex pumila* was collected at the mouth of Parihākoakoa Stream 'on a sandy outwash at the top of the storm beach . . . washed away in a flood in May 1960'.[9] The other two sand-related records are from over a century ago. There is also the local presence of sand kānuka (*Kunzea amathicola*) at Te Maraeroa. Taken collectively these four records might suggest that a sandy area (perhaps a small estuary) may once have existed on the southwestern coast by one of the stream mouths. The salt-marsh ribbonwood (*Plagianthus divaricatus*) recorded at Te Tītoki Point in 1984 could also have been part of such a dune/estuarine association as well as the estuarine seagrass (*Zostera muelleri*) collected in the Shakespear period.

## The naturalised flora

The naturalised flora include some 201 species — less than one-third of the total flora. The process of naturalisation began very slowly but has steadily been increasing (see Fig. 5). Only a small number of these would have reached the island unaided by humans. Eighty per cent of these naturalised species are, or were, known only near habitation at Te Maraeroa. Of those found elsewhere on the island, only two are woody: prickly hakea (*Hakea sericea*) (possibly taken to the island by gumdiggers and now a weed species that is being managed) and kiwifruit (*Actinidia chinensis* var. *deliciosa*) (a single stem about 2 metres long, established from a toilet stop up by the Ōrau Hut). The main source of the naturalised species has been either purposeful introductions or as contamination of pasture species, clothing, footwear, machinery, etc. One of the sources has been the cultivated plants in the rangers' garden and environs; most of those have now been eradicated.

Most of the naturalised species are plants of the open areas. Only six have established unaided by humans in the forest: climbing asparagus (*Asparagus scandens*) and inkweed (*Phytolacca octandra*), both of which have fruit that is dispersed by birds; and mist flower (*Ageratina riparia*), Mexican devil (*A. adenophora*), Brazilian fireweed (*Erechtites valerianifolia*) and fleabane (*Erigeron sumatrensis*), all with windblown seed. The open coastal slopes and cliffs are habitat for the hardy species: the worst infestations are the pampas grasses (*Cortaderia* spp.), which are managed, and a few herbaceous species. The main environmental weeds that have been or are being managed on the island (24 species) are covered in Chapter 4. Since 2011 much stricter biosecurity and quarantine checks for all visitors have virtually eliminated humans as the vector for new introductions. However, new species, both indigenous and exotic, will slowly continue to establish by seed or spore as a result of being blown, floated (including plant propagules), ingested by or attached to birds, or from illegal landings. There are no biosecurity checks of a kererū arriving on Hauturu from the mainland loaded with seed from monkey apple (*Syzygium smithii*), loquat (*Eriobotrya japonica*) or bangalow palm (*Archontophoenix cunninghamiana*).

---

*Clockwise from top left*: Koromiko (*Hebe pubescens* subsp. *sejuncta*). (EC) The trunk of pukatea (*Laurelia novae-zelandiae*). (LW) *Archeria racemosa* (Ericaceae), an erect shrub/tree of the upper slopes, with attractive summer flowers. (JB) *Metrosideros parkinsonii* flower buds from a newly discovered population on the summit ridge. (RW)

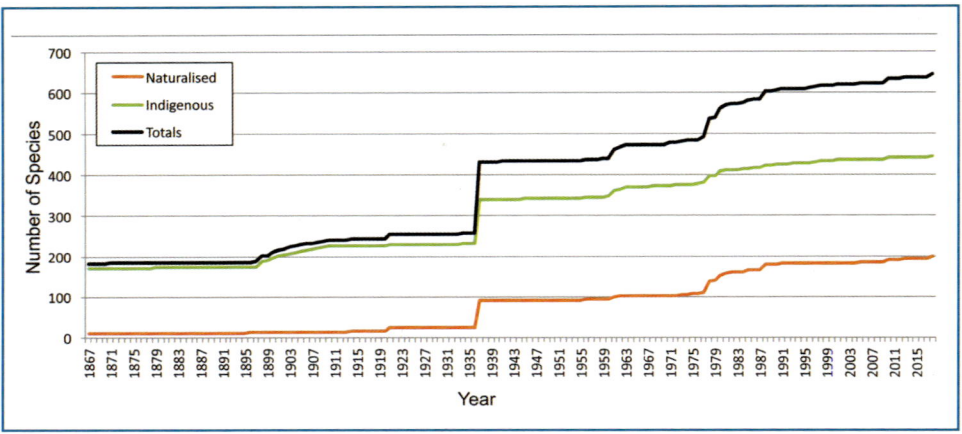

*Fig. 5:* Cumulative naturalised and indigenous vascular plant species records over time, based on the first literature record or herbarium voucher (whichever is the earlier). Note the spikes in 1937 and 1961. When the first-record date is uncertain, the median of the suspected range is plotted.

## Nationally and regionally threatened species on Hauturu

Remarkably, there are 60 national and 56 regional threatened and uncommon species recorded for Hauturu (see Appendix 6).[10] Of these 16 (14 per cent) haven't been recorded for at least 40 years and are presumed extinct. For several of these threatened species Hauturu has the only known population in the Auckland region, including: *Carmichaelia williamsii*, *Coprosma neglecta*, *Dactylanthus taylorii*, *Fuscospora solandri*, *Leptopteris superba*, *Peraxilla tetrapetala*, *Pimelea acra* and *Thismia rodwayi*. It is also the regional stronghold for several other species.

The māwhai (*Sicyos mawhai*), which is listed as nationally at risk, is present in at least two coastal sites on the island. This trailing or climbing plant — the only native member of the cucumber family — has small, spiny, dry fruit with reflexed barbs. Over the last 70 years the southern range of this endemic New Zealand species has been severely reduced, and it has gone from the inner Hauraki Gulf and the mainland as well. The reason for this reduction appears to be its susceptibility to cucumber mosaic virus; this is why no cucurbits are cultivated on the island and visitors are not permitted to bring cucumber sandwiches.

The nationally at risk mida (*Mida salicifolia*), a hemiparasite that attaches onto the roots of hosts, is a small erect tree to around 6 metres tall. It is frequent in the regenerating forest under kānuka and in the kauri–hard beech forests. Its extremely variable leaf size, which often mimics maire species (*Nestegis* spp.), and its variation of oppositely and alternately arranged leaves on the same plant can make it a challenge to identify for those unfamiliar with it.

The nationally at risk *Ranunculus urvilleanus* is a locally common upright herb to 60 centimetres tall, scattered around the wetter areas of Te Maraeroa. This is the best Auckland regional population.

Kirk's daisy (*Brachyglottis kirkii*) is taxonomically problematic on Hauturu. The variety *kirkii* (leaves 2–4 centimetres across, flowers November to February, often epiphytic) is a threatened taxon, and the variety *angustior* (leaves 1–2 centimetres across, flowers March to June, usually terrestrial) is not threatened. However, the two varieties on Hauturu are 'somewhat intermediate',[11] with flowers present in March, June and October; leaves can be from 2 to 6.8 centimetres across and plants terrestrial. More observations and recording are required.

## Two remarkable disjunct distributions and some interesting additions since 1961

*Metrosideros parkinsonii* was discovered in 1959 around the summit of Parkins Knoll at 620 metres asl, with a total of 20–30 spindly, multi-stemmed shrubs reaching the lower canopy.[12] It was already known to be present on the summit area of Aotea. In 2017 Richard Walle could locate only five plants on Parkins Knoll, mostly single-stemmed and 6–8 metres tall — perhaps being shaded out of the regenerating canopy. In November 2017 Liam Walle located a new population of 2–3 small shrubs on the summit ridge, 350 metres south of Parkins Knoll. Remarkably, the nearest population to these two Barrier Island populations is 540 kilometres away in the northern South Island near Collingwood. Perhaps a species of petrel accidentally transported some of the dust-like seed.

In 1978 Dick Veitch discovered a single stand of about 30 black beech trees, 30–50 centimetres diameter at breast height (dbh), on Hauturu. The trees were on a northern ridge above Ōrau Gorge at c. 220 metres asl. This discovery extends the northern range for the species some 250 kilometres.[13] How it reached Hauturu remains a mystery.

Other interesting discoveries include *Raukaua simplex* (1980), *Pomaderris kumeraho* (1986), *Thismia rodwayi* (1987), *Blechnum colensoi* (1993), *Melicytus lanceolatus* (1993) and *Danhatchia australis* (2001). Three of these are, like the black beech, at their northern geographical limit (see page 184).

## Absentees and surprisingly small populations of some species

We now know that kiore have an enormous impact on vegetation as well as on fauna, by eating seeds, seedlings, pollinators and seed dispersers. Campbell and Atkinson showed that on New Zealand's northern offshore islands the recruitment of 11 woody species was depressed by kiore: coastal karamū, taupata, kohekohe, coastal māhoe, coastal maire (*Nestegis apetala*), parapara, karo, tawāpou, houpara, nīkau and tūrepo or large-leaved milktree (*Streblus banksii*).[14] From seedling plots monitored before and after the Hauturu kiore eradication Campbell showed that 13 woody species increased after eradication. They are (in descending order): parapara, coastal māhoe, tāwari, rewarewa, nīkau, tānekaha (*Phyllocladus trichomanoides*), white maire (*Nestegis lanceolata*), kahikatea (*Dacrycarpus dacrydioides*), kareao or supplejack, pigeonwood (*Hedycarya arborea*), kohekohe, haekaro and karaka. Combining these results shows that kiore decrease the recruitment of at least 20 northern coastal species. These species are all present on Hauturu except for coastal maire and tūrepo; the presence of kiore until 2004 may be the reason why these two coastal species are absent. Perhaps the oddly small populations of rimu (*Dacrydium cupressinum*), tītoki (*Alectryon excelsus*), ngaio and tawāpou are also in part due to the now-historic effects of kiore.

Hamilton and Atkinson noted the discovery of two rimu seedlings: one was transplanted to the rangers' garden and is now about 15 metres tall; the other one died.[15] In 1981 Bruce Hayward discovered a 6-metre-tall rimu at about 260 metres asl on the ridge south of Te Hue Stream, and more recently two rimu (20 and 50 centimetres diameter) have been discovered close together at a similar altitude on the same ridge.

The absence of parataniwha (*Elatostema rugosum*) and putaputawētā or marbleleaf (*Carpodetus serratus*) is difficult to explain. There are other species that could be expected to occur on Hauturu because of their nearby presence, but that have not been found there. For instance, the hydrangea-like subshrub *Lobelia physaloides* reaches as far south as Rakitū off the east coast of Aotea but is not known on either of the Barrier islands; and *Celmisia adamsii* var. *adamsii* is on Castle Rock to the south-southwest of Hauturu, *C. adamsii* var. *rugulosa* is to the northwest on Mount Manaia on the Whangārei Heads peninsula and, at lower altitudes, *C. major* is to the east on Aotea — yet the genus is absent from Hauturu. Perhaps the reason is just a lack of chance dispersals.

## Large-leaved coastal forms

Hamilton and Atkinson listed six woody species that showed a tendency to have larger-leaved forms on Hauturu compared with the same species on the adjacent mainland, and they gave the example of taurepo (*Rhabdothamnus solandri*), one of the more extreme cases.[16] They included the very large-leaved nīkau (they recorded one frond as measuring 6.7 metres long, with leaflets 1.2 metres long × 76 millimetres across), but surprisingly they did not include tawaroa, the wide-leaved form of tawa, which is the only form observed on the island. Ross Beever looked at many examples of large-leaved forms on New Zealand's northern offshore islands, including some from Hauturu. His graph for taurepo showed that the leaves from Hauturu, Taranga and Mayor islands stood out as more than twice the size of their mainland counterparts.[17] However, a more recent herbarium collection (AK 206855) from Mount Karioi on the west coast of the North Island, south of Raglan, bucks the trend as it also has large leaves (average leaf 53.2 millimetres × 41.0 millimetres). Beever suggested that these large-leaved forms are possibly ancestral, dating back to preglacial times, and that their smaller-leaved relatives evolved in more southern refugia.

## Geographical limits

Changing geographical limits can be an indication of climate change, and therefore they are important to monitor: Are they stable, retreating or expanding? Sixteen vascular plant species appear to have their geographical limits on Hauturu; seven of these are shared with Aotea/Great Barrier Island (noted as +GBI).

**Northern limits**: *Archeria racemosa* (+GBI), *Blechnum colensoi*, *Cardamine chlorina*, *Carex horizontalis*, *Coprosma dodonaeifolia* (+GBI), black beech, *Lindsaea viridis* (+GBI), *Melicytus lanceolatus*, *Metrosideros parkinsonii* (+GBI), *Peraxilla tetrapetala* (the more northern Tutamoe Range population died out in 1995), *Pseudopanax colensoi* var. *colensoi*, *Pseudopanax discolor* (+GBI), *Raukaua simplex* var. *simplex* (+GBI).

**Southern limits**: *Coprosma neglecta*, *Hebe pubescens* subsp. *sejuncta* (+GBI), *Pimelea acra*.

## Comparison of the floras of three Hauraki Gulf high peaks

In 1973 Lucy Moore made the first comparison of the floras of the upper areas of Hauturu (722 metres asl), Aotea (627 metres) and Moehau on the Coromandel (892 metres).

We now have a better knowledge of what is present. Some 19 montane taxa have their northern geographical limit on the taller Moehau and do not extend to either of the Barrier islands. Many species occur on all three peaks but some are unexpectedly absent from one or two peaks, as shown in the following list, which suggests that dispersal mechanisms are important limiting factors.

- Plants on Hauturu high peaks but not recorded from Aotea or Moehau: *Blechnum colensoi, Melicytus lanceolatus, Pimelea acra*.
- Plants on Hauturu and Aotea high peaks, but not on Moehau: *Archeria racemosa, Metrosideros parkinsonii, Nertera villosa, Pseudopanax discolor*.
- Plants on Aotea and Moehau high peaks but not recorded for Hauturu: *Lepidothamnus intermedius, Pseudowintera colorata*.
- Plants on Aotea high peaks but not recorded from Hauturu or Moehau: *Dracophyllum patens, Epacris sinclairii, Halocarpus kirkii, Kunzea sinclairii, Manoao colensoi, Olearia allomii*.
- Plants on Hauturu and Moehau high peaks but not on Aotea: *Carex horizontalis, Dracophyllum traversii, Griselinia littoralis, Peraxilla tetrapetala, Pseudopanax colensoi* var. *colensoi*.
- Plants on Moehau high peaks but not recorded for Hauturu and Aotea: *Androstoma empetrifolia, Aporostylis bifolia, Carpha alpina, Celmisia incana, Coprosma colensoi, C. foetidissima, Halocarpus biformis, Hymenophyllum villosum, Juncus novae-zelandiae, Kelleria dieffenbachii, Libocedrus bidwillii, Luzuriaga parviflora, Oreobolus pectinatus, Ourisia macrophylla* ssp. *macrophylla, Pentachondra pumila, Phyllocladus alpinus, Poa colensoi, Podocarpus laetus* × *P. nivalis, Rytidosperma setifolium*.
- Plants on all three peaks: *Alseuosmia macrophylla, Ascarina lucida, Astelia microsperma, A. trinervia, Brachyglottis kirkii, Coprosma dodonaeifolia, Corokia buddleioides, Gahnia pauciflora, Hymenophyllum lyallii, Ixerba brexioides, Libertia micrantha, Metrosideros albiflora, M. umbellata, Myrsine salicina, Notogrammitis billardierei, N. pseudociliata, Phyllocladus toatoa, Pittosporum kirkii, Pseudowintera axillaris, Quintinia serrata, Raukaua simplex* var. *simplex, Trichomanes strictum, Weinmannia sylvicola*.

## The future

Many of the changes to the vegetation of Hauturu will be subtle because the kiore depredated species should now slowly increase at the expense of some of the non kiore depredated species. The most obvious changes will be with the regeneration of Te Maraeroa. Karo, houpara, taupata, coastal māhoe and others should continue to increase along the coastal fringe. Forest will continue to creep over rank pasture from the edges; wetland species currently absent might re-establish, for example kahikatea, pukatea; and with time species such as cabbage tree, raupō, *Ranunculus amphitrichus* and *Schoenoplectus tabernaemontani*, whose seeds were plentiful in the peat sample from Te Maraeroa, might return. Another obvious change is the kānuka collapsing as trees die off due to age and are slowly replaced by a vigorous broadleaf and conifer canopy.

The absence of mammalian predators allows Hauturu to have the best fully functioning indigenous ecosystem in the region. For example, the bird-pollinated shrub taurepo is able to set abundant fruit on Hauturu but not on the mainland,[18] and the same applies to the abundant shrub toropapa. A different example is the Hauturu short-tailed bats that pollinate the *Dactylanthus taylorii* — this occurs nowhere else in the region.

Frequent and regular surveys need to be carried out to monitor new plant species arrivals and local extinctions, ideally every five years. Good management relies on knowing what is present. There will no doubt be discussions about possible 'reintroductions', but our preference would be to concentrate on removing the targeted exotics, and hopefully the 'missing' native plants will turn up by themselves.

## Table 1: History of vascular plant recording for Hauturu

| Botanist/ collectors | Fieldwork period | Publications and herbarium vouchers (herbaria in parentheses) |
|---|---|---|
| Thomas Kirk | 1867–68, 1871 (S and SE coast) | Recorded c. 182 indigenous and 7 naturalised spp. (Kirk 1869) and 9 herbarium specimens (AK and WELT) |
| Thomas Cheeseman | 1896–1908 (6 visits) | 110 herbarium specimens (AK) |
| Edith Smith and Frances Shakespear | 1897–1910 (lived on the island) | 498 herbarium specimens (mainly AK) |
| James Adams | Jan 1901 (with T. Cheeseman) | 62 herbarium specimens (AK) |
| Walter R. B. Oliver | 1921–32 | 310 herbarium specimens (WELT) |
| W. M. Hamilton | 1932–37 (surveyed most of the island) | Recorded 351 indigenous and 83 naturalised spp. (Hamilton 1937) |
| W. M. Hamilton and Ian Atkinson | 1938–61 | Revised and expanded Hamilton's (1937) vegetation and flora account recording 368 indigenous and 92 naturalised spp. (Hamilton and Atkinson 1961) 91 herbarium specimens (mainly CHR) |
| P. Alison Lush | 1949 | 158 herbarium specimens (WELT) |
| Ronald Melville et al. | 1962 | 111 herbarium specimens (AK and CHR) |
| C. (Phyllis) Hynes | 1963 | 163 herbarium specimens (AK) |
| Ross Beever, Alan Esler, Maureen Young and Ewen Cameron | 1967–2012 (multiple visits to the island) | Revised and expanded Hamilton and Atkinson's list (Beever et al. 2012) to: 428 indigenous and 198 naturalised spp. and 349 herbarium specimens (AK and CHR) |
| Patrick Brownsey | 1974, 1984 | 56 herbarium specimens (mainly WELT) |
| Anthony Wright | 1981, 1990 | 78 herbarium specimens (AK) |
| W. (Bill) Sykes | 1985 | 125 herbarium specimens (CHR) |
| Ewen Cameron and Maureen Young | 2018 | 54 herbarium specimens (AK) |
| **Total collections** |  | **2950** = 1608 (AK), 617 (CHR), 725 (WELT) |

## Table 2: Hauturu vegetation zones

| After Hamilton & Atkinson (1961) | Our system after Atkinson & Stanley (2018) | Area (ha)* |
|---|---|---|
| 1. Paspalum pasture (40 acres) | 1. Pasture–fernland/sedgeland | 16 |
| 2. Cyperus/Carex (4 acres) | | |
| 3. Leptospermum (2480 acres) | 2. Kānuka/kanuka–mānuka forest | 910 |
| 4. Muehlenbeckia communities (10 acres) | 3. Coastal forest and scrub | 262 |
| 5. Open cliff communities (140 acres) | | |
| 6. Coastal scrub | | |
| 7. Pōhutukawa forest<br>Forest on cliffs and talus (300 acres)<br>Pua Mataahu forest (3 acres)<br>Hingaia rock forest (26 acres) | | |
| 8. Pōhutukawa/broadleaf forest (235 acres) | | |
| 9. Kauri forest (1000 acres)<br>With miro as co-dominant<br>With beech and rātā as important elements<br>With rātā as co-dominant | 4. Kauri and kauri–hard beech forests | 373 |
| 10. Beech forest (20 acres) | | |
| 11. Rātā/tawa forest (2270 acres) | 5. Northern rātā/pūriri–tawaroa forest | 832 |
| 12. Tawhero/tawa forest (780 acres) | 6. Tōwai–tawaroa forest | 359 |
| 13. Quintinia/Ixerba/southern rātā (170 acres) | 7. Quintinia–tāwari–southern rātā forest and scrub | 65 |
| 14. Summit scrub (7 acres) | | |
| **Total (ha)** | | **2817** |

* adjusted to fit the present known total area of Hauturu

## Table 3: Totals of wild vascular species and hybrids

| Vascular plant group | Indigenous | Naturalised | Totals (% naturalised) |
|---|---|---|---|
| **Lycophytes** | 5 | 1 | **6 (17%)** |
| **Ferns** | 109 | 0 | **109 (0%)** |
| **Conifers** | 10 | 0 | **10 (0%)** |
| **Angiosperms — dicots** | 212 | 142 | **354 (40%)** |
| **Angiosperms — monocots** | 102 | 58 | **160 (36%)** |
| **Totals** | 438 | 201 | **639 (32%)** |

# Ferns and lycophytes

True maidenhair (*Adiantum aethiopicumi*) (LW)

Common maidenhair (*Adiantum cunninghamii*) (LW)

Small maidenhair (*Adiantum diaphanum*) (LW)

Rosy maidenhair (*Adiantum hispidulum*) (LW)

Jointed fern (*Arthropteris tenella*) (LW)

Hen and chickens fern (*Asplenium bulbiferum*) (LW)

Hanging spleenwort (*Asplenium flaccidum*) (LW)

Coastal spleenwort (*Asplenium haurakiense*) (LW)

*Asplenium lamprophyllum* (LW)

# Ferns and lycophytes

Shining spleenwort (*Asplenium oblongifolium*) (LW)

Sickle spleenwort (*Asplenium polyodon*) (LW)

*Blechnum blechnoides* (LW)

*Blechnum chambersii* (LW)

Crown fern/piupiu (*Blechnum discolor*) (LW)

Thread fern (*Blechnum filiforme*) (LW)

*Blechnum fraseri* (LW)

*Blechnum nigrum* (LW)

A new kiokio frond (*Blechnum novae-zelandiae*) (LW)

# Ferns and lycophytes

*Blechnum procerum* (LW)

Black tree fern/mamaku (*Cyathea medullaris*) (LW)

Rasp fern (*Doodia australis*) (LW)

Carrier tangle fern/waewae kākā (*Gleichenia microphylla*) (LW)

*Hymenophyllum dilatatum* (LW)

Kidney fern/raurenga (*Hymenophyllum nephrophyllum*) (LW)

*Hypolepis ambigua* (LW)

*Lastreopsis glabella* (LW)

Hairy legs (*Lastreopsis hispida*) (LW)

# Ferns and lycophytes

*Lindsaea linearis* (LW)

*Lindsaea viridis* (OB)

*Lindsaea trichomanoides* (LW)

*Lycopodium deuterodensum* (LW)

Climbing club moss/waewae-koukou (*Lycopodium volubile*) (LW)

Bushmen's mattress/mangemange (*Lygodium articulatum*) (LW)

Hounds tongue (*Microsorum pustulatum*) (SF)

Fragrant fern/mokimoki (*Microsorum scandens*) (LW)

*Notogrammis heterophylla* (LW)

# Ferns and lycophytes

*Paesia scaberula* (LW)

*Pellaea rotundifolia* (LW)

Tassel fern/iwituna (*Phlegmariurus varius*) (LW)

Gully fern/pākau (*Pneumatopteris pennigera*) (LW)

Shield fern (*Polystichum wawranum*) (LW)

Bracken/rārahu (*Pteridium esculentum*) (LW)

*Pteris comans* (LW)

Turawera (*Pteris tremula*) (LW)

*Pyrrosia eleagnifolia* (LW)

# Ferns and lycophytes

*Rumohra adiantiformis* (LW)

Bifurcated comb fern (*Schizaea bifida*) (LW)

Fan fern (*Schizaea dichotoma*) (LW)

Umbrella fern/tapuwae kōtuku (*Sticherus cunninghamii*) (LW)

Chain fern (*Tmesipteris elongata*) (LW)

# Gymnosperms

Kauri (*Agathis australis*) (AA)

Kauri (*Agathis australis*) (LW)

Kahikatea (*Dacrycarpus dacrydioides*) (LW)

Rimu (*Dacrydium cupressinum*) (LW)

Toatoa (*Phyllocladus toatoa*) (LW)

Tānekaha male cones (*Phyllocladus trichomanoides*) (LW)

Tānekaha (*Phyllocladus trichomanoides*) (LW)

Hall's tōtara (*Podocarpus laetus*) (LW)

Miro (*Prumnopitys ferruginea*) (LW)

## Angiosperms — dicotyledons

Toropapa (*Alseuosmia macrophylla*) (SF)

Shore celery (*Apium prostratum*) (LW)

Archeria racemosa (SF)

Wineberry/makomako flowers (*Aristotelia serrata*) (LW)

Wineberry/makomako seedling (*Aristotelia serrata*) (LW)

Hutu (*Ascarina lucida*) (LW)

Taraire (*Beilschmiedia tarairi*) (LW)

Tawaroa (*Beilschmiedia tawaroa*) (LW)

Kirk's daisy/kohurangi flower (*Brachyglottis kirkii*) (NF)

**Vegetation and vascular flora**

## Angiosperms — dicotyledons

Kirk's daisy/kohurangi (*Brachyglottis kirkii*) (LW)

Rangiora (*Brachyglottis repanda*) (LW)

*Calystegia sepium* ssp. *roseata* (SF)

Shore bindweed (*Calystegia soldanella*) (LW)

Climbing convolvulus (*Calystegia tuguriorum*) (LW)

Native broom (*Carmichaelia australis*) (LW)

Native broom flower (*Carmichaelia australis*) (LW)

Giant-flowered broom (*Carmichaelia williamsii*) (WMH)

*Centipida minima* (DR)

# Angiosperms — dicotyledons

Berry saltbush (*Chenopodium triandrum*) (LW)

Yellow clematis (*Clematis cunninghamii*) (LW)

Clematis seed (*Clematis paniculata*) (LW)

Māmāngi (*Coprosma arborea*) (LW)

Kanono (*Coprosma autumnalis*) (LW)

Shining karamū (*Coprosma lucida*) (LW)

*Coprosma macrocarpa* × C. *propinqua* (LW)

Taupata (*Coprosma repens*) (LW)

*Coprosma rhamnoides* (LW)

**Vegetation and vascular flora**

# Angiosperms — dicotyledons

Karamū (*Coprosma robusta*) (LW)

Tutu (*Coriaria arborea*) (LW)

Korokia (*Corokia buddleioides*) (AW)

Karaka fruit (*Corynocarpus laevigatus*) (SF)

Flowers of Hades/pua o Te Rēinga (*Dactylanthus taylorii*) (LW)

Mercury Bay weed (*Dichondra repens*) (LW)

New Zealand ice plant (*Disphyma australe*) (LW)

Akeake (*Dodonaea viscosa*) (LW)

*Dracophyllum latifolium* (LW)

# Angiosperms — dicotyledons

*Dracophyllum traversii* (LW)

Sundew (*Drosera auriculata*) (SF)

Kohekohe flower (*Dysoxylum spectabile*) (LW)

Kohekohe seedling (*Dysoxylum spectabile*) (LW)

Kohekohe fruit (*Dysoxylum spectabile*) (LW)

Sand milkweed (*Euphorbia glauca*) (AW)

Kōtukutuku (*Fuchsia excorticata*) (LW)

Tawhairaunui/hard beech in flower (*Fuscospora truncata*) (LW)

Tawhairaunui/hard beech (*Fuscospora truncata*) (LW)

**Vegetation and vascular flora**

# Angiosperms — dicotyledons

Snowberry (*Gaultheria antipoda*) (LW)

Hangehange (*Geniostoma ligustrifolium*) (LW)

*Gonocarpus incanus* (LW)

Puka (*Griselinia lucida*) (LW)

Koromiko flower (*Hebe macrocarpa* var. *latisepala*) (SF)

Koromiko (*Hebe pubescens* subsp. *sejuncta*) (LW)

Pigeonwood/porokaiwhiri (*Hedycarya arborea*) (LW)

Niniao (*Helichrysum lanceolatum*) (LW)

Tāwari (*Ixerba brexioides*) (LW)

# Angiosperms — dicotyledons

Rewarewa flowers (*Knightia excelsa*) (LW)

Dwarf mistletoes on kānuka (*Korthalsella salicornioides*) (LW)

Kānuka (*Kunzea robusta*) (NF)

Papatāniwhaniwha (*Lagenophora pumila*) (LW)

Pukatea (*Laurelia novae-zelandiae*) (LW)

Mairehau (*Leionema nudum*) (SF)

*Leptinella tenella* (LW)

Mānuka (*Leptospermum scoparium*) (LW)

Mingimingi fruiting (*Leucopogon fasiculatus*) (LW)

**Vegetation and vascular flora**

# Angiosperms — dicotyledons

Mangeao (*Litsea calicaris*) (LW)

Shore lobelia (*Lobelia anceps*) (LW)

Ramarama (*Lophomyrtus bullata*) (LW)

Wharangi (*Melicope ternata*) (LW)

Coastal māhoe (*Melicytus novae-zelandiae*) (LW)

Māhoe flower (*Melicytus ramiflorus*) (LW)

Māhoe seedling (*Melicytus ramiflorus*) (LW)

White rātā/akatea (*Metrosideros albiflora*) (LW)

Carmine rātā (*Metrosideros carminea*) (LW)

## Angiosperms — dicotyledons

Pōhutukawa (*Metrosideros excelsa*) (LW)

Pōhutukawa flowers (*Metrosideros excelsa*) (LW)

Akakura, a rare yellow form (*Metrosideros fulgens*) (KB)

Akakura (*Metrosideros fulgens*) (LW)

Small white rātā/akatea (*Metrosideros perforata*) (LW)

Northern tree rātā (*Metrosideros robusta*) (LW)

Southern rātā (*Metrosideros umbellata*) (LW)

Maire-taike mida (*Mida salicifolia*) (LW)

Wire vine (*Muehlenbeckia complexa*) (LW)

## Angiosperms — dicotyledons

Māpou (*Myrsine australis*) (LW)

Toro (*Myrsine salicina*) (LW)

*Nertera depressa* (LW)

Akepiro (*Olearia furfuracea*) (LW)

Heketara (*Olearia rani*) (LW)

New Zealand jasmine (*Parsonsia capsularis*) (LW)

*Peperomia urvilleana* (LW)

Mistletoe flower (*Peraxilla tetrapetala*) (WMH)

Mistletoe (*Peraxilla tetrapetala*) (RG)

# Angiosperms — dicotyledons

Pinātoro (*Pimelea urvilleana*) (DR)

Kawakawa (*Piper excelsum*) (LW)

Bird catcher/parapara (*Pisonia brunoniana*) (LW)

Tāwhiri karo (*Pittosporum cornifolium*) (LW)

Karo (*Pittosporum crassifolium*) (LW)

*Pittosporum kirkii* (LW)

*Pittosporum kirkii* fruit (AW)

Haekaro (*Pittosporum umbellatum*) (SF)

Haekaro fruit (*Pittosporum umbellatum*) (LW)

Vegetation and vascular flora 205

# Angiosperms — dicotyledons

Tawāpou (*Planchonella costata*) (LW)

*Pomaderris amoena* (LW)

Five finger/whauwhaupaku (*Pseudopanax arboreus*) (LW)

*Pseudopanax discolor* (AW)

Horopito (*Pseudowintera axillaris*) (LW)

Quintinia/tāwheowheo (*Quintinia serrata*) (LW)

Raukawa (*Raukaua edgerleyi*) (LW)

Taurepo (*Rhabdothamnus solandri*) (NF)

Bush lawyer (*Rubus cissoides*) (LW)

# Angiosperms — dicotyledons

Patē seedling (*Schefflera digitata*) (LW)

Native cucumber/māwhai (*Sicyos mawhai*) (AW)

Native cucumber/māwhai fruit (*Sicyos mawhai*) (LW)

*Solanum americanum* (LW)

Poroporo flower (*Solanum aviculare*) (SF)

Poroporo (*Solanum aviculare*) (LW)

Sea spurrey (*Spergularia tasmanica*) (LW)

Native spinach (*Tetragonia implexicoma*) (LW)

Toru flower (*Toronia toru*) (AW)

**Vegetation and vascular flora**

## Angiosperms — dicotyledons

Toru (*Toronia toru*) (AW)

Pūriri (*Vitex lucens*) (LW)

Pūriri tree (*Vitex lucens*) (LW)

Tōwai (*Weinmannia silvicola*) (LW)

## Angiosperms — monocotyledons

Pixiecap orchid (*Acianthus sinclairii*) (LW)

Rengarenga lily (*Arthropodium cirratum*) (LW)

Coastal astelia (*Astelia banksii*) (LW)

# Angiosperms — monocotyledons

Bush lily/kakaha (*Astelia fragrans*) (LW)

Kakaha flower (*Astelia hastata*) (LW)

Kakaha (*Astelia hastata*) (SF)

*Astelia microsperma* (LW)

*Astelia trinerva* (CRV)

Toetoe (*Austroderia splendens*) (LW)

Pygmy orchid (*Bulbophyllum pygmaeum*) (LW)

Lady's fingers (*Caladenia chlorostyla*) (LW)

*Carex solandri* (LW)

# Angiosperms — monocotyledons

Bush cabbage tree/tī ngahere (*Cordyline banksii*) (LW)

Dwarf cabbage tree/tī rauriki (*Cordyline pumilio*) (LW)

Spider orchid (*Corybas acuminatus*) (LW)

Helmet orchid (*Corybas cheesemanii*) (LW)

Spider orchid (*Corybas rivularis*) (CRV)

Spider orchid (*Corybas trilobus*) (CRV)

Giant umbrella sedge (*Cyperus ustulatus*) (LW)

Gnat orchid (*Cyrtostylis oblonga*) (CRV)

Winikā (*Dendrobium cunninghamii*) (DR)

## Angiosperms — monocotyledons

Tūrutu blueberry (*Dianella nigra*) (LW)

*Drymoanthus adversus* (SF)

Autumn orchid (*Earina autumnalis*) (LW)

Spring orchid (*Earina mucronata*) (LW)

Kiekie (*Freycinetia banksii*) (LW)

Native iris (*Libertia grandiflora*) (LW)

*Libertia micrantha* (LW)

Onion orchid (*Microtis unifolia*) (CRV)

*Morelotia affinis* (LW)

**Vegetation and vascular flora**

# Angiosperms — monocotyledons

Horned orchid (*Orthoceras novae-zeelandiae*) (CRV)

Flax/harakeke (*Phormium tenax*) (LW)

Mountain flax/wharariki (*Phormium cookianum*) (LW)

Small greenhood orchid (*Pterostylis alobula*) (CRV)

Small greenhood orchid (*Pterostylis trullifolia*) (CRV)

Greenhood orchid (*Pterostylis graminea*) (CRV)

Nīkau (*Rhopalostylis sapida*) (RG)

Supplejack (*Ripogonum scandens*) (LW)

Sun orchid (*Thelymitra longifolia*) (CRV)

*Thelymitra longifolia* (CRV)

# 6.6 Mosses

— JESSICA BEEVER

**All plants need water, but mosses have circumvented the need for roots, instead absorbing moisture over their entire leafy surface. Carpeting the ground, tree trunks and branches, their luxuriance in the 'moss forest' of Hauturu is due to the benign moisture conditions provided by the island's frequent cloud cap, high rainfall, high humidity and low evapotranspiration, all with exposure to relatively high light. Accounts of the forest vegetation of Hauturu have referred to the richness of small plants on the island's upper slopes and ridges, the so-called 'moss forest' or 'cloud forest'.[1] The terminology 'moss forest' is, however, rather inaccurate botanically, as the spectacular swards of small plants are composed of bryophytes, that is, of both mosses and their relatives the liverworts.**

It is not only the high elevations of Hauturu that have intrigued bryologists — the island supports a richness of mosses throughout. The ecological distribution of these plants is determined largely by their microenvironment, rather than being related to the broad vegetation types defined by dominant vascular plants (see Chapter 6.5). Factors such as the precise substrates that are available for colonisation (the porosity and chemical composition of the supporting surface), the availability of light and the moisture regimes that prevail will be primary factors controlling distribution of species on the island. Thus the many deep valleys and dark streamsides differ in their suite of mosses from the well-lit, drier regenerating forests on the lower slopes. Mosses found on the shoreline are species that flourish in conditions of high light, and there the presence of salt spray may also be important. The highly disturbed vegetation, which has been subjected to considerable recent human activity on the flat of Te Maraeroa, has supported certain moss species — in pasture, mown lawns, vehicle tracks, the bare ground adjacent to buildings and even on the concrete of manmade structures. Here, if light exposure is high and moisture is largely dependent on direct rainfall, the substrate becomes the primary controlling factor. Some of the same species have been recorded away from habitation, where sites of natural disturbance provide similar microenvironments.

All mosses found on the island today must have found their own way there or hitchhiked with visitors — avian, human or with domestic animals. Mosses are on the

*Dawsonia superba*, the world's tallest self-supporting moss, growing on Thumb Track, Hauturu. (DR)

whole highly vagile organisms, with a wide range of specialised asexual structures, as well as spores produced in capsules in a sexual cycle. Moss spores are small and light, so are readily dispersed in air currents, and some are highly resistant to desiccation, freezing and exposure to ultraviolet light and are thus able to be dispersed even long-distance on a global scale. In addition, mosses are totipotent, so that plant fragments can be units of dispersal. By such means the present-day moss flora of Hauturu has been assembled.

Mosses have important roles in our native ecosystems: they act as a living mulch, absorbing moisture when it rains and slowly releasing it, thereby protecting against rapid runoff, and prolonging periods of high humidity. These small plants can provide a suitable seedling bed for vascular plants, and many invertebrates find a home among them.

## History of moss studies

Early Pākehā residents of the island, Edith Smith and Frances Shakespear, made some collections of mosses, but it was not until W. M. Hamilton took an interest in the island's flora and explored the terrain extensively that a short list of 14 conspicuous mosses was published in 1937. These included *Ptychomnion aciculare* and *Rosulabryum subtomentosum*. Numerous other botanists, including T. C. Chambers, J. M. Dingley, L. B. Moore, F. J. Newhook and I. B. Wormald, have collected mosses on Hauturu as part of other projects, and specialist bryophyte studies have been undertaken by Jessica Beever and John Braggins with P. J. Brownsey and B. Polly, primarily over the decade 1980–90 but also on a three-day visit by Beever in January 2018. Today numerous moss specimens are lodged in the nation's major herbaria, namely AK, CHR, WELT and MPN. The details of the localities and habitat of specimens are available on the databases of those institutions, and voucher specimens can be verified as the need arises. Collectively, records now comprise 158 species of moss; and four species (*Distichophyllum crispulum*, *Fissidens tenellus*, *Hypnum cupressiforme* and *Weissia controversa*) each have two varieties recognised on Hauturu.

## Habitats

The bush on the summit ridge, where conditions of high light and low evapotranspiration prevail, supports dense masses of bryophytes as epiphytes perched on other plants and as ground-dwellers. Among the mosses here are spectacular large forms of *Dicranoloma robustum*, including many with long leaves tinged vinous red. These were formerly considered to belong to an endemic species, *Dicranoloma cylindropyxis*. Three other mosses — *Rhizogonium pennatum*, *Campylopus purpureocaulis* and *Austrohondaella limata* — are known on Hauturu only from above 600 metres asl, and from similar elevations elsewhere in northern New Zealand.

'Pipe-cleaner moss' (*Ptychomnion aciculare*) (LJ)

*Rosulabryum subtomentosum* (LJ)

*Dicranoloma robustum* (cylindropyxis form), summit ridge (JB)

*Fissidens taxifolius* (TL)

A quite different suite of mosses has been recorded on Mount Hauruia (Bald Rock) and its footstool. From exposed andesitic rock outcrops between c. 500 and 575 metres asl there are records of *Papillaria flavolimbata, Racomitrium crispulum, R. lanuginosum* and *Polytrichum commune*. Another polytrichaceous moss, *Polytrichum juniperinum*, has also been found here, but associated with some soil. This moss is a typical species of the cleared land around trig stations on many of northern New Zealand's high points, for example Moehau and Tutamoe. On Hauturu it has been found at the modified summit of the island as well as at the natural site on Hauruia.

A further group of mosses are found both in cloud forest at high elevations and in the deep-cut valleys where there is similarly low evapotranspiration, but these species are tolerant of low light intensities; examples include members of the families Daltoniaceae (*Distichophyllum crispulum, D. microcarpum, D. rotundifolium* and *Calyptrochaeta cristata*) and Hookeriaceae (*Achrophyllum dentatum*). *Mittenia plumula* has been recorded from only one site, in a dense sheet on slightly overhung wet rock in thick tōwai–tawaroa forest at 580 metres. This unusual moss is classified in its own family, and is noted for its luminescence, a result of its ability to focus light when light is available only at very low intensities.[2]

Another substrate that is sometimes colonised by mosses in wet, dark forest is fronds of certain ferns: *Blechnum nigrum, Trichomanes elongatum* and *T. endlicherianum*. Three species of the moss genus *Distichophyllum* have been observed growing as epiphylls on ferns on Hauturu: *D. crispulum* var. *adnatum, D. pulchellum* and *D. rotundifolium*. Other mosses observed growing as fern epiphytes are *Cyathophorum bulbosum, Dicranoloma menziesii, Lopidium concinnum, Ptychomnion aciculare, Rhaphidorrhynchium amoenum* and *Wijkia extenuata*. The latter was also observed as an epiphyte on the fern ally *Tmesipteris*. Even mosses have become substrate for other mosses in this habitat: *Achrophyllum quadrifarium* and the beautiful umbrella moss *Dendrohypopterygium filiculaeforme* were seen supporting shoots of *Distichophyllum rotundifolium*. Epiphyllous mosses have been found only in moist, densely shaded sites where evapotranspiration would be low at all times, but even at quite low elevations on the island, for example at 50 metres asl on a south-facing slope in the lower Haowhenua catchment, and at 15 metres asl on the wall of the canyon carved by the Parihākoakoa Stream.

Moss epiphytes have been recorded on the trunks of trees when sufficient light penetrates down into the forest, as is often the case in the kānuka/mānuka, kauri–hard beech and rātā/pūriri–tawaroa forests. Here several species of *Macromitrium* creep on bark, namely *M. gracile, M. longipes* and *M. ligulaefolium*, as well as *Cladomnion ericoides, Hypnum chrysogaster, Leratia obtusifolia, Orthorrhynchium elegans* and *Weymouthia cochlearifolia*. Fallen epiphytes from higher in the canopy, found on branches lying on the ground, include *Holomitrium perichaetiale, Dicranoloma menziesii* and *Dicnemon calycinum*.

Humus on the forest floor is a favoured substrate for the 'pipe-cleaner moss' *Ptychomnion aciculare* and one of the larger 'Bryum' mosses, *Rosulabryum subtomentosum*. Rotting wood is often colonised by *Hypnum chrysogaster* and *Rhaphidorrhynchium amoenum*.

Sites of natural disturbance in forest provide a fresh soil surface for primary colonisers.

When large landslips occur in the forest a partial canopy gap can be opened up, and thus higher light intensities are suddenly present on the forest floor. Such newly available bare soil can provide substrate for a suite of early colonisers: *Campylopodium medium*, *Oligotrichum tenuirostre* and *Polytrichadelphus magellanicus*. Soil on the upended root base of wind-thrown trees, or in fresh soil cavities so created, provides similar microhabitats but usually with much less light; here *Dicranella vaginata*, *Ditrichum difficile*, *Fissidens pallidus*, *Hypnodendron arcuatum*, *Pogonatum subulatum* and *Wilsoniella blindioides* have been recorded. And small patches of bare earth that have been created by natural erosion on the dark forest floor are often colonised by the minute plants of *Fissidens tenellus*.

The unpolluted forest streams of Hauturu provide habitat for *Fissidens rigidulus*, a moss found throughout New Zealand in fast-flowing oligotrophic waters. A form of *Fissidens tenellus* var. *tenellus* with the tagname *Fissidens* 'microaquatic' occurs on Hauturu but has been detected only when stream levels have been very low. In March 1987 the Awaroa Stream had a dry bed even at 100 metres elevation, and here dozens of minute plants of this moss were noted barely 3 millimetres tall and producing copious spores. On the same visit this tiny *Fissidens* was also found at several sites in the Tirikakawa Stream, on rock that would usually be submerged. *Sematophyllum jolliffii* is a moss that evidently requires at least frequent submergence, and has been recorded in the Awaroa, Tirikakawa and Te Waikohare streams. Species that are tolerant of submergence and are found on rock in creek beds are *Echinodium umbrosum*, *Fissidens asplenioides* and *F. blechnoides*, and *Ctenidium pubescens* is a coloniser on flood-scoured soil banks. *Camptochaete pulvinata* is a common species on rocks at low elevations, especially in streambeds.

A high-light environment, with perhaps a salt influence, is present on the coast of Hauturu. Here, on solid rock where it is exposed at the base of the sea cliff, mosses such as *Ischyrodon lepturus*, *Leptostomum macrocarpum*, *Macromitrium brevicaule*, *Ptychomitrium australe*, *Sematophyllum homomallum*, *Syntrichia laevipila*, *S. papillosa* and *Tortella flavovirens* have been recorded. Numerous mosses occur where there is some shade from a light canopy of pōhutukawa or other coastal vegetation (such as *Phormium tenax*, *Peperomia urvilleana*, *Microsorum pustulatum* or *Astelia banksii*), often accumulating humus among their shoots: *Campylopus clavatus*, *C. introflexus*, *Holomitrium perichaetiale*, *Macromitrium gracile*, *M. prorepens*, *Rosulabryum subtomentosum*, *Sematophyllum contiguum* and *Thuidium sparsum*. On more porous substrates such as beach boulders comprised of soft conglomerate material *Ditrichum difficile* may be present, and on coastal landslips, early colonisers *Bryum dichotomum* and *Didymodon australasiae* can occur. The fissured bark of exposed pōhutukawa roots can support *Campylopus pyriformis* and *Rosulabryum capillare*. On more-or-less stable boulders at the back of the beach *Hypnum cupressiforme* has occasionally established.

*Calyptrochaeta apiculata* is an uncommon coastal moss in northern New Zealand, and Hauturu has the northernmost records. It has been found only twice on the island, first by Lucy Moore in 1941 in a 'salt meadow' at 'Awaroa Point', and more recently at a stream mouth in that vicinity.

On the highly modified habitats of Te Maraeroa a range of mosses has been found that have been recorded from that area alone on Hauturu. Seven have been found on bare soil, either adjacent to buildings, on tractor tracks or in the rangers' garden, or — only from surveys made in 1990 or earlier — on bare patches in pasture: *Ceratodon purpureus*, *Chenia leptophylla**, *Eurhynchium praelongum**, *Pleuridium subulatum*, *Tortula truncata**, *Weissia austrocrispa* and *W. controversa* var. *controversa*. On other substrates the following have been recorded: *Barbula unguiculata**, *Bryum argenteum*, and *Tortula muralis* on concrete, and *Bryum duriusculum* — a species that colonises old fire sites — was found growing on both soil and burnt wood at a former bonfire site. Another suite of mosses has been recorded both from this highly modified area and from more natural sites elsewhere on the island where similar microenvironments are available, albeit in quite different vegetation. These are *Bryum sauteri*, *Campylopus pyriformis*, *Fissidens curvatus* and *Funaria hygrometrica*.

Some of the mosses recorded on Te Maraeroa (marked * above) are classified as exotic in the recently published *Checklist of the New Zealand Flora*.[3] However, it is often not possible to ascertain whether particular species are native to New Zealand or exotic as many mosses are naturally widespread in the world, even cosmopolitan. Some may be truly native but also inhabit human-induced sites if the microenvironment provided there is suitable. Future research using genetic tools may better clarify the native versus exotic status of individual taxa. However, for one moss species on Hauturu, *Fissidens taxifolius*, its exotic status in New Zealand is well established.[4] This moss, known colloquially as 'rogue Fissidens moss' is an invasive species that was well established on Hauturu at Te Maraeroa by January 2018, and has been recorded on the Thumb Track to at least 275 metres elevation.

In January 2018 no mosses were recorded on soil in the retired pasture at Te Maraeroa; with the dense growth of vascular plants there is now no bare soil present. Mosses may be expected to establish in the future, though, as the vegetation succession progresses.

## Department of Conservation classified mosses

Three mosses recorded on Hauturu are 'classified species'[5] according to the Department of Conservation's New Zealand Threat Classification System.[6] These are *Tortella cirrhata*, *Ischyrodon lepturus* and *Thuidium cymbifolium*, all of which are classified as 'naturally uncommon'. *Tortella cirrhata* is known on Hauturu from a single site, growing on soil with another pottiaceous halophyte, *Tortella flavovirens*, in Ōrau Cove on the northern coast of the island. This moss bears the qualifier 'range restricted'[7] as it is known only from a few coastal localities in northern parts of the country, including Rangitoto Island in the inner Hauraki Gulf, where the moss was initially discovered by Lucy Moore.

*Ischyrodon lepturus* is likewise recorded on Hauturu from one site only, a lightly shaded east-facing, sloping rockface at the base of the sea cliff, c. 2 metres asl at the southern end of Hingaia. Other halophytes — *Macromitrium brevicaule* and *Tortella flavovirens* — grew nearby, with a range of other rock-dwelling species that are not confined to coastal habitats: *Ptychomitrium australe*, *Syntrichia laevipila*, *S. papillosa* and *Leptostomum macrocarpum*. *Ischyrodon lepturus* was also first collected in New Zealand by Moore, in this case on Māuitaha in the Hen and Chicken Islands. There are now several records for this moss around Cook Strait, and well-developed colonies occur on the consolidated boulder banks at the northern end of Kāpiti Island. The species has the qualifier 'sparse',[8] reflecting its scattered occurrence over a wide range. It may well be that further searching of the Hauturu coastline will show the species to be more abundant there, since there would seem to be no lack of suitable habitat in places where the boulder beach is stable at its landward edge.

Hauturu is an important refuge for *Tortella cirrhata* and *Ischyrodon lepturus*. Elsewhere they are threatened by kikuyu grass (*Cenchrus clandestinus*), the serious coastal weed of mainland northern New Zealand. This invasive grass has been recorded on Hauturu on the helicopter landing area and around the boathouse.[9] Eradication has been attempted but is not yet confirmed.

The third classified moss, *Thuidium cymbifolium*, is essentially a tropical species. In New Zealand it has occasional records from the Hunua Ranges northwards. On Hauturu it is recorded from three low-elevation sites: on muddy breccia on a stream bank at East Cape, from soil at the edge of Valley Track at c. 10 metres asl, and from compacted soil on the Shag Track.

## Northernmost limits

Available data indicate that three moss species are at their northernmost New Zealand limit on Hauturu. Two inhabit only the highest elevations, above 600 metres — *Campylopus purpureocaulis* and *Rhizogonium pennatum*; the latter has been recorded at a similar latitude on Aotea. The third northernmost record is for *Calyptrochaeta apiculata*, which has been recorded close to sea level on Hauturu, as it is in all its mainland sites; it reaches higher elevations only on Rakiura/Stewart Island and the subantarctic islands. Future monitoring of localities for these taxa may give useful information relevant to climate change.

## Conclusion

Some 30 per cent of New Zealand's total moss flora (159/520 species), spread over more than half New Zealand's moss families (40/73 families), have been recorded on Hauturu. These include three species classified by the Department of Conservation as 'naturally uncommon'. A similar proportion of New Zealand's liverwort and hornwort flora are also recorded here (see Chapter 6.7). Thus the island is an extremely valuable sanctuary for the bryophytes of northern New Zealand. Its value is enhanced by the fact that incursion by invasive groundcover weeds has been minimal, apart from the rogue Fissidens moss, *Fissidens taxifolius*. Existing strict quarantine measures need to be maintained to mitigate against the arrival of other threats, and kikuyu grass must be eliminated, in order to sustain the mana and the mauri of Te Hauturu-o-Toi and all of its taonga.

# 6.7
# Liverworts and hornworts

— JOHN E. BRAGGINS

**The liverwort flora of Te Hauturu-o-Toi/Little Barrier Island is rich and varied, with 195 taxa identified in 71 genera and including four species in which more than one variety is present on the island. The herbarium collections include 66 specimens that have not yet been determined to species level because the taxonomy of the New Zealand taxa in those genera is as yet poorly understood. With around 200 taxa the Hauturu liverwort flora represents about 28 per cent of the estimated 700 species in New Zealand. This level of variety may in part reflect the potential for dispersal available to such spore-bearing plants.**

Traditional textbook illustrations of liverworts show a predominance of thalloid (leafless) forms. On Hauturu those forms are particularly sparse, with 18 such species. The only common one is a small *Metzgeria*; most have few records, or just a single record. Leafy taxa predominate, and the most common of these is *Bazzania adnexa*, with over 30 specimens recorded. At least two species of liverwort were first described from Hauturu and are now known from a few records elsewhere in New Zealand: *Lembidium longifolium* and *Kurzia fragifolia* (described as *Telaranea fragifolia*). There has been no subsequent record of them on the island.

The size of the flora on Hauturu and the nature of the liverwort species found there make for an interesting comparison with those of Rangitoto, an island of similar area in the inner Hauraki Gulf but much younger and at lower altitude. On Rangitoto only 70 species of liverwort have been recorded.[1]

The richness of the liverwort flora of Hauturu reflects the range of microhabitats, particularly the much older forest, the existence of more-or-less permanent streams and moist, often cloud-covered 'moss forest' areas. The richest and most noticeable development of larger liverworts is in this high-altitude summit forest; and, as is often the case elsewhere in New Zealand, the 'moss' may in fact be mainly liverworts. On Hauturu these are the areas represented by vegetation types: tōwai/tawaroa forest and Quintinia–tāwari–southern rātā forest and scrub (Chapter 6.5). Among these are many of the big, visually obvious species of liverworts, and it is these that are generally missing from Rangitoto. The larger, moisture-loving taxa include three species of *Schistochila*, *Gottschea* (1), *Isotachis* (1), *Lepicolea* (1), *Tetracymbaliella* (1), *Anastrophyllopsis* (1), *Heteroscyphus* (1), *Lepidozia* (2), *Plagiochila* (2), *Temnoma* (1) and *Lepidolaena* (1).

The range of species at lower altitudes and in drier zones may also be quite wide.

*Chandonanthus squarrosus* growing on a beech tree. (JB)

Most are confined to streamside locations, and others grow on exposed bark; the latter are often tiny species that are adapted to survive drought cycles.

The collections are mostly from the southern half of the island, with a smaller number from the northern slopes. Liverworts are generally more abundant than mosses and lichens, especially in the higher and cooler, wetter sites, so that even this limited selection probably gives a good indication of the species present. The frequently occurring liverworts on Hauturu are generally also common on the mainland. Based on the collections, they are a mixture of larger, more visible species and other small species that are either usually mixed in with other bigger specimens and detected later under a microscope — such as *Metalejeunea cucullata*, which is not usually detected in the field — or grow in conspicuous patches, especially on bark — such as *Frullania rostellata* and *Thysananthus anguiformis*, both of which are able to survive dry conditions. Other species that are not always evident in the field include *Radula strangulata*, which is by far the commonest of the many *Radula* species on the island. Large, obvious species include three *Bazzania* species; the commonest of these is *B. adnexa*, which was also the most frequently found liverwort in the forest survey for the carbon monitoring survey in the South Island. Two other forest *Bazzania* species — *B. hochstetteri* and *B. tayloriana* — although less prevalent, still had 20 or more records from Hauturu. Most of the frequently collected species are widely distributed at a range of altitudes and, thus, in a variety of forest types. A number of the species represented by only one or two collections are from the higher and wetter tops, while others are scattered elsewhere. Those not collected often include a number of small, insignificant species, even by liverwort standards. Other common species are *Chiloscyphus lentus*, which is usually a low epiphyte on trunks or on fallen bark or wood and also on soil or rock, and *Cuspidatula monodon*, which is also one of the most common species in similar sites on Rangitoto.

Some of the more colourful species are best represented in more open areas where they take on a stronger colouration, often in the red–brown or purple–red colour range. Conspicuous examples are *Cuspidatula monodon*, *Syzygiella colorata*, *Lepidolaena taylorii*, *L. clavigera* and *Anastrophyllopsis subcomplicata*. Some epiphytes on tree bark may also be dark-coloured where there is good light exposure. Species of *Thysananthus*, *Spruceanthus* and *Lopholejeunea* can be almost black, and several *Frullania* species and *Goebeliella* are often red–brown. In contrast, forest-floor species tend to be shades of green but may form substantial clumps and so are conspicuous, for example *Bazzania*, *Lepidozia*, *Schistochila* and *Gottschea* species, and the similar but smaller *Neolepidozia* and *Tricholepidozia*.

The smallest species are mostly Lejeuneaceae, and even small fragments of bark may have stems of several species on them. It is characteristic of both liverworts and mosses to grow in complex mixtures, especially when epiphytic in well-lit, damp areas, so that collections often resemble a miniature forest rather than a discrete individual collection.

Some 57 species are known from single collections only. This is in part due to the extremely small size of the plants and their habit of growing among other, often larger species. The very small plants, consisting sometimes only of minute threads, may not be detected until specimens are examined later with a microscope. Of these, collected by accident as bycatch, a few turn out to be unusual and rare records. The frequency of single specimen records can also be attributed to the few collecting trips to the island by

*Bazzania hochstetteri* (JB)

*Chiloscyphus semiteres* (JB)

*Echinolejeunea papillata* (JB)

*Leiomitra lanata* (JB)

*Lepidolaena clavigera* (JB)

*Schistochila glaucescens* (JB)

*Schistochila nobilis* (JB)

*Thysananthus anguiformis* (JB)

specialist collectors who are able to recognise novelties when they encounter them. The historical collections, the oldest of which date from the end of the nineteenth century, include the only record not in the AK collections — an *Asterella* collected by T. Kirk in 1893. Others from 1939 and 1945 are generally comprised of a few large conspicuous taxa, particularly from near the summit, and have little habitat information. In the case of tiny species, these may be more prevalent than appears, as others have been overlooked in the field. The nature of intensive collecting means that, in general, the number of specimens reflects how common a species was on the island.

Almost all the liverworts found on Hauturu are native to New Zealand; the only exception is the widespread weedy species *Lunularia cruciata*.

## Hornworts

The hornwort flora of the island is impressive, with 13 collections in four genera. These represent four of the 13 species native to New Zealand. Three of the species are terrestrial and the fourth and most common, *Dendroceros granulatus*, is epiphytic. *Phaeoceros carolinianus* is known from only a single collection, although it is widespread elsewhere in New Zealand.

## History of collections

The list of liverworts and hornworts recorded from Hauturu is based on collections housed at the herbarium of the Auckland Museum (AK) — in total about 1200 specimens. Most voucher specimens result from John Braggins' collecting expeditions in 1980 and 1984 in conjunction with Jessica Beever. These visits were based at the ranger station at the southwest corner of the island. Samples were collected from the main Summit Track and the Thumb Track to the summit of Mount Hauturu, East Cape to Pōhutukawa Flat (Hingaia) via Mount Hauruia (Bald Rock), Mount Whēkauwhēkau, Mount Kiriraukawa and Kauri Ridge. The Valley and Shag tracks near the ranger station were also explored, as were the lower reaches of the Waipawa, Awaroa and Parihākoakoa streams and a ridge leading from the Awaroa towards Hauruia. A few earlier, historic collections were available in the herbaria, and other important specimens collected later by E. K. Cameron, M. A. M. Renner and R. M. Bellingham added useful records. R. M. Schuster made a substantial collection in 1962 but these are unidentified and are not yet available for study. There are 75 specimens from Hauturu in other New Zealand herbaria, some of which are duplicates of AK specimens; of these only one is a new record.

This list updates an earlier one[2] as a number of species or varieties (16 in total) have since been recognised. In addition there have been numerous name changes as a result of taxonomic and nomenclatural research, often made possible by advances in study techniques. The current list (see Chapter 9) attempts to update names where possible, with some of the most recent changes indicated. Some of these changes conflict with Engel and Glenny.[3]

## Conclusion

The species list, together with that provided by Beever for mosses (see Chapter 9), is a statement of what we currently know of the bryoflora of Hauturu. Later taxonomic decisions will affect it, but as all records are based on voucher specimens, the original identifications can be checked. Part of the value of such lists is that attention is drawn to things that can be expected to be there or might be there, and it is hoped that this encourages careful searching for other species in the future.

# 6.8 Lichens

— BRUCE W. HAYWARD

**Two hundred and sixty-nine lichen species in 104 genera are recorded here from Hauturu. This represents about 16 per cent of the presently known lichen flora of New Zealand. The species list in Chapter 9 includes all the more common lichens and the majority of the more distinctive foliose and fruticose species, but undoubtedly there are many crustose and rarer species not yet found. A complete lichen flora for Hauturu might be nearer 400 species. Approximately 20 per cent of the Hauturu lichen species are endemic to New Zealand, 30 per cent are cosmopolitan species and 25 per cent have an Australasian distribution. This is the most diverse flora so far recorded from any northern New Zealand offshore island, slightly more than the 247 species so far recorded from Aotea/Great Barrier Island. Aotea's crustose lichens need more study, and it can be expected to have a greater lichen diversity than Hauturu.**

Lichens that are rarely recorded from New Zealand[1] but have been found on Hauturu include only the second New Zealand records of the Australian *Calicium robustellum*, the pantropical foliose *Hypotrachyna immaculata*, the southern hemisphere crustose *Arthothelium fusconigrum* and the southern hemisphere foliose *Heterodermia isidiophora* and endemic foliose *Sticta livida*; the third New Zealand records of the cosmopolitan crustose *Arthonia cinnabarina*, the cosmopolitan crustose foliicolous *Coenogonium zonatum* and the endemic crustose *Lopadium monosporum*; and the fourth records of the endemic crustose *Megalospora bartletti*, *Melaspilea subeffigurans* and *Phaeographis inusta* and the Australasian fruticose *Usnea molliuscula*.

The list of lichens from Hauturu includes the northernmost records from New Zealand of at least 41 species (asterisked in the species list). Some of these northernmost lichen records are because Hauturu and central Great Barrier have the northernmost occurrences of high-altitude, high-humidity forest and scrub in New Zealand. Some of the northernmost records from Hauturu are shared with Aotea, as the high peaks of Hauturu (722 metres) and Hirakimata/Mount Hobson (627 metres), the summit of Aotea, are at the same latitude.

*Pseudocyphellaria dissimilis*, a foliose lichen that prefers the moist cloud forest habitat of Hauturu. (DR)

## Lichen associations and altitudinal zonation

In tandem with the vascular vegetation, the lichen flora exhibits a general altitudinal zonation. The black lichens *Lichina confinis* and *Verrucaria microporoides* are the only two lichens that grow on, and are restricted to, intertidal rocks. There is a high diversity in lichens that grow on rocks and soil in the maritime zone just above high tide, dominated by the foliose *Parmotrema*, *Xanthoparmelia*, *Heterodermia* and *Pannaria elixii*, the crustose genera *Buellia*, *Lecanora*, *Pertusaria* and the fruticose *Ramalina celastri* and *R. australiensis*. Coastal scrub has a distinctive association of yellow *Caloplaca*, *Xanthoria* and *Teloschistes*, grey-green *Physcia caesia* and white crustose *Opegrapha intertexta*. A diverse lichen flora grows on the gnarled branches of coastal pōhutukawa; it is dominated by foliose *Hypotrachyna*, *Parmotrema*, *Parmelinopsis* and clumps of *Usnea inermis*.

Kānuka (*Kunzea robusta*) and kānuka–mixed forest below about 300 metres elevation have the richest lichen floras, dominated by foliose forms including nine species of *Pseudocyphellaria* (especially *P. carpoloma*, *P. coriacea* and *P. montagnei*), eight species of *Pannaria*, plus *Coccocarpia*, *Heterodermia* and *Sticta latifrons*. There is a diverse flora of ground-dwelling fruticose lichens on Hauturu, mainly *Cladonia confusa*, *Cladia aggregata*, *Stereocaulon ramulosum* and nine other species of *Cladonia*.

Kauri and mixed-forest types between 200 and 500 metres elevation also have rich epiphytic lichen floras, dominated by foliose lichens such as *Pseudocyphellaria glabra*, *P. multifida*, *Pannaria*, *Heterodermia* and *Menegazzia*, but crustose *Megalospora* and *Megalaria grossa* are also common. The unusual fruticose lichen *Metus conglomeratus* and the foliose *Pseudocyphellaria dissimilis* are commonly found growing on the forest floor.

Lichens in the high-humidity mixed forest and scrub above 500 metres are dominated by the distinctive stipitate foliose *Sticta filix* with subdominant foliose lichens *Pseudocyphellaria multifida*, *P. faveolata*, *P. glabra*, *P. coronata*, *Bundophoron australe* and fruticose *Cladia aggregata*. *Lefidium tenerum*, *Siphula gracilis* and the script lichen *Leiorreuma exaltata* are abundant above 650 metres. Lichens that are restricted to these higher elevations above 500 metres on Hauturu and elsewhere in northern New Zealand include the fruticose *Siphula decumbens*, *S. gracilis*, *Cladia sullivanii* and *Cladonia nudicaulis*, the foliose *Sticta filix*, *Lefidium tenerum*, *Nephroma plumbeum*, *Peltigera nana*, *Pseudocyphellaria faveolata*, *P. rubella* and *Xanthoparmelia mougeotina*, and crustose *Coccotrema cucurbitula* and *Leiorreuma exaltata*. Taxa that were not found below 300 metres altitude include the foliose *Pseudocyphellaria glabra*, *P. multifida*, *Bundophoron australe* and *B. patagonicum*, orange-fruited crustose *Miltidea ceroplasta* and bright yellow-fruited crustose *Stirtoniella kelica*.

## The collections

The lichens of Hauturu were extensively collected and identified by Anthony Wright, Glenys Hayward and Bruce Hayward during two week-long field trips to the island by the Auckland University Field Club in 1981 and by the Offshore Islands Research Group in 1990, published in the Auckland Museum Records in 1991.

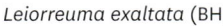
*Coccotrema cucurbitula* (BH)

*Sticta filix* (BH)

*Leiorreuma exaltata* (BH)

*Bundophoron australe* (BH)

*Xanthoparmelia australasica* (BH)

*Menegazzia neozelandica* (BH)

# 6.9 Fungi

— PETER BUCHANAN

**Fungi are important components of forest ecosystems but are often overlooked. Their functions include assisting plant nutrition through mutually beneficial relationships with plant roots; nutrient cycling through decomposition of dead plants, fungi and animals; plant protection through the presence within plants of beneficial endophytes; parasitism of plants, fungi and animals; and as a food source for invertebrates and possibly for some vertebrates. Knowledge of the fungi of Te Hauturu-o-Toi/Little Barrier Island is based on relatively few collecting expeditions, with 815 specimens of fungi, mostly native, from Hauturu in the national collection (Fungarium PDD, Manaaki Whenua), representing 450 species determined to at least genus level. A few other records have been gathered from literature, from iNaturalist reports and from contributed images of Hauturu biota.**

On Hauturu fungi are most visible during autumn when declining temperatures and increasing moisture levels stimulate development of fungal fruitbodies such as the fleshy mushrooms and other 'macrofungi' — i.e. those with clearly visible reproductive structures. Most mushrooms that develop under tawhai (beech, *Fuscospora* spp.), mānuka (*Leptospermum scoparium*) and kānuka (*Kunzea robusta*) are directly connected to the roots of these trees — the so-called 'mutualist' fungi. These include many mushrooms of the genera *Cortinarius*, *Russula* and *Hebeloma*, and truffle-like fungi such as *Rossbeevera pachydermis*. They form a beneficial (ectomycorrhizal) relationship with the root that assists it to absorb water and minerals. In return the fungus gains sugars produced through photosynthesis in the leaves and translocated downwards to the roots. The fine, branched, thread-like filaments (hyphae) of these fungi permeate a large volume of soil surrounding associated roots, enabling increased uptake by both fungus and root.

Fungi that form mushrooms on fallen wood (for example *Cyptotrama* sp., *Hypholoma australianum* and *Hygrocybe* sp.) or on the forest floor away from tawhai, mānuka, and kānuka (for example *Pluteus velutinornatus* and the coral fungus *Clavulinopsis sulcata*), function mainly as 'decomposers' and control the ecosystem process of nutrient cycling.

Werewere-kōkako (*Entoloma hochstetteri*), summit ridge, Hauturu. (LW)

These mushrooms develop from a hidden feeding stage of the fungus within the wood, causing the wood to decay and visually rot, or within the soil. Invertebrates also assist the eventual conversion of decomposing wood to humus on the forest floor. Other wood-decay fungi form crust-like or bracket fruitbodies (*Pycnoporus coccineus*, *Podoserpula petalodes* and *Stereum versicolor*) that vary from flexible and short-lived to perennial brackets; the latter may be firmer than the underlying wood (*Ganoderma* sp., *Fomes hemitephrus*). Among other decomposers on Hauturu are the earthstars (*Geastrum* sp.) and bird's nest fungi (*Nidula niveotomentosa*).

Several introduced fungus-like and fungal parasites have been recorded on Hauturu. *Phytophthora cinnamomi*, a pathogen on many native and exotic trees in Aotearoa and elsewhere, was isolated by Newhook from soil underneath nine or 10 kauri trees sampled on rarely visited northern slopes.[1] Newhook suggested that the pathogen may have spread from two former Māori settlements on the island;[2] it may have originally accompanied kūmara brought by early voyaging Māori. The related *P. agathidicida*, cause of the serious kauri dieback disease, has to date not been detected in island-wide surveys of Hauturu;[3] nor has the wind-borne invasive myrtle rust (*Austropuccinia psidii*) that threatens a broad range of hosts in Myrtaceae.

*Aspergillus fumigatus* is a soil-borne fungus that causes the respiratory fungal disease aspergillosis in hihi/stitchbird (*Notiomystis cincta*). This fungus is present on Hauturu,[4] but it has been detected in lower concentrations there and causes a lower rate of mortality of hihi than on Tiritiri Matangi Island and in two mainland locations. This is fortunate, as Hauturu is home to the only self-sustaining population of hihi.

Slime moulds are fungus-like organisms that are no longer considered to belong in the fungal kingdom, but are often encountered when looking for fungi. They feed on bacteria and become visible to humans only when they suddenly coalesce to form a collective 'plasmodium'.

## Some distinctive fungi of Hauturu

Short lists of the macrofungi from Hauturu (i.e. fungi with visible reproductive structures such as mushrooms and brackets) were first published by Aiken, and Aiken and Shreeves.[5] The earlier list included 15 fungi, though only seven were identified to species level — by Shirley Baker at PDD, the fungarium at what was then the Department of Scientific and Industrial Research (DSIR). Most of the 57 macrofungi recorded in the 1954 list were wood-decay fungi identified by John Gilmour at the Forest Research Institute and Joan Dingley at PDD.

Of the 450 fungal species recorded on Hauturu, four have *hauturu* in their species names and their type (name-bearing) specimen(s) were collected on the island. None of these four, however, is now known to be confined to Hauturu.

The leaf-spot fungus *Phyllachora hauturu* subsp. *hauturu* was described by Johnston and Cannon as having black, rounded fruitbodies less than 1 mm diameter, forming on blackened regions of living leaves of *Myrsine australis* or māpou.[6] Although this host is common, the fungus is known from only two collections — on Hauturu in 1945 and on

Rangitoto in 2002. They described two other subspecies of *P. hauturu*, both on other *Myrsine* species — from the Chatham Islands and from subantarctic islands.

Another leaf-spot fungus, *Lophodermium hauturuanum*, is also known from Northland, Auckland, Coromandel and Buller. It is characterised by small elongate pustulate fruitbodies (Fig. 1) in yellowish areas of dead leaves of *Gahnia* spp.

Dingley described *Nectria hauturu* on *Carmichelia* sp. (broom), on which it forms tiny (less than 0.5 millimetres diameter) salmon-pink, flask-shaped fruitbodies in small groups.[7] Subsequently it was found to occur on a range of hosts in several North Island locations and overseas, and the name is now considered a synonym of *Clonostachys ralfsii*.

Dingley also described *Cordyceps hauturu* (Fig. 2), a 'vegetable caterpillar' that is parasitic on larvae of a hepialid moth.[8] The fungus is known from only two collections, both pre-1953; the other is from Auckland. *Cordyceps hauturu* is similar in overall appearance to the comparatively common *Ophiocordyceps robertsii* or āwheto, which has also been recorded on Hauturu. Māori collected and traded āwheto as a source of powdered charcoal to produce an intense black pigment for tā moko (tattooing). In Tibet, *Ophiocordyceps sinensis* — a fungus that resembles *C. hauturu* and *O. robertsii* — is collected and traded as an expensive herbal medicine. The related *Cordyceps kirkii*, previously recorded on the Cook Strait giant wētā (*Deinacrida rugosa*) and the ground wētā (*Hemiandrus maculifrons*), has been newly recorded on Hauturu on dead wētāpunga (*Deinacrida heteracantha*). Although it is parasitic this native *Cordyceps* is unlikely to negatively affect wētāpunga at a population level.

## Māori knowledge and uses of fungi on Hauturu

Māori likely used other fungi on Hauturu as a resource, too. Pūtawa (*Laetiporus portentosus*) is one of the most conspicuous fungi on the island where it grows as a large, though short-lived, hoof-shaped bracket on living tawhairaunui (*Fuscospora truncata*). Once dried, it was valued by Māori for fire-starting and fire-carrying. *Armillaria novaezelandiae* (harore) — a close relative of *A. limonea* — and hakeke (*Auricularia cornea*) are common edible mushrooms that grow on fallen wood. Hakeke was also widely collected and dried for export as a food to China in the late 1800s/early 1900s, although it is not known whether Hauturu forests were harvested for this purpose.

The fly-dispersed — and hence smelly — stinkhorn fungi *Aseroe rubra* (puapuatai) and *Ileodictyon cibarium* (matakupenga), and the puffballs or pukurau (for example *Morganella compacta*) are other fungi of Hauturu that may have been eaten in their immature stages by Māori. The sky-blue werewere-kōkako (*Entoloma hochstetteri*) is a surprising sight when seen against the greens and browns of the forest. This mushroom is featured on the NZ$50 banknote, illustrating the Māori story that the kōkako (*Callaeas wilsoni*) rubs its cheek against the mushroom to gain the characteristic blue colour of its wattle.

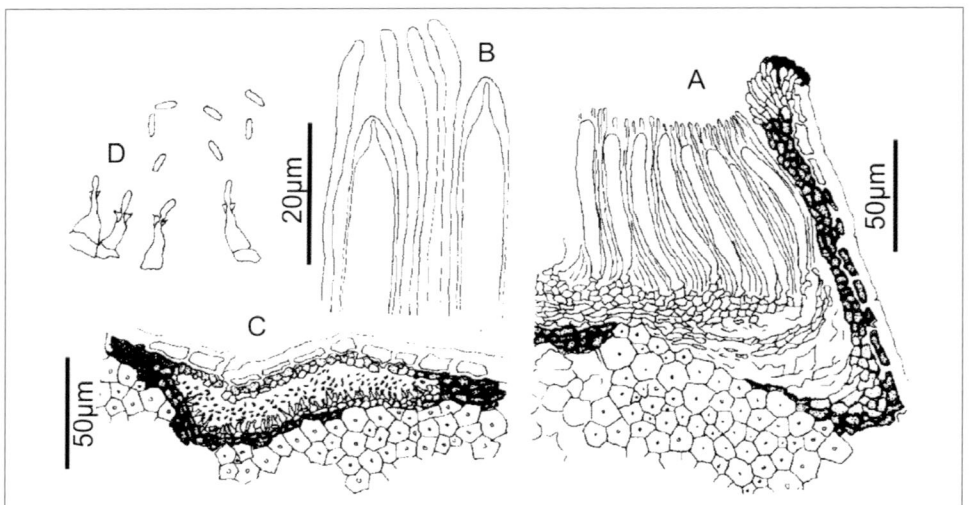

*Fig. 1: Lophodermium hauturuanum* (Johnston 1989): **A** Margin of ascocarp (sexual fruitbody) in vertical section. **B** Apex of asci and paraphyses within ascocarp; ascospores not shown within asci. **C** Pycnidium (asexual fruitbody) in vertical section. **D** Conidiogenous cells and conidia (single-celled asexual spores) from pycnidium.

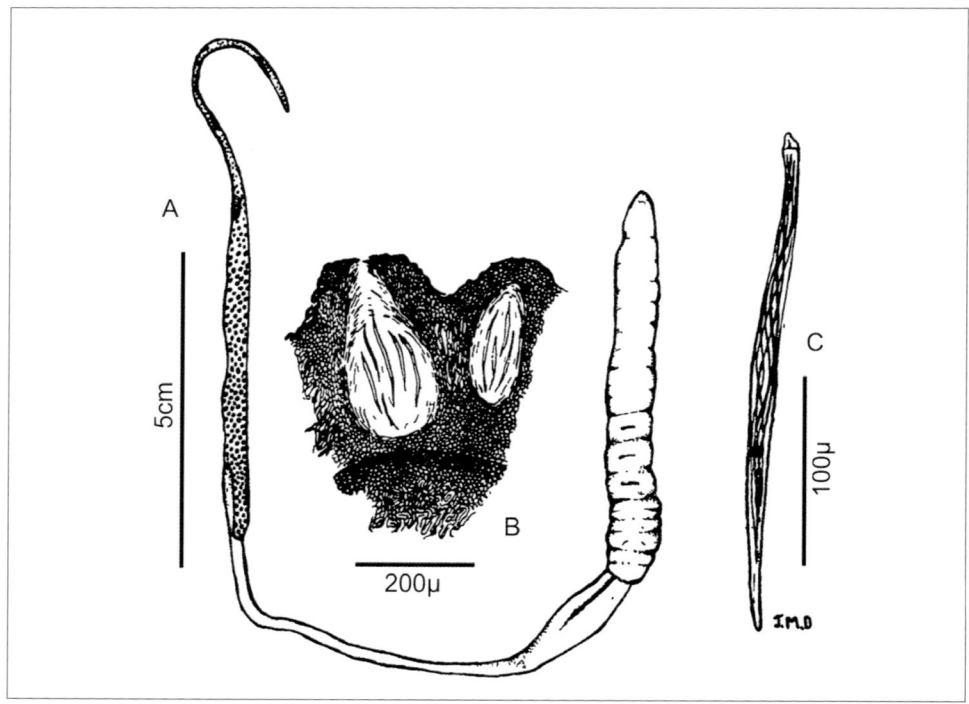

*Fig. 2: Cordyceps hauturu* (Dingley 1953): **A** Mummified caterpillar (right) and aerial fungal fruitbody (left), though typically linear in form with caterpillar basal and buried. **B** Section through fertile part of fungal fruitbody showing flask-shaped perithecia. **C** Ascus from within a perithecium, containing eight narrow, elongate ascospores that are twisted together.

## Fungi first described from Hauturu

Altogether 33 fungal species were described as new, based on type specimens from Hauturu. Recognition of many of these species is thanks to certain key mycologists who visited and undertook extensive collecting on Hauturu: Bryce Kendrick, Canada (four new species of microfungi), Egon Horak, Austria (11 new species of mushrooms), and Peter Johnston, Manaaki Whenua, Auckland (eight new species of microfungi). Of the 33, four of Horak's mushroom species (*Cortinarius gemmeus*, *Entoloma blandiodorum*, *E. consanguineum* and *Tubaria perplexa*) are known only from Hauturu, along with the similarly restricted wood-decay fungus *Hymenochaete magnahypha*. Another wood-decay fungus, *Dichomitus newhookii*, is known only from Hauturu and the remote Three Kings Islands.

## Live cultures and fungal DNA from Hauturu

Hauturu fungi are poorly represented in culture collections: only four live fungal cultures are held in the International Collection of Microorganisms from Plants (ICMP) at Manaaki Whenua. However, an important DNA resource for Hauturu fungi awaits detailed analyses after altitudinal eDNA sampling of soil-borne fungi as a component of overall soil biodiversity.[9] Sequence identification to fungal species level has yet to be undertaken, but detection of 2715 fungal operational taxonomic units (DNA-distinguished entities) from sampled soils already indicates a high diversity of fungi in Hauturu soils — as expected. Meantime, eDNA analyses by Andrew Dopheide have documented similarity between community patterns of soil fungi and those of above-ground plants, which is evidence of the level of interactions between fungi and plants.[10] An altitudinal influence on fungal community composition was also demonstrated in the Hauturu study.

*Armillaria limonea* (LB)

Puapuatai/stinkhorn fungus (*Aseroe rubra*) (SF)

Hakeke/wood ear (*Auricularia cornea*) (LB)

*Clavulinopsis sulcata* (SF)

*Cordyceps kirkii* (DS)

*Cortinarius* or *Hebeloma* sp. (LB)

*Cyptotrama* sp. (SF)

Werewere-kōkako (*Entoloma hochstetteri*) (SF)

*Fomes hemitephrus* (SF)

Orange pore fungus (*Favolaschia calocera*) (LB)

*Ganoderma* sp. (SF)

*Geastrum* sp. and *Clavulinopsis* sp. (LB)

*Hygrocybe* sp. (LB)

*Hypholoma australianum* (LB)

Matakupenga/basket fungus (*Ileodictyon cibarium*) (SF)

Pūtawa/puku tawai (*Laetiporus portentosus*) (SF)

*Morganella compacta* (LB)

*Nidula niveotomentosa* (LB)

*Pluteus velutinornatus* (SF)

*Podoserpula pusio* (LB)

*Pycnoporus coccineus* (SF)

*Rossbeevera pachydermis* (LB)

*Russula* sp. (LB)

Slime mould plasmodium (LW)

*Stereum versicolor* (LB)

# 6.10
# Stream vertebrates
— LYN WADE

**Te Hauturu-o-Toi/Little Barrier Island has many streams that have eroded steep and sinuous gorges through the fragmented volcanic rock, particularly on the northern and eastern coasts, with many treacherous waterfalls. Most streams are single or low-order, and a number of the southern and western streams have occasional sidestreams and wider streambeds with more alluvium. During drier spells many carry little or no water, and this along with periodic flood events impacts on the survival of the stream fauna. The majority of streams consist of boulders, cobbles and coarse gravel with a variable organic component, and exit to the sea by running over cliffs or under the substantial boulder beaches.**

A survey of aquatic vertebrates carried out by Mike McGlynn from the National Institute of Water and Atmospheric Research (NIWA) and other researchers in 2000 showed the presence of longfin eel (*Anguilla dieffenbachii*), banded kōkopu (*Galaxias fasciatus*) and redfin bully (*Gobiomorphus huttoni*).[1] Redfin bully were present in only one stream. A later survey conducted by Lyn Wade in 2013 covered several more of the island's streams, including at least one stream from each quarter of the island.[2] This survey confirmed the presence of longfin eel and banded kōkopu. Several of the western and southern streams contained both species in moderate numbers. The banded kōkopu measured in the 2013 survey ranged in size from 88 millimetres to 220 millimetres. Juvenile galaxiids were observed in Tirikakawa Stream but were not measured. The largest longfin eel measured was 500 millimetres. Small elvers were observed but not measured. Both eel and kōkopu are diadromous — they spend a brief time in the ocean as a normal part of their life cycle. Considering the disturbed regime of the island streams and the difficult access across boulder beaches from the ocean it is possible that the banded kōkopu may breed within the stream systems. Several species of South Island galaxiid are known to have made the switch from migratory to non-migratory breeding patterns.[3] That these species are strong climbers is important for their survival in the rugged and harsh environment of Hauturu's intermittent streams. In a dry summer high numbers of banded kōkupu congregate in the few remaining pools.

The latest threat ranking for New Zealand freshwater fishes places longfin eel and redfin bully as declining and banded kōkopu as not threatened.[4]

Hauturu forest stream. (NF)

# 6.11 Seaweeds

— MIKE WILCOX

**Boulder beaches all along the southern and eastern sides of Te Hauturu-o-Toi/Little Barrier Island and in bays on the north and west sides extend from above high tide down into the subtidal zone. The intertidal boulders that are rolled about by the sea are bare of prominent algal growth, but the sublittoral fringe down into the sublittoral has a rich algal flora, doubtless enhanced by the clear and sediment-free water adjacent to a bush-clad island. This type of shoreline is unusual in the Hauraki Gulf. Apart from the boulder shores, Haowhenua Point, Te Hue Point and Ngatamahine Point are solid rock headlands on the very exposed northeast and northern coasts, and there are numerous small rock stacks (for example Crayfish Rock) and some solid rock platforms.**

The seaweeds of Hauturu have been investigated by several researchers. Dellow, Trevarthen, Bergquist and Dromgoole have recorded species they found there and described the pattern of algal zonation,[1] while Morton mentions some of the seaweeds and highlights the hazardous instability of the boulder beaches as a habitat for intertidal marine life.[2] However, the earliest known records of seaweeds are the collections made there by Miss Edith M. Smith in the early 1900s and forwarded by Robert Shakespear to Kew Royal Botanic Garden collections, and now held at the British Museum of Natural History, London. Cotton mentions that Edith Smith collected *Liagora harveyana*, *Perithalia capillaris* and *Xiphophora chondrophylla* on Hauturu.[3] Important collections were made by Alison Lush of the Dominion Museum (now the Museum of New Zealand Te Papa Tongarewa), who was there for two weeks in October 1949 with Vivienne Dellow.[4] Assisted by Lyn Wade, Mike Wilcox made collections and observations in March 2010.[5]

## Red algae

The only obvious red alga on mid-intertidal boulders is the encrusting *Apophlaea sinclairii*, which is particularly abundant behind Hingaia Reef. It is succeeded at lower levels by *Catenellopsis oligarthra*, *Caulacanthus ustulatus*, *Centroceras clavulatum*, *Gelidium caulacantheum*, *Gigartina minuta*, *Hildenbrandia*, *Nothogenia pulvinata*,

*Ecklonia radiata*, a common brown algae that prefers to cling to stable boulders or solid rock. (LWh)

*Phacelocarpus labillardierei, Polysiphonia scopulorum, Rhizopogonia asperata* and non-geniculate coralline 'pink-paint' ('Lithothamnia').

Records of red algae in the high-tide zone are *Bostrychia intricata, Catenella nipae* and *Pyropia plicata*. Dellow has recorded '*Bangia*' sp. (as *Bangia atropurpurea*) and *Rhodochorton purpureum*.[6] In March 2010 *Bostrychia intricata* was found growing associated with *Catenella fusiformis* at the shaded base of large mid-tidal boulders near the West Landing.

In March 2010, low-tide boulders on the western shore had a short, mixed turf of red algae comprising *Aphanocladia delicatula, Caulacanthus ustulatus, Chondria lanceolata, Dipterosiphonia dendritica* and *Gayliella flaccida*.

Epiphytes recorded are *Abroteia suborbiculare, Dasya subtilis, Dasyclonium ovalifolium, Laurencia distichophylla, Lomentaria caespitosa, Nancythalia humilis* and *Pleurostichidium falkenbergii*. *Vertebrata aterrima* was found in March 2010 growing epiphytically at the base of *Carpophyllum plumosum*.

## Brown algae

These are the physically dominant seaweeds on Hauturu. The only brown algae of the middle intertidal boulders are the crust species *Ralfsia verrucosa* and *Hapalospongidion saxigenum*. In addition, *Petalonia binghamiae, Scytothamnus australis* and *Splachnidium rugosum* have been recorded,[7] and *Hormosira banksii* has been seen in mid-littoral pools.[8]

The sublittoral fringe (near to or just below the shore, visible at low spring tides) is dominated by *Cystophora torulosa* and *Xiphophora chondrophylla*, together with *Dictyota kunthii* and *Spatoglossum* sp. This last species (previously called *S. chapmanii*) is a notable seaweed on Hauturu; the only other Auckland occurrence is at the Mokohinau Islands. The upper sublittoral has abundant *Carpophyllum maschalocarpum, C. plumosum, Colpomenia peregrina, Cystophora retroflexa, Dictyota kunthii, Ecklonia radiata, Halopteris virgata, Sargassum sinclairii, Spatoglossum* sp., *Zonaria aureomarginata* and *Z. turneriana*. As Dromgoole observed, the algae of the upper sublittoral show a pattern of occurrence related to boulder size: *Carpophyllum maschalocarpum* and *Ecklonia radiata* occur only on the larger, more stable boulders, and on the smaller boulders a thick canopy of *Carpophyllum plumosum, Cystophora retroflexa* and *Dictyota kunthii* dominates, with an undercover of *Halopteris virgata, Zonaria turneriana* and *Spatoglossum* sp.[9]

The lower sublittoral, from depths of 3 metres down to 20 metres, is dominated by *Carpophyllum plumosum* and *Ecklonia radiata*, together with *Carpophyllum flexuosum, Cystophora retroflexa, Sargassum sinclairii* and *Zonaria turneriana*. *Ecklonia radiata* occurs at very high densities around Crayfish Rock. The filamentous brown algae *Bachelotia antillarum* and *Feldmannia mitchelliae* were found in March 2010 growing on the mooring buoy rope in front of the south landing.

*Above*: Boulder beach, Hauturu. (MW) *Below*: Pacific sea lettuce (*Ulva australis*), and the brown crustose alga *Ralfsia verrucosa* on the outer fringe of the boulder beach. (LW)

## Green algae

Hauturu has a meagre green algal flora. Intertidal species recorded are *Bryopsis plumosa*, *Caulerpa geminata*, *Codium convolutum*, *Derbesia novae-zelandiae*, *Lychaete herpestica*, *Microdictyon mutabile*, *Ulva procera* and *U. australis* — this last by far the most conspicuous green alga on the island. *Ulva australis* — the common sea lettuce on Hauturu — was abundant in March 2010 on the southern and western boulder shores in the lower intertidal zone and down to a depth of 5 metres. It is present year-round. Sea lettuce on intertidal rocks is a fast-growing, short-lived pioneer that quickly colonises available sites, particularly when the water has become enriched with nutrients, but usually becomes ousted by other algae on clean sites. The fast and constant water motion may also be conducive to sea lettuce growth.

## Other seaweeds

The cyanobacterium *Isactis plana* commonly occurs as circular greenish crusts on mid-low tide boulders, while the filamentous marine diatom *Melosira moniliformis* was observed in a brackish sea pond on the southern shore, forming a brownish fuzz that was smothering branch debris and cast thalli of *Carpophyllum maschalocarpum*.

*Spatoglossum* sp. is a very significant and unusual brown algae in Auckland waters, with Hauturu being the main locality. (KC)

# Chapter seven
# Seas around Hauturu

— ROGER GRACE

**The location of Te Hauturu-o-Toi/Little Barrier Island in the mid to outer Hauraki Gulf means the coastlines around the island are bathed in clear subtropical waters of the East Auckland Current from the north, but also occasionally by more turbid coastal waters of the inner Hauraki Gulf. The influence from the north is generally dominant, with some subtropical marine species at Hauturu similar to some at the Poor Knights and Mokohinau islands. Storm waves are more influential on the northern and eastern shores; the southern and western shores are a little more sheltered from the bigger ocean swells. This influences the distribution of marine life in the intertidal zone around the island, as many marine organisms are adapted to a limited range of wave action.**

Boulder beaches are the main shore type around most of the island, with several hard rock shores, mainly on headlands in the north. Each of these two rock shore types support a different array of intertidal marine life, as the boulder shores are mobile in stormy conditions and boulder beaches are inhospitable when the rocks are rolling around in the waves, creating high risk of crushing marine life — and a very noisy clatter. There are some specialist marine animals that can cope with these conditions: they either have very strong shells, such as the black nerita (*Nerita atromentosa*), or a very low profile so they can avoid crushing, such as the fragile limpet (*Atalacmea fragilis*) and the radiate limpet (*Cellana radians*). Other animals avoid harm by being particularly agile and hiding deeper among the boulders as they crash around, such as some shore crabs living on boulder beaches. Despite these adaptations it is inevitable that there is some mortality when waves move the boulders on these shores. Algae, which are of course unable to move, are virtually absent on the mobile boulders in the intertidal zone.

Intertidal marine life attached to the hard rock shores is more typical of those on mainland and other island rocky shores where there is moderate to high wave action. Several species of barnacles are found on the rock shores, with different species occurring at different levels on the shore and in different degrees of wave action. The tiny periwinkle (*Littorina unifasciata*) occurs high on the shore above the barnacle zone. Around mid-tide

A leatherjacket (*Meuschenia scaber*) with *Ecklonia radiata* against a colourful background of algae and corals. (LWh) *Previous*: Land and sea meet, boulder beach, Hauturu. (NF)

level a wide range of grazing snails, limpets and chitons occur, where they feed on a short fuzz of attached algae. Colonies of the Pacific oyster (*Crassostrea gigas*) are found in more sheltered spots, and they even occur in sheltered pockets at the northern end of the island, where they are small and stunted in their growth because they do not do well in heavy wave action.

Around low-tide level and below, algae soon become the visually dominant life forms (see Chapter 6.11). The hard rock shores typically have a wider range of algal species than the mobile boulder shores. The commonest of the larger brown seaweeds is the flapjack (*Carpophyllum maschalocarpum*), which forms dense masses of fronds constantly moving in the sea, sloshing around near low tide and dominating the shallow mixed-weed zone.

Once we go a little deeper around most of the Hauturu shallow reefs, the scene is a dramatic contrast from the successful wildlife and vegetation restoration story on the land. The last 50 years or more of heavy fishing pressure around the island has had a devastating effect on the health of its reefs, leaving a desolate wasteland of urchin or kina barrens, depleted fish and other kaimoana stocks, and the loss of previously productive kelp forests (*Ecklonia radiata*). What we have now is a mere shadow of the rich and diverse marine ecology of yesteryear.

Just below the shallow mixed-weed zone of flapjack and other seaweeds, around most of the island the rocky bottom is largely bare rock or covered in a thin film of encrusting pink coralline seaweed and a few small turfing species. Large numbers of kina or sea urchins (*Evechinus chloroticus*) are dotted over the bottom of either solid rock or boulders. The depth range of the kina barrens varies but is typically from about 3 metres to about 8 or 10 metres. It is a dynamic zone with kina densities ranging from about one per square metre up to patches that may contain 30 or more urchins in a square metre.

Other grazing species that can survive in this zone are the limpets (*Cellana stellifera*) in shallower water, and the large solid-shelled Cook's turban (*Cookia sulcata*). The giant chiton (*Eudoxochiton nobilis*) may also be found in this zone.

Heavy grazing by kina prevents the re-establishment of larger seaweeds and many of the smaller turfing species. Kina have multiplied to 'plague' proportions because fishing has removed most of their predators, such as snapper (*Chrysophrys auratus*), crayfish (*Jasus edwardsii*) and some of the reef fish; there are no longer sufficient numbers of these predators to carry out their ecological function of keeping kina numbers in check. The resulting kina barrens have little biodiversity and productivity compared with the original rich kelp forest that used to occupy this zone.

The deeper reefs around Hauturu are still in a reasonably healthy state apart from lower reef-fish and crayfish numbers. At the bottom edge of the kina barrens, the large kelp forms dense beds with individual plants up to a metre or so tall.

*Above*: A diverse mixture of brown, red and green seaweeds typical of the shallow mixed-weed zone on stable boulders on the west side of Hauturu. The pale bare rock at the top is intertidal. A red moki (*Cheilodactylus spectabilis*), one of the long-lived keystone reef species, cruises among the boulders. Depth about 1 metre. (LWh) *Below*: Kina barrens can develop on solid rock or boulders. Here solid rock is covered in encrusting coralline algae and small turfing species, dominated by grazing sea urchins or kina, which have eaten the kelp forest. Seasonal green alga adjacent is *Ulva australis*. Depth about 3 metres. Te Ananuiarau Bay, northwestern Hauturu. (KS)

The forest supports a wealth of invertebrate and algae life by way of the shelter offered by the structure of the kelp and its canopy, and the crevice-like openings among the stable boulders. Numerous encrusting sponges, bryozoans, ascidians, anemones, polychaete worms and hydroids smother the rock surfaces and provide a rich diversity of life now absent from the kina barrens above.

Nestled among the boulders may be brown and yellow feather stars (*Comanthus novaezelandiae*), their feather-like fronds bearing little superficial resemblance to the seastars and urchins to which they are related. Small numbers of mostly small crayfish may be found hiding deep in boulder crevices, though their numbers and size are nothing like they would have been in the middle of last century. Of similar colour and greater abundance is the red rock crab (*Plagusia capensis*), running rapidly sideways to the nearest rock crevice to escape danger.

As we approach deeper water around 30 metres, the density of the kelp forest diminishes because these plants are reaching the lower limit of their light tolerance; like all plants they require light to photosynthesise and grow. In this part of the Hauraki Gulf the kelp does not penetrate deeper than around 30 metres, although several smaller, less light-demanding red algal species may go a bit deeper.

Below the level of the bottom of the kelp forest, the rocky bottom is a kaleidoscope of colourful sponges and dozens of species of invertebrate life, though the colours of these species may be lost to a diver without a torch. Many of these species are attached permanently to the bottom, such as the sponges, hydroids and bryozoans, so they are obliged to feed on plankton drifting to them in the water column.

The rocky bottom around Hauturu reaches a depth of over 40 metres off the northeast of the island, though around most of the island it is truncated by the change to sandy seabed at a much shallower level, particularly off the western and southwestern shores.

The sandy seabed supports an entirely different array of marine life from the hard rocky bottom. A few species can be seen on the sand surface, such as scallops (*Pecten novaezelandiae*) and the comb star (*Astropecten polyacanthus*), but most of the life on the sandy bottom is buried beneath the surface. Beds of the morning star shell (*Tawera spissa*) sometimes reach densities of well over 1000 per square metre and constitute an important food resource for some fish species. The large dog cockle (*Tucetona laticostata*) will be present in more current-swept, coarser sediments, as well as the purple clam (*Venericardia purpurata*). Myriad polychaete worms, some with sensitive feathery tentacles above the surface, many species of tiny crustaceans and a shell collector's treasure trove of micromolluscs make the sandy bottom anything but the visual desert some unobservant divers complain about.

---

*Above*: Tall kelp forest (*Ecklonia radiata*) off northern Hauturu. A few small kingfish (*Seriola grandis*) and mado (*Atypicthys strigatus*) swim above the kelp. Depth about 15 metres. (LWh) *Below*: Below the kelp forest a kaleidoscope of colourful sponges, bryozoans and other filter-feeding invertebrates dominate the deep reef. In the photo can be seen a large tube sponge (*Siphonochalina latituba*), orange soft coral with white polyps (*Alcyonium aurantiacum*), and some small cup corals (*Culicea rubeola*) among a yellow encrusting sponge. A leatherjacket (*Meuschenia scaber*) nibbles at the sponges and encrusting life, helping to maintain the high diversity of the deep reef sponge garden. Depth about 25 metres, northern Hauturu. (LWh)

Fishing pressure around Hauturu is increasing as the population of the Auckland region rises and more Aucklanders see Hauturu as a prime recreational fishing destination. Their access to the fish around the island is improving with better boats, high-tech fish-finding gear, GPS — and more leisure time. Although the fish life around Hauturu is basically similar ecologically to what was there before, with a prominent subtropical element in the species mix, the abundance and size of many of these species are way below what would be natural for this area.

The most abundant reef fish by far is the common spotty or paketi (*Notolabrus celidotus*). Too small to be a target species for recreational fishers, the spotty is now 'top of the population' by default as nobody is interested in catching it. Other species with subtropical affinities but now in severely reduced numbers include sandager's wrasse (*Coris sandeyeri*), banded wrasse (*Notolabrus fucicola*) and red pigfish (*Bodianus unimaculatus*).

A wide range of other reef-fish species makes up the rest of the fish list. These include the plankton-feeders blue maomao (*Scorpis violaceous*), sweep (*Scorpis lineolata*) and demoiselle (*Chromis dispilus*) and, in deeper water, butterfly perch (*Caesioperca lepidoptera*) and pink maomao (*Caprodon longimanus*). Often reef-dwelling but also in more open water are yellowtail jack mackerel (*Trachurus novaezelandiae*) and juvenile trevally (*Pseudocaranx georgianus*). Down among the rocks will be red moki (*Cheilodactylus spectabilis*), marblefish (*Aplodactylus arctidens*) and, in shallow water, kelpfish or hiwihiwi (*Macronemus marmoratus*). Nibbling at reef life or attacking larger plankton is the leatherjacket (*Meushenia scaber*). Over the reefs and feeding on seaweeds are the herbivorous parore (*Girella tricuspidata*) and sometimes quite large silver drummer (*Kyphosus sydneyanus*). Sliding through the kelp and leaving telltale circular bite marks on the kelp fronds is the butterfish or greenbone (*Odax pullus*). Skirting the edge of the reef will be occasional porae (*Nemadactylus douglasii*) and often quite abundant goatfish (*Upeneichthys lineatus*), their barbels under the chin probing the sand for worms and crustaceans. Cruising around over the reef or resting on the sand is the eagle ray (*Myliobatus tenuicaudatus*), which crunches up shellfish it digs up from the sand, or smashes its favourite Cook's turban shell on the shallow reef, using its strong vice-like jaws. The snapper (*Chrysophrys auratus*) is by far the most popular fish for recreational fishers, followed by the pelagic kingfish (*Seriola lalandi*) and kahawai (*Arripis trutta*) when the snapper aren't biting.

The Hauraki Gulf supports a number of resident or regular visitor marine mammals, many of which are seen close to Hauturu. Fur seals (*Arctocephalus forsteri*) are slowly increasing in numbers in the Gulf after being hunted to very low numbers and retreating to the south. A resident pod of bottlenose dolphins (*Tursiops truncatus*) hangs around close to the shores of Hauturu and Aotea/Great Barrier Island. Sometimes large aggregations of over 100 common dolphins (*Delphinus delphis*) are seen in the Gulf, but not usually close

*Above*: Pilchards (*Sardinops neopilchardus*) are a critical part of the food chain of the pelagic environment of the Hauraki Gulf. (DT) *Below*: Near the bottom of the pelagic food chain of the Hauraki Gulf shrimp-like krill (*Nictiphanes* sp.) provide seasonal food for fish and for seabirds, some of which rely on krill to feed their chicks. (DT)

to the islands. Killer whales or orca (*Orcinus orca*) are now frequently seen in the Hauraki Gulf. Historically, considerable numbers of humpback whales (*Megaptera novaeangliae*) seasonally migrated either north or south through the Gulf, often passing close to Hauturu, though they are much less common now. A resident pod of Bryde's whales (*Balaenoptera edeni brydei*) seems to live mostly in the Hauraki Gulf and they are often seen close to the island. In recent years there have been visits from southern right whales (*Eubalaena australis*) and pigmy blue whales (*Balaenoptera musculus brevicauda*).

## Relationships between land and sea

The importance of interactions between animals and plants on the land and those in the adjacent seas is often overlooked. Many of these relationships have been lost from mainland coasts because of the massive ecological changes brought about by clearance of the land, agricultural practices and urban development, and loss of nesting seabirds because of habitat changes and predators. On some of New Zealand's conservation islands, including Hauturu, these relationships are still largely intact.

Seabirds nesting on the island feed in the ocean nearby, as well as further afield. They come back to the island to feed their chicks, and many of them develop in burrows well within the forest canopy. Over thousands of years the faeces of the chicks and adults, rich in nutrients from fish and krill, have fertilised the forest and undergrowth of the islands. This has been part of the nutrient cycle of coastal forests for millennia. Unfortunately, nesting seabirds are now rare in the mainland forests, and the coastal forests may be suffering from a lack of nutrients from this source.

It is very important that feeding opportunities in the waters near the islands are maintained, and not depleted by inappropriate commercial fishing practices. The relatively recent development of a fishery for pilchards in the Gulf is potentially damaging to the ecology of the open waters of the Hauraki Gulf.

Pilchards (*Sardinops neopilchardus*) are a critical part of the food chain of the pelagic environment. They form a substantial component of the food of several seabird species, other fish and marine mammals and also the big feeding frenzies of trevally, gannets, dolphins and sometimes Bryde's whales. In turn the seasonal abundance of shrimp-like krill (*Nictiphanes* sp.) provides food not only for the pilchards and some other fish, but directly for seabirds such as red-billed gulls. Depletion of local food sources for seabirds, particularly during the breeding season, means the birds have to travel much longer distances to find food, keeping them away from their chicks for longer periods and reducing the likelihood of breeding success.

## Historical changes

Few people realise the massive scale of changes to the ecology of the Hauraki Gulf through the late 1800s and 1900s and continuing into this century, brought about largely by industrial-scale fishing, both commercial and recreational. Changes to the ecology of shallow rock-reef systems around Hauturu are typical of similar changes around other offshore islands and around the mainland coasts.

The situation around the mainland coasts and inner Hauraki Gulf is further exacerbated by changes in land-use practices — farming, forestry, urban development — leading to serious impacts from sediment runoff from the land. Fortunately most of our offshore islands, including Hauturu, are remote from this particular impact.

The development of a trawler fleet working out of Auckland in the mid-1900s was the beginning of a continuing onslaught on the ecology of the Gulf. The presence of pockets of biogenic reefs — 'oasis' communities — scattered throughout the Gulf posed a great nuisance to early trawling operations. These reefs had developed over probably thousands of years, growing entirely on skeletons of their ancestors, with no solid rock substrates involved at all. Black coral trees metres high and hundreds of years old, as well as giant sponges, 'coral beds' of bryozoans and fields of horse mussels clogged and ripped the trawl nets and reduced the efficiency of the fishing operation. The fishermen's answer to this problem was to drag a heavy chain between two trawl vessels, flattening everything in its path and clearing the way for the trawl nets to have an easy run at the fish.

The original 'oasis' communities would have been extremely rich biodiversity hotspots, and would have provided refuge for huge numbers of fish and, particularly, for juveniles of important recreational and commercial species such as snapper. They may also have been important 'signposts' for migrating fish through the Gulf. In contrast, the seafloor of the Gulf today is like a barren bowling green. Research is now documenting the importance of upright benthic structures such as sponges, bryozoans, horse mussels (*Atrina zelandica*) and, in shallow waters, seagrass (*Zostera* sp.) as nursery areas for juvenile fish. In fact, recent studies by NIWA have indicated that lack of such benthic structure is a major bottleneck, hindering recovery of snapper populations from many years of overfishing.

Commercial fishing trawlers continue to work the waters of the Hauraki Gulf, including those near Hauturu, and maintain the featureless mud bottom of extensive areas of the Gulf. Recovery of any biological structure of value as juvenile fish habitat is impossible with regular bottom-impacting trawling.

In the 1970s another form of fishing was introduced that impacted particularly on the reef ecosystems. Around all the islands of the Gulf, gill-netting over reef areas became a popular commercial fishing activity. Although this method is relatively short-lived because the fish are 'cleaned out' from a reef area quite quickly, serial depletion of fish from reef after reef was a serious problem for several years.

The introduction of the Quota Management System (QMS) to fisheries management in 1986 brought with it some serious and unanticipated ecological impacts throughout the Gulf including around Hauturu. In their wisdom the then Ministry of Agriculture and Fisheries (MAF) decided that in order to achieve the maximum sustainable yield (MSY) from most fisheries, the population should be reduced to only 20 per cent of the original unfished biomass. This meant that the remaining fish had little competition from their peers, and could grow at the maximum possible rate. The annual weight increment of the fishery was the harvestable stock — so snapper were rapidly fished down to only 20 per cent of the original stock. Unfortunately this policy took no account of impacts on the rest of the ecology: fishery managers only took an interest in each single species — in this case snapper — with no regard for any concept of an ecosystem approach to fishery management, as has been talked about more recently.

Snapper is a key species in the marine ecology of the Hauraki Gulf, and big changes to its population impact hugely on the rest of the ecology. Snapper is a major predator on the kina or sea urchin of shallow reefs, and removal of 80 per cent of the snapper had a huge impact on the kina population. Kina thrived with the drastic reduction in predation. This is an artificial situation causing seriously degraded reef ecology brought about by not leaving sufficient snapper in the system to carry out their normal ecological function of controlling the kina population.

In recent years the successors of MAF — the Ministry for Primary Industries (MPI) and, now, Fisheries NZ — have changed their mind about the desirable standing biomass of snapper and are now trying to push the snapper population up to 40 per cent of its pre-fished biomass. This is a great goal and would improve the ecological health of the Gulf, possibly leading to a reduction in kina barrens and recovery of kelp forests. But it will take a long time and the pathway to achieve this is unclear.

Adding greatly to the kina barren problem is the situation with crayfish in the Hauraki Gulf and the rock lobster management area known as CRA2 (Hauraki Gulf/Bay of Plenty). Crayfish are also important predators of kina and their effects on kina populations are the same as those of snapper. For many years MPI have monitored the decline of crayfish catches in CRA2; they have fiddled around with quotas from time to time but done nothing effective to arrest the decline. The crayfish situation is now so dire in CRA2 that there have been serious calls to close the fishery. Catch per unit effort is by far the lowest in the country and recruitment of juveniles into the stock is virtually zero. Once abundant crayfish stocks around Hauturu are now reduced to a situation where crays are quite rare, and it would be barely economic for commercial fishers to pursue them.

In the open waters around Hauturu and other offshore islands, huge boil-ups of mainly trevally and kahawai were historically spectacular. Although some still occur, their size and frequency are greatly diminished, and the individual fish in the schools are much smaller than in the past. Industrial purse-seining, particularly in the 1970s, drastically reduced trevally populations throughout the Hauraki Gulf, Bay of Plenty and the Northland east coast. Although there has been little study of this, it is likely it will have had an impact on island-breeding birds of the Gulf because of the lack of trevally to push krill to the surface where the birds can catch them to feed their young. This particularly affects the red-billed gulls, which do not dive and can only catch krill in the top 50–100 millimetres of water.

Few people are aware that hāpuku (*Polyprion oxygeneios*) used to be a common shallow-water reef fish, and would have occurred around the coasts of Hauturu. In the middle of last century large hāpuku were caught off the rocks at Cape Rodney, and on the reefs around Tiritiri Matangi, where the reefs are no deeper than around 30 metres. We now think of hāpuku as a deepwater fish that is occasionally caught on rock pinnacles offshore in 100–200 metres of water. They have disappeared in diving depths, including around Hauturu. We may never know what their ecological function was on our shallow reefs. Conventional marine reserves will not bring them back: a more drastic regional fisheries approach would be required, and a lot of time.

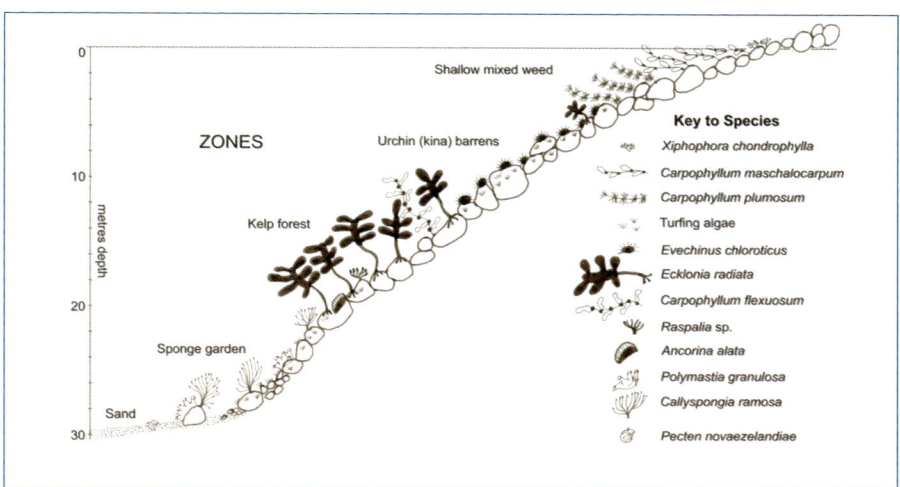

*Fig. 1*: Generalised underwater profile and zones on the boulder bottom of western Hauturu. The kina barrens zone is an artifact of fishing. Prior to snapper and crayfish being fished down to unsustainably low levels, the kelp forest extended right up to the lower edge of the shallow mixed-weed zone. A marine reserve is the best way to restore snapper and crayfish numbers and sizes to a point where they can eat the kina and allow the kelp forest, with its high ecological value and diversity, to recover. Diagram by Roger Grace.

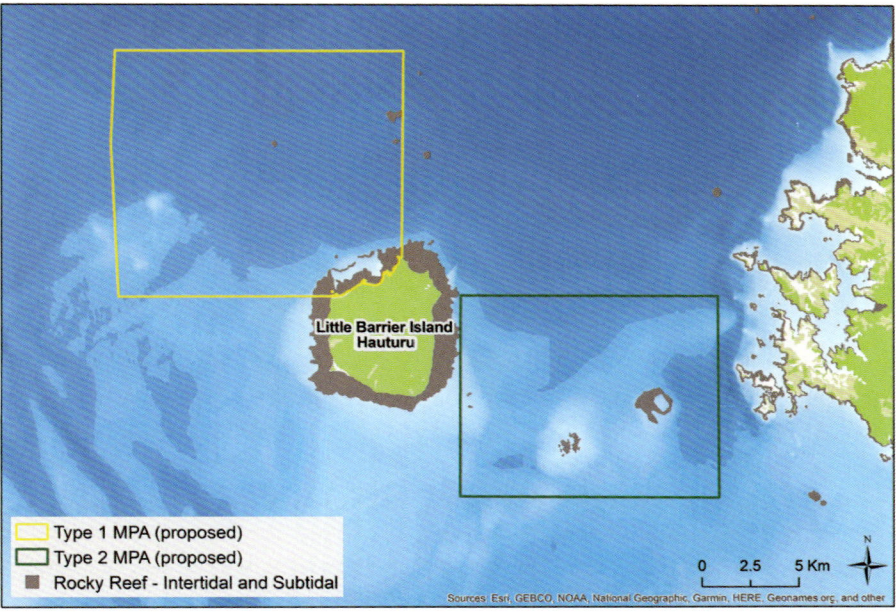

*Fig. 2*: The Marine Spatial Plan presented in the Seachange document released in 2016 included 15 recommended Marine Protected Areas (MPA) in the Hauraki Gulf. Close to Hauturu, a large Type 1 MPA to the northwest of the island would protect marine life from the surface to the seabed. To the east of Hauturu, a large Type 2 MPA would protect the seabed from damaging activities such as trawling and dredging. Dark brown areas on the map are rocky reefs. The benefits to marine biodiversity would be profound. Map courtesy of Irene Pohl/Department of Conservation.

## Hopes for the future

Despite the depleted nature of the waters and reefs around Hauturu, there is hope for the future if there is understanding and interest from the public and a political will to arrest the decline and restore the Gulf towards its earlier richness. We will never get back to how it was in the early part of last century, however — there has been too much long-term irreversible damage.

To restore marine health to the Hauraki Gulf and the waters around Hauturu a two-pronged approach is required. First, a network of no-take zones located in key areas and habitats throughout the Hauraki Gulf is needed, and of sufficient area to protect at least 10 per cent of the area of the Gulf. Second, a drastic change in fisheries management is needed to allow recovery of seriously depleted species like snapper and crayfish, and to allow recovery from habitat-damaging fishing practices such as trawling and dredging. Fishery changes should head towards an ecosystem approach, where impacts of the removal of one species are assessed in terms of their relationship to other species in the ecosystem.

The recent Seachange process, whereby government and local authority agencies and a widely representative selection of interest groups and NGOs got together to assess the problems of the Hauraki Gulf and to think about how these could be reversed, could go a long way toward a solution to the declining health of the Gulf. There is, however, no clear way forward to implement the resulting Marine Spatial Plan, and without a substantial push from the public and uptake by agencies nothing is likely to happen. The Minister of Conservation has recently announced the establishment of a Seachange Ministerial Advisory Group to further progress the process.

Tools for marine protection include the Marine Reserves Act 1971, customary provisions in the Fisheries Act 1986, controls on methods, seasons and areas in, among others, the Fisheries Act, special legislation, a new Marine Protected Areas Act in the pipeline, and an innovative approach using the Resource Management Act. Potential use of these provisions is evolving rapidly.

An example of potential protection around Hauturu, taken from the Marine Spatial Plan, is shown in Figure 2. If this was implemented the benefits to marine biodiversity at Hauturu would be profound.

There is great public interest in restoration and conservation efforts on the land at Hauturu, and this is particularly evident at Tiritiri Matangi further to the south. The same level of concern, however, is lacking when it comes to the marine environment, probably largely because the underwater world is 'out of sight, out of mind' to most people. They are simply not aware of the seriously degraded state of the marine aspects of the Hauraki Gulf and Hauturu. Integrated protection and recovery of land and sea habitats around some of our Gulf islands is desirable and makes a lot of sense. Only with public understanding of the issues, and a political will from authorities, will we see an improvement in the sea as we are experiencing on land.

A scuba diver uses a transect line and underwater writing board to count fish on a 50 x 10-metre transect south of Te Maraeroa. (DT)

# Chapter eight
# The future
— DAVE TOWNS AND MATT RAYNER

**W. M. Hamilton and his co-workers were incredibly privileged to gain access to Te Hauturu-o-Toi/ Little Barrier Island, and to describe a piece of primeval New Zealand. However, the reasons why the island holds such extraordinary biodiversity are only now becoming apparent. Furthermore, rather than degrading over time, as is evident on much of the mainland, native ecosystems on Hauturu are undergoing renewal following the recent removal of the last invasive predators.**

The Database of Island Invasive Species Eradications hosted by Island Conservation in 2018 listed 925 successful pest eradications from islands. Hauturu stands out among these because of the size of the island and the complexity and integrity of its ecological systems. The previous chapters demonstrate that the effects of kiore and cats were selective, at times additive, and often subtle. For some groups of organisms, particularly seabirds, assemblage structure was severely compromised by these introduced predators, with potential knock-on effects through changes to the soil environments. But for other elements of the biota, composition of their assemblages was not affected by invasive mammals but has been strongly influenced by other invasions. For example, the flora now contains over 200 species of exotic plants, some of which are serious environmental weeds. On the other hand, the chapter on vegetation identifies at least 20 species of native plants that were previously suppressed by kiore and are now likely to become much more widespread.

A complex interplay now exists between recovery of native species, reset ecosystems such as those driven by seabirds, constant attempts by exotics to establish and a rapidly changing climate. How the island's ecosystems evolve will thus be of great interest to conservation biologists and the public of New Zealand for generations to come. The section in this book on forest birds indicates that not all species are winners when these communities readjust after invasive predators are removed. In fact, familiar species such as silvereyes and fantails have declined in abundance while others such as bellbirds and robins have increased.

The international significance of Hauturu should not be underestimated. The island is the largest landmass in a chain of nature reserves extending along the northeastern coastline of New Zealand from the Three Kings to the Aldermen Islands, identified to UNESCO by the New Zealand government in 2007 as a potential serial island World

Forest trail, Waipawa Track. (LW) *Previous*: Summit ridge, looking northeast towards Mount Ōrau from Mount Hauturu (722 metres). (AA)

Heritage site called Whakarua Moutere. If these islands eventually gain this listing they will join other archipelagos such as the Galápagos as sites of 'outstanding universal value'. To New Zealanders, gaining World Heritage listing may be irrelevant. We simply need to understand and treat these islands, and Hauturu in particular, for what they are: the repositories of a disproportionate amount of our biological and cultural wealth.

Beyond cultural and political considerations, Hauturu has a regional and international role as a nesting site and refuge for a growing number of very mobile species of birds. Some, such as kererū, kākā and tūī, are known to travel large distances over land and water. The periodic appearance of bellbirds on the mainland and islands adjacent to Hauturu indicates that these species will fly many kilometres over water. This mobility was spectacularly confirmed in 2004 when the eradication of mammal pests from the Tāwharanui open sanctuary, 23 kilometres from Hauturu, saw the immediate establishment of bellbirds in its protected forests with song types that matched those of birds from Hauturu. These transboundary effects mean that birds from Hauturu can provide services such as seed dispersal (kererū) and pollination (bellbirds and tūī) far from the island and then return at will. With an increasing network of protected islands and mainland habitats from the Hauraki Gulf to Auckland city, the positive 'footprint' of Hauturu is now extending far beyond its shores.

The regional transboundary effects are minor compared with the distances travelled by burrowing seabirds. For example, birds from the huge (and growing) colonies of Cook's petrels migrate from Hauturu to the North Pacific Ocean, then return each year to breed. Black petrels undertake similar migrations, but to the productive waters of the Humboldt Current off the west coast of South America. During their migration and residence in far-off oceans, these seabirds are a vital component in marine food webs: they feed on zooplankton and fish and provide a rich guano fertiliser that feeds phytoplankton and thus fires up and drives the entire marine food chain. Of the land-based migrants, the shining cuckoo and long-tailed cuckoo play an ecological role as predators and consumers in far-off island chains of the South Pacific. It is important to remember that conservation efforts on a small island at the bottom of the world can have important implications far beyond the island, and even beyond New Zealand's boundaries.

Of course, these features of Hauturu could once again be placed under threat if rodents are allowed to reinvade or new, insidious threats such as diseases become established. Its distance offshore may help to shield Hauturu from these problems, but the size of the island and its extreme ruggedness mean that dealing with any pest or disease incursions will be particularly challenging. Vigilance and biosecurity are now more important than ever.

## Insidious threats

Biosecurity may seem like an inconvenient delay when making a trip to a nature reserve island such as Hauturu. But care with the way people, goods and equipment are

*Above*: Kōkako (*Callaeas wilsoni*). (AA) *Below*: Cook's petrel/tītī (*Pterodroma cookii*). (EW)

transported to these islands has paid off: incursions by unwanted species such as mice, rats and invasive invertebrates are rare for islands beyond swimming range from the mainland. When rodent incursions have happened, 'hit squads' of people aided by rodent detector dogs and an array of toxins and traps have so far successfully neutralised them. Of course, not every arrival is a pregnant female, which has further helped with avoiding full-blown invasions. However, most incursions have been to relatively small islands. The challenge presented by Hauturu is its size and terrain, which means border control is essential. With an increasing human population in the Auckland region and consequent growth in boat traffic around the island, new monitoring methods are needed to ensure species introductions don't accidentally occur under our noses.

Until recently the deployment of remote monitoring technologies such as satellites and drones seemed a far-fetched method for coastal surveillance of nature reserve islands. But as these technologies become more reliable and affordable they can increasingly be seen as realistic tools for coastal monitoring during periods of high boat presence and traffic.

A powerful technique for assessing invasion risk on the islands has been borrowed from forensic genetic methods used by the police. By sampling rodents and stoats that reach islands or are found on boats, it is possible to track their origins. For example, ship rats in the Bay of Islands were found, using microsatellite DNA markers, to have originated from outside the area rather than the adjacent mainland. For this to happen the rats must have stowed away on a boat. Similarly, a stoat found on Rangitoto Island after the species had been eradicated from the island lacked the markers found in the original island population and in those on neighbouring Waiheke Island. The genetic fingerprint of this animal indicated that it most likely originated from near Musick Point on the mainland, which suggests that it swam to Rangitoto, possibly via Motukorea/Browns Island. These genetic advances are now being complemented by technical advances, such as traps and tunnels that produce an electronic signal when they have been activated. But however sophisticated these devices might become, there is unlikely to be a time when we can relax about potential reinvasions or new invasions.

Perversely, rodent incursions may be more easily managed than insidious problems caused by invasive invertebrates such as wasps and ants, the establishment of weeds, and the arrival of plant pathogens. When commuting birds such as kererū or exotic species such as starlings move from island to mainland, they can bring unwelcome visitors on their return: the seeds of invasive plants. Japanese honeysuckle, climbing asparagus, privet and even Moreton Bay fig can transform native plant communities and are now a risk on Hauturu. Windborne plants such as pampas and mothplant — which has become an expensive problem on islands such as Great Mercury — can reach Hauturu.

Windborne diseases such as myrtle rust are likely the most difficult invasion to deal with. Isolation from the mainland and care with biosecurity should, however, insulate the forests of Hauturu from arguably the greatest threat to northern forests: the spread of kauri dieback. If the kauri forests of Hauturu remain free of this disease, the significance of the island for native biodiversity would be hugely important, should kauri disappear elsewhere.

## Future: Expectations, mysteries and tough questions

When Hauturu was first established as a nature reserve in the 1890s it was seen as an island haven for birds disappearing from the mainland, but also as a glimpse of the past. Over the last 50 years, beginning with the eradication of cats, it has been changing into something else. Seabird numbers are increasing and so is the number of species nesting — in particular with the unexpected reappearance of the New Zealand storm petrel. Now the island's ecosystems have become a showcase because of their rich diversity. For international observers Hauturu is an inspiration about what can be achieved for biodiversity conservation. For New Zealanders in general, we now have a benchmark against which we can measure the effects of invasive species elsewhere in the country.

The island may eventually provide lessons, too, about the resilience of species and ecosystems in the face of a rapidly changing climate. As the Conservation Management Plan produced for the island in 2017 states, 'Te Hauturu-o-Toi is the gold standard for predator-free New Zealand and New Zealand's jewel in the crown for conservation.'[1] For Ngāti Manuhiri, the Management Plan reveals the multi-layered significance of the island: as a place where people can reconnect with nature in a form lost everywhere else; a place for rekindling traditions and legends; a wānanga or place of learning; and a place where they can exercise kaitiakitanga (guardianship) for the benefit of their descendants and all other New Zealanders.

The species and ecosystems present on Hauturu appear to have a secure and assured future. However, as species on the island continue their recovery, other questions will need to be given considerable thought. The first set of questions involve value judgements about what wildlife managers bring to and take from the island in the form of translocations; how we value the marine environment surrounding Hauturu; and how new initiatives might encompass our increasing knowledge of the marine–terrestrial connection. These questions should not be taken lightly because they could affect the principles outlined by Ngāti Manuhiri in the Conservation Management Plan. The second group of questions reflect our lack of knowledge about the long-term effects of biological changes under way within the ecosystems on Hauturu.

## Translocations

It is unknown how many species disappeared after kiore reached Hauturu. Two species of large nocturnal lizards have survived only on islands that have never been invaded by rats. Both of these species, robust skink and McGregor's skink, are often associated with areas heavily burrowed by seabirds. Robust skinks in particular use the seabird burrows as shelter, emerging from them at night to prey on invertebrates among the leaf litter. A local population of robust skinks has survived on an islet in the Mokohinau group, and a local population of McGregor's skinks exists on an islet near Taranga/Hen Island. Both species have scattered distributions on islands as far south as Cook Strait, and their remains have been found in the North Island in caves and sand-dune deposits. The previous dryland connection between Hauturu and the last surviving island populations of these two species

indicates a likely continuous distribution that once included Hauturu. Should such species be reintroduced?

More tangible evidence in the form of a museum specimen reveals Hauturu as one of the last places in the world that the North Island snipe called home before its extinction in the late nineteenth century. This species was extremely vulnerable to introduced predators and disappeared rapidly from Hauturu after the introduction of cats in the 1870s, but was probably already uncommon on the island as a result of rat predation. Although its predators are now removed from Hauturu, the endemic North Island snipe is globally extinct. Could an analogue snipe species be introduced to fulfil the snipe's ecological role as a consumer of soil-dwelling invertebrates? Related species are found on southern offshore islands and in the Chatham Islands. And if snipe were to be reintroduced to Hauturu, which species should it be?

Finally, as seabird populations increase on Hauturu, several studies indicate that the number of species able to recolonise is probably limited by its distance from other breeding populations. A few species, such as ōi/grey-faced petrels and fluttering shearwaters, seem willing to colonise newly available sites. However, other species, such as Pycroft's petrel, little shearwater, flesh-footed shearwater and white-faced storm petrel inhabit nearby archipelagos but are unknown on Hauturu. None of these species seems able to colonise new sites more than a few hundred metres over water from existing colonies. Should some or all of these species be reintroduced to Hauturu? And if they were, would there be risks to the expanding population of New Zealand storm petrels?

New Zealand has a long history of translocations and Hauturu was initially established as a reserve to which birds would be translocated. However, there are always risks of unforeseen consequences when species are moved outside their known range. Even emergency translocations deemed to be in the national interest can carry risks — this is why they are sometimes called planned invasions.

## Connecting and protecting moana and whenua

Stand where the tide laps Hauturu's boulder beaches and you are confronted with conservation hypocrisy that is grounded in New Zealanders' value judgements about the environment and how we exploit and protect it. Take a few steps up onto the land and you enter a 'nature reserve', an area of complete nature protection where you can be prosecuted for harming any native wildlife — though before you get far a ranger will likely appear requesting your paperwork, as entry is by permit only. Take a few steps down into the marine realm, however, and you enter New Zealand's last great frontier of environmental exploitation, where there is little protection for native biodiversity beyond 'bag limits' for larger, more preferred species. This terrible dichotomy is based on our bias towards the land, and on a historical cultural perspective that what happens beneath the sea does not matter, and that such resources are inexhaustible. Increasingly,

Moss forest, summit ridge. (AA)

however, we are learning that the ocean and its resources are not infinite and that there is a complex web of relationships between land and sea. In the case of island conservation, in particular, science is telling us that protecting the terrestrial world of our island nature reserves without protecting surrounding marine habitats makes no sense at all.

Again, seabirds feature in this story. We know nutrients are transferred from the sea to the land in the form of seabird guano, but we have only recently discovered that these nutrients also flow back into the sea to enrich the coastal marine habitats surrounding our precious island reserves. This relationship was first revealed in a study from the reefs off Palmyra Atoll in the North Pacific, where researchers studying the distribution of manta rays inadvertently discovered that the rays like feeding off coastlines supporting unmodified forest that contained seabird colonies. The researchers went on to show that where seabirds are present their guano enriches coastal reef ecosystems, and this enhances the growth of phytoplankton and the zooplankton that the manta rays like to feed on. Another study found that the reef diversity of tropical seas was lower in areas surrounding islands where rats — and therefore no seabirds — were present. Where seabirds flourished in the absence of rats, the reef ecosystems abounded with life. Closer to home, in the Hauraki Gulf, recent research has confirmed this relationship: it has revealed higher algal diversity in seaweed forests surrounding islands with a higher density of seabirds. The take-home message: the land and sea are deeply connected, with seabirds acting as the ecological link. Given this interconnection, and the strengthening of nutrient pathways off Hauturu with growing seabird populations, it is essential that the nature reserve's footprint be extended beyond its terrestrial boundaries. Marine as well as terrestrial protection would see a strengthening of these relationships — for example, the return of large snapper and crayfish, which are predators of the algae-eating kina, would see the restoration of what are now kina barrens to beautiful undersea kelp forests, and an increase in the amount of storm-cast seaweed and detritus on Hauturu's coastline. Such material in turn supports a whole community of plants, invertebrates, reptiles and birds. Hauturu, with its pristine terrestrial communities, presents a rare laboratory where such sea–land ecological relationships could be studied in the context of restoring the marine environment.

## New initiatives

Whether the extraordinary biodiversity on Hauturu and other islands in the Whakarua Moutere World Heritage proposal should be celebrated through formal nomination needs further careful consideration. There are advantages and disadvantages to designation as a World Heritage site. The advantages would likely include global recognition of the biological and cultural importance of the islands involved, their significance to the long-term future of a host of distinctive species of plants and animals, and their role in the development of pest-eradication technologies. Eight of the nine islands or island groups proposed for the listing are now free of all invasive mammals. From a national perspective,

Sunset at Te Tītoki Point. (NF)

a listed site is required to have consistent and appropriate management; indeed, UNESCO can fine governments that fail to meet management obligations. One could argue that since all of the sites are nature reserves, management of access and biosecurity risk are already covered by legislation. However, there is plenty of recent evidence that offshore reserves are easy targets for budget cuts. Perhaps World Heritage listing would provide secure funding and clearer guidelines for priority tasks.

One characteristic of these islands identified by the World Heritage proposal is the almost total lack of marine reserves to complement the high levels of terrestrial protection. The only exception is the Poor Knights, with its globally recognised marine ecosystems. As shown earlier, there is a stark contrast between the innovative management and protection of terrestrial ecosystems and the destruction of marine ecosystems. Even if the World Heritage proposal is shelved, the question of how to reconnect the land with the sea around islands such as Hauturu must be addressed.

## Tough questions

Our knowledge of Hauturu has increased substantially in the decades since the work of W. M. Hamilton and his colleagues, as this book confirms. But there is still much we do not understand about the island, particularly about how its ecosystems might function and change in the long-term as a result of our various management actions. We now know that the fate of Hauturu is inextricably linked to the ocean, and no doubt a major driver of change on the island will come from the sea. Currently, seabird populations are just beginning a process of recovery after a century of decline in the face of introduced mammals. Given the massive historic populations of species such as Cook's and black petrels, it is unlikely that these species are food limited; and with the huge amount of breeding habitat on the island, they could increase substantially. What effect will this have on Hauturu?

Similar large island ecosystems elsewhere are much more modified and often overrun by even more damaging introduced species than those on Hauturu. Other locations where the effects of seabirds on islands have been well studied are at least an order of magnitude smaller than Hauturu. We know that small, low-altitude islands can be heavily modified by burrowing seabirds and also by the effects of salt-laden winds. Hauturu, however, because of its size and elevation, provides a completely different type of seabird island and a different range of possibilities for how it might change. For example, will dense populations of seabirds under old-growth kauri forest on Hauturu function as they do on small islands with young coastal broadleaf forest? We know that kauri can change soil chemistry and structure, but so do seabirds. How will these two interact? Will dense burrowing under old-growth forest promote a change in the structure of the forest to a higher density, lower stem-diameter canopy? If such habitat changes do occur they will most certainly affect other forest dwellers, or even the entire food web, possibly reducing the suitability of the island for some taxa while making it more suitable for others. Could such changes alter the suitability of the island for translocated species, or increase its susceptibility to invasive weeds?

These questions may take decades or even longer to produce definitive answers because the rate of change will often be subtle and slow. The same can be said of some conservation efforts. The reproductive output of tuatara is so low and they are so slow to reach sexual maturity that they will likely remain a rarely encountered component of the island's ecosystem for many decades. The time taken for this species to reach the kind of densities found on other islands may well need to be measured in centuries. Of course, in parallel with this gradual increase, the global climate is also changing. So the final question is: Will the growing diversity and complexity of Hauturu's ecosystems provide insights into their resilience against external stresses, such as those likely to stem from climate change?

To understand such issues we need to keep monitoring Hauturu's ecosystems and species. As described in the preceding chapters, we already have useful baseline information for some of these. However, there are other areas where our knowledge is patchy or lacking and these need further study. Our findings on Hauturu could have global application, as communities elsewhere grapple with a changing climate and other anthropogenic effects on their ecosystems and species.

# Chapter nine
# Species lists

# Aquatic and terrestrial invertebrates

| ORDER/Family | Species |
|---|---|
| **CHILOPODA (CENTIPEDES)** | |
| **LITHIBIOMORPHA (STONE CENTIPEDES)** | |
| Ballophillidae | Ballophilus hounselli |
| Scolopendridae | Cormocephalus rubriceps (hura, hara, giant centipede) |
| **ANNELIDA (EARTHWORMS)** | |
| Acanthodrilidae | Dinodriloides beddardi |
| | Rhododrilus similis |
| Lumbricidae | Allolobophora caliginosa |
| | Octoclasium cyaneum |
| Megascolecidae | Anisochaeta antarcticus |
| | Anisochaeta gigantea |
| | Anisochaeta shakespeari |
| | Diporochaeta obtusa |
| | Megascolides irregularis |
| | Tokea maorica |
| | Tokea reptans |
| **COLLEMBOLA (SPRINGTAILS)** | |
| | Holacanthella duospinosa |
| **ONYCHOPHORA (VELVETWORMS)** | |
| Peripatopsidae | Peripatoides sympatrica (ngāokeoke/peripatus) |
| **NEMATOMORPHA (NEMATODES)** | |
| **PLATYHELMINTHES (FLATWORMS)** | |
| | Caenoplana coerulea (black flatworm) |
| | Planaria sp. |
| **TERRESTRIAL MOLLUSCA (SLUGS AND SNAILS)** | |
| Athoracophoridae | Anthoracophorus bitentaculatus rufovenosus |
| | Pseudaneitea sp. (puṭokoropiropi/leaf-veined slug) |

| ORDER/Family | Species |
|---|---|
| | Vomanus maemoreus |
| Charopidae | Cavellia serpentinula |
| | Cavellioropa microrhina |
| | Charopa chrysaugeia |
| | Charopa coma |
| | Charopa pilsbryi |
| | Charopa pseudanguicula |
| | Flammocharopa costulata |
| | Flectola colensoi |
| | Flectola caputspinulae |
| | Flectola roseveari |
| | Flectola sterkiana |
| | Huonodon hectori |
| | Ptychodon tau |
| Clausiliidae | Heliodiscus singleyanus |
| Flammulinidae | Allodiscus planulatus |
| | Allodiscus urquharti |
| | Flammulina chiron |
| | Flammulina feredayi |
| | Flammulina perdita |
| | Phenacohelix ponsonbyi |
| | Serpho kevi |
| | Suteria ide |
| | Thalassohelix zelandiae |
| | Therasia decidua |
| | Therasiella celinde |
| Hydrocenidae | Hydrocena purchasi |
| Laomidae | Laoma leimonias |
| | Laoma pirongiaensis |
| | Laoma poecilosticta |
| | Paralaoma lateumbilicata |
| | Paralaoma serratocostata |
| | Paralaoma sp.n.1 |
| | Paralaoma sp.n.6 |

*Previous:* Awaroa Stream. (OB)

| ORDER/Family | Species |
|---|---|
| | Paralaoma sp.n.29 |
| | Paralaoma sp.n.30 |
| | Paralaoma sp.n.32 |
| | Paralaoma sp.n.33 |
| | Paralaoma sp.n.38 |
| | Phrixgnathus allachroidus |
| | Phrixgnathus ariel |
| | Phrixgnathus cheesemani |
| | Phrixgnathus conella |
| | Phrixgnathus erigone |
| | Phrixgnathus glabriasculus |
| | Phrixgnathus māriae |
| | Phrixgnathus microreticulatus |
| | Phrixgnathus moellendorffi |
| | Phrixgnathus sublucidus |
| | Phrixgnathus sp.n.40 |
| | Phrixgnathus sp.n.59 |
| **Liareidae** | Liarea carinella |
| | Liarea turriculata |
| **Otoconchidae** | Otoconcha dimidiata |
| **Oxychilidae** | Oxychilus alliarius |
| **Pupinidae** | Cytora cytora |
| | Cytora pallida |
| | Cytora torquillum |
| **Rhytididae** | Delos coresia |
| | Delos jeffreysiana |
| **ISOPODA** | |
| **Armadillidae** | Sp.1 |
| **AMPHIPODA** | |
| | Parorchestia tenuis |
| | Puhuruhuru aotearoa |
| **EPHEMEROPTERA (MAYFLIES)** | |
| **Ameletopsidae** | Ameletopsis perscitus (yellow dun) |
| **Ichthybotidae** | Ichthybotus hudsoni |
| **Leptophlebiidae** | Acanthophlebia cruentata |

| ORDER/Family | Species |
|---|---|
| | Arachnocolus phillipsi |
| | Austroclima sepia |
| | Deleatidium fumosum |
| | Deleatidium lillii |
| | Deleatidium myzobranchia |
| | Isothraulus abditus |
| | Mauiulus luma |
| | Neozephhlebia scita |
| | Zephlebia borealis |
| | Zephlebia cruentata |
| | Zephlebia dentata |
| | Zephlebia inconspicua |
| | Zephlebia nebulosa |
| | Zephlebia spectabilis |
| | Zephlebia versicolor |
| **Siphlonuridae** | Coloburiscus humeralis |
| | Nesamaletus ornatus |
| **MEGALOPTERA (DOBSONFLIES)** | |
| **Corydalidae** | Archichauliodes diversus |
| **NEUROPTERA (LACEWINGS)** | |
| **Hemerobiidae** | Drepanacra binocula |
| | Micromus tasmaniae |
| **Myrmeleontidae** | Weeleus acutus |
| **ODONATA (DRAGONFLIES, etc.)** | |
| **Corduliidae** | Antipodochlora braueri |
| | Procordulia smithii |
| **Coenagrionidae** | Xanthocnemis zealandica |
| **PLECOPTERA (STONEFLIES)** | |
| **Austroperlidae** | Austroperla cyrene |
| **Eustheniidae** | Stenoperla prasina |
| **Gripopterygidae** | Acroperla trivacuata |
| | Holcoperla angularis |
| | Zelandobius confusus |
| | Zelandobius pallidus |
| | Zelandobius sp.n. |
| | Zelandoperla agnetis |
| **Notonemouridae** | Spaniocerca zelandica |

| ORDER/Family | Species |
|---|---|
| **TRICHOPTERA (CADDISFLIES)** | |
| Chathamiidae | Philanisus plebeius |
| | Sp. 1 |
| Heliopschidae | Helicopsyche zealandica |
| | Helicopsyche sp. |
| Hydrobiosidae | Hydrochorema crassicaudatum |
| | Psilochorema mimicum |
| | Psilochorema sp. |
| Leptoceridae | Triplectides obsoletus |
| Philopotamidae | Dolophilodes mixta |
| | Hydropsyche fimbriata |
| | Oeconesus maori |
| | Olinga feredayi |
| | Olinga sp. |
| | Oxyethira albiceps |
| | Plectrocneunia sp. |
| | Polyplectopus puerilis |
| | Pycnocentria sp. |
| **ARANEAE (SPIDERS)** | |
| Agelenidae | Orepukia sp.nr geophila |
| | Sp.n. |
| Amphinectidae | Aorangia sp. |
| | Paramamoea sp. |
| | Reinga media |
| Anapidae | Paranapis insula |
| | Zealanapis sp. |
| Clubionidae | Clubiona consensa |
| Dictynidae | Paradictyna rufoflava |
| Gnaphosidae | Hypodrassodes dalmasi |
| | Hypodrassodes maoricus |
| | Hypodrassodes sp.1 |
| | Hypodrassodes sp.2 |
| Hexathelidae | Hexathele hochstetteri (tunnelweb spider) |
| | Porrhthele antipodiana (tunnelweb spider) |
| | Porrhthele quadrigyna (tunnelweb spider) |
| Huttoniidae | Huttonia sp. |

| ORDER/Family | Species |
|---|---|
| Linyphiidae | Dunedinia denticulata |
| Lycosidae | Anteropsis hilaris |
| | Artoria hospita |
| | Lycosa sp. |
| Malkaridae | Sp.n. |
| Mecysmaucheniidae | Zearchaea clypeata |
| Micropholcommatidae | Taphiassa sp. |
| Migidae | Migas insularis (trapdoor spider) |
| Mimetidae | Australomimetus sp. |
| Nemesiidae | Stanwellia hapua (trapdoor spider) |
| Orsolobidae | Sp.n. |
| Pararchaeidae | Pararchaea sp. |
| Pisauridae | Dolomedes aquaticus |
| | Dolomedes minor (nursery-web spider) |
| Salticidae | Helpis minitabunda |
| | Trite planiceps |
| Segestriidae | Sp.n. |
| Stiphididae | Cambridgea ambigua |
| | Cambridgea foliata |
| | Cambridgea ramsayi |
| Synotaxidae | Pahoroides sp. |
| Theridiidae | Episinus sp |
| Thomisidae | Sidymella angulata |
| | Sidymella angularis |
| Zodariidae | Forsterella faceta |
| Zoropsidae | Uliodon sp.1 |
| | Uliodon sp.2 |
| **OPILIONES (HARVESTMEN)** | |
| Neopilionidae | Forsteropsalis pureora |
| | Sp.n. |
| Pettalidae | Rakaia granulosa |
| **PSEUDOSCORPIONS** | |
| Chernetidae | Apatochernes sp. |
| | Sathrochthonius taraha |

| ORDER/Family | Species |
|---|---|
| **ACARI (MITES)** | |
| Acaridae | Tryophagus vanheurui |
| **PROTURA (CONEHEADS)** | |
| Eosentomidae | Eosentomon dawsoni |
| Acerentomidae | Gracilentulus gracilus |
| **DIPTERA (FLIES)** | |
| Anisopodidae | Sylvicola sp. |
| Asilidae | Neoitamus lascus |
| Bibionidae | Dilophus neoinsolitus |
| Brachystomatidae | Ceratomerus aquilonius |
| | Ceratomerus brevifurcatus |
| | Ceratomerus sp. (Hauturu endemic) |
| Calliphoridae | Pollenia nigrisquama |
| Canthyloscelididae | Canthyloscelis sp. |
| Chathamiidae | Chathamia integripennis |
| Chironomidae | Sp.1 |
| Chloropidae | Gaurax mesopleuralis |
| | Hippelates insignificans |
| | Lasiopleura quadriseta |
| Coelopidae | Chaetocoelopa littoralis |
| Culicidae | Aedes antipodeus |
| | Aedes notoscriptus |
| | Culex asteliae |
| | Culex pervigilans |
| | Culex quinquefasciata |
| | Maorigoeldia argyropus |
| | Opifex fuscus |
| Cypselomatidae | Heloclusia aristata |
| Dixidae | Nothodixa campbelli |
| | Paradixa fuscinervis |
| | Paradixa harrisi |
| | Paradixa neozelandica |
| | Paradixa sp. |
| Dolichopodidae | Australachalus sp. |
| | Parentia aotearoa |
| | Parentia calignosa |
| | Parentia gemmata |

| ORDER/Family | Species |
|---|---|
| | Parentia recticosta |
| | Parentia restricta |
| | Parentia titirangi |
| Drosophilidae | Drosophila kirki |
| Empididae | Chelfera sp. |
| | Hemerodromia radialis |
| Ephydridae | Hydrellia tritici |
| Fanniidae | Australofannia sp. |
| Heleomyzidae | Allophylina albitarsis |
| | Xeneura picata |
| Heteromyzidae | Aneuria angusta |
| | Fenwickia affinis |
| Hippoboscidae | Ornithoica stipituri |
| | Ornithomya variegata (avian parasitic fly) |
| Huttoninidae | Huttonia brevis |
| Keroplatidae | Arachnocampa luminosa |
| Lauxaniidae | Sapromyza neozelandica |
| | Trypaneoides costata |
| Limoniidae | Limonia wiseana |
| Lonchopteridae | Lonchoptera bifurcata |
| Mytocetophilidae | Mytocetophila sp. |
| | Zygomyia sp. |
| Mystacinobiidae | Mystacinobia zelandica (New Zealand batfly) |
| Orthocladiidae | Sp.1 |
| Sciaridae | Bradysia impatiens |
| Sciomyzidae | Eulimnia milleri |
| Simuliidae | Austrosimulium australense |
| Sphaceroceridae | Howickia omamari |
| | Howickia sp. |
| | Phthitia thomasi |
| Stratiomyidae | Zealandoberis violacea |
| Syrphidae | Eristalis tenax |
| Tabanidae | Scaptia adrel |
| Tachinidae | Trigonospila brevifacies |
| Tanypodidae | Sp.1 |
| Teratomyzidae | Teratomyza neozelandica |

| ORDER/Family | Species |
|---|---|
| Therevidae | Anabarhynchus longipennis |
| | Ectinorhynchus cupreus |
| Tipulidae | Sp.1 |
| **HEMIPTERA (BUGS)** | |
| Acanthosomatidae | Rhopalimorpha obscura |
| Anthocoridae | Sp.1 |
| Aphidoidae | Sp.1 |
| Aphrophoridae | Carystoterpa fingens |
| | Carystoterpa minor |
| | Carystoterpa vagans |
| Aradidae | Acaraptera myersi |
| | Adenocoris spiniventris |
| | Aneurus brouni |
| | Calisius zealandicus |
| | Carventaptera sp. |
| | Leuraptera zealandica |
| | Neadenocoris ovatus |
| | Neocarventus angulatus |
| | Woodwardiessa quadrata |
| Ceratocombidae | Ceratocombus novaezelandiae |
| Cercopidae | Carystoterpa fingens |
| | Carystoterpa minor |
| | Carystoterpa vagans |
| | Carystoterpa sp. |
| | Pseudaphronella jactator |
| Cicadellidae | Anzygina zealandica |
| | Arawa sp. |
| | Maiestas vetus |
| | Novothymbris sp. |
| | Zelopsis nothofagi |
| Cicadidae | Amphipsalta cingulata (clapping cicada) |
| | Amphipsalta zelandica (chorus cicada) |
| | Kikihia cutora cutora (northern snoring cicada) |
| | Kikihia ochrina (April green cicada) |

| ORDER/Family | Species |
|---|---|
| | Notopsalta sericea (clay bank cicada) |
| Cixiidae | Aka finitima |
| | Cermada hybrid1 |
| | Cermada inexspectata |
| | Cermada punctimargo |
| | Cermada sp. |
| | Cixius inexspectatus |
| | Cixius punctimargo |
| | Huttia nigrifrons |
| | Koroana rufifrons |
| | Oliarus atkinsoni |
| | Oliarus oppositus |
| | Tiriteana clarkei |
| Coccidae | Aphenochiton pubens |
| | Aphenochiton subtilis |
| | Crystallotesta leptospermi |
| | Crystallotesta ornatella |
| | Ctenochiton viridus |
| | Kalasiris perforata |
| | Plumichiton pallicinus |
| Coelostomidiidae | Coelostomidia wairoensis |
| | Coelostomidia zelandica |
| Cydnidae | Macroscytus australis |
| Diaspididae | Aspidiotus nerii |
| | Hemiberlesia rapax |
| Delphacidae | Sulix tasmani |
| | Ugyops pelorus |
| | Ugyops rhadamanthus |
| Derbidae | Eocenchrea maorica |
| Flatidae | Anzora unicolor |
| Lygaeidae | Rhypodes clavicornis |
| | Rhypodes sp. |
| Miridae | Chaetedus sp. |
| | Chinamiris acutospinosus |
| | Chinamiris indeclivis |
| | Diomocoris maoricus |
| | Felisacus elegantulus |
| | Halticus minutus |

| ORDER/Family | Species |
|---|---|
| | *Sidnia kinbergi* |
| | *Wekamiris auropilosus* |
| | *Zanchius ater* |
| **Myerslopiidae** | *Pemmation aspera* |
| **Pemphigidae** | *Pemphigus* sp. |
| **Pentatomidae** | *Cermatulus nasalis nasalis* |
| | *Cuspiona simplex* |
| | *Dictyotus caenosus* |
| | *Glaucias amyoti* |
| | *Monteithiella humeralis* |
| | *Nezara viridula* |
| | *Oechalia schellenbergii* |
| | *Oncacontias vittatus* |
| | *Philapodemus australis* |
| | *Rhopalimorpha obscura* |
| **Pseudcoccidae** | *Balanococcus gahniicola* |
| **Rhyparochromidae** | *Dieuches notatus* |
| | *Plinthisus* sp. |
| | *Remaudiereana inornata* |
| | *Targarema stali* |
| | *Targarema* sp. |
| | *Tomocoris ornatus* |
| | *Truncala insularis* |
| | *Trypetocoris separatus* |
| **Ricaniidae** | *Scolypopa australis* |
| **Tingidae** | *Tanybyrsa cumberi* |
| **Veliidae** | *Microvelia macgregori* |
| **HYMENOPTERA (BEES, WASPS, ANTS)** | |
| **Apidae** | *Bombus terrestris* (bumble bee) |
| **Bethylidae** | *Goniozus jamiei* |
| | *Sierola lucyae* |
| **Braconidae** | *Aphareta aotea* |
| | *Asobara albiclara* |
| | *Asobara persimilis* |
| | *Aspilota* sp. |
| | *Aucklandella pyrastis* |
| | *Aucklandella ursula* |

| ORDER/Family | Species |
|---|---|
| | *Meteorus pulchricornis* |
| | *Meteorus novazealandicus* |
| | *Schauinslandia* sp. |
| | Sp.1 |
| **Colletidae** | *Leioproctus imitatus* |
| | *Leioproctus pango* |
| **Diapriidae** | *Betyla fulva* |
| | *Diphoropia sinvosa* |
| | *Maoripria verticillata* |
| | *Paramesius* sp. |
| **Formicidae** | *Austroponera* sp. |
| | *Heteroponera microps* |
| | *Huberia brounii* |
| | *Megaspilus fuscipennis* |
| | *Monomorium antarcticum* |
| | *Monomorium smithii* |
| | *Prolasius* sp. |
| | *Strumigenys chiricahua* |
| **Gasteruptiidae** | *Pseudofoenus* sp. |
| **Halicitidae** | *Lasioglossum sordidum* |
| **Ichneumonidae** | *Aclastus* sp. |
| | *Ascogaster bicolorata* |
| | *Casinaria* sp.n. (Hauturu) |
| | *Mesochorus* sp. |
| | Sp.1 (Campopleginae) |
| | Sp.2 (Cryptinae) |
| | Sp.3 (Ichneumoninae) |
| | Sp.4 (Tersilochinae) |
| **Mymaridae** | *Australomymar* sp. |
| **Pompilidae** | *Priocnemis monachus* (large black hunter wasp) |
| | *Sphictostethus nitidus* (golden hunter wasp) |
| **Pteromalidae** | *Aphobetus maskeli* |
| | *Neapterolelaps* sp. |
| | *Ophelosia charlesi* |
| **Sphecidae** | *Pison spinolae* |
| **Vespidae** | *Polistes chinensis* |

| ORDER/Family | Species |
|---|---|
| | Polistes humilis |
| | Vespula germanica |
| | Vespula vulgaris |
| **BLATTODEA (COCKROACHES)** | |
| Blattidae | Celatoblatta undulivitta |
| | Celeriblattina minor |
| | Maoriblatta novaeseelandiae |
| | Maoriblatta sp. |
| | Ornatiblatta maori |
| **ORTHOPTERA (WĒTĀ)** | |
| Anostotomatidae | Deinacrida heteracantha (wētāpunga/giant wētā) |
| | Hemiandrus pallitarsus (ground wētā) |
| | Hemideina thoracica (Auckland tree wētā) |
| Rhaphidophoridae | Gymnoplectron acanthocera (Auckland cave wētā) |
| Tettigoniidae | Caedicia simplex |
| **SIPHONAPTERA (FLEAS)** | |
| Ceratophyllidae | Ceratophyllus petrochelidoni |
| **THYSANOPTERA (THRIPS)** | |
| Merothripidae | Merothrips floridensis |
| Thripidae | Anaphothrips dubius |
| | Apterothrips secticornis |
| | Chirothrips manicatus |
| | Heliothrips haemorrhoidalis |
| | Hercinothrips bicinctus |
| | Microcephalothrips abdominalis |
| | Pseudanaphothrips achetus |
| | Thrips obscuratus |
| | Thrips simplex |
| **PHASMATODEA (STICK INSECTS)** | |
| Phasmatidae | Acanthoxyla geisovii (prickly stick insect) |

| ORDER/Family | Species |
|---|---|
| | Argosarchus horridus (bristly stick insect) |
| | Asteliaphasma naomi |
| | Clitarchus hookeri (common stick insect/ smooth stick insect) |
| | Spinotectarchus acornatus |
| **PHTHIRAPTERA (LICE)** | |
| Menoponidae | Menacanthus eurysternus |
| Philopteridae | Naubates fuliginosus |
| | Pectinopygus varius |
| | Rallicola rodericki |
| **LEPIDOPTERA (MOTHS AND BUTTERFLIES)** | |
| Crambidae | Culladia cuneiferellus |
| | Deana hybreasalis |
| | Eudonia colpota |
| | Eudonia cymatias |
| | Eudonia dochmia |
| | Eudonia philerga |
| | Eudonia steropaea |
| | Eudonia submarginalis |
| | Eudonia zophochlaena |
| | Glaucocharis chrystochyta |
| | Hygraula nitens |
| | Mnesictena flavidalis |
| | Orocrambus flexuosellus |
| | Orocrambus ramosellus |
| | Scoparia minuculalis |
| Elachistidae | Circoxena ditrocha |
| | Elachista archaeonoma |
| | Elachista gerasmia |
| Erebidae | Nyctemera annulata |
| Geometridae | Austrocidaria bipartita |
| | Austrocidaria parora |
| | Azelina venustula |
| | Cleora scriptaria |
| | Declana floccosa |

| ORDER/Family | Species |
|---|---|
| | Declana junctilinea |
| | Epyaxa venipunctata |
| | Helastia cinerearia |
| | Ischalis gallaria |
| | Pasiphila inducata |
| | Poecilasthena schistaria |
| | Poecilasthena pulchraria |
| | Pseudocoremia dugdalei |
| | Pseudocoremia melinata |
| | Pseudocoremia suavis |
| Heliozelidae | Heliozela catoptrias |
| Hepialidae | Aenetus virescens |
| Lycaenidae | Lycaena rauparaha |
| | Lycaena salustius |
| Micropterigidae | Zealandopterix zonodoxa |
| | Sabatinea chalcophanes |
| Momphidae | Zapyrastra callipharia |
| Noctuidae | Mythimna separata |
| | Proteuxoa tetronycha |
| Nympalidae | Danaus chrysippus |
| | Dodonidia helmsii (forest ringlet butterfly/ Helm's butterfly) |
| | Hypolimnas bolina nerina (blue moon butterfly) |
| | Vanessa kershawi (Australian painted lady) |
| Oecophoridae | Izatha epiphanes |
| | Izatha mesoschista |
| | Izatha peroneanella |
| | Tingera armigerella |
| Pieridae | Pieris rapae |
| Pyralidae | Patagonoides farinaria |
| Tortrictidae | Argyroploce chlorosaris |
| | Argyroploce lautana |
| | Capua semiferana |
| | Ctenopseustis fraterna |
| | Epalxiphora axenana |
| | Gryptospasma querula |
| | Pyrgotis calligypsa |

| ORDER/Family | Species |
|---|---|
| | Strepsicrates ejectana |
| **COLEOPTERA (BEETLES)** | |
| Aderidae | Xylophilus brouni |
| | Xylophilus coloratus |
| Anthicidae | Sapintus pellucidipes |
| Athribidae | Arecopais spectabilis |
| | Cacephalus huttoni |
| | Dysnocryptus inflatus |
| | Eugonissus conulus |
| | Hoplorhaphus spinifer |
| | Liromus pardalis |
| | Notochoragus crassus |
| | Pleosporius bullatus |
| Brentidae | Lasiorhynchus barbicornis (giraffe weevil) |
| | Neocyba metrosideros |
| Cantharidae | Asilis reflexa |
| Carabidae (ground beetles) | Anchomenus sulcitarsus |
| | Anomotarsus variegatus |
| | Aulacopodus calathoides |
| | Clivini vagans |
| | Ctenognathus bidens |
| | Ctenognathus cardiophorus |
| | Ctenognathus lucifugus |
| | Ctenognathus novaezelandiae |
| | Demetrida nasuta |
| | Dichrochile maura |
| | Duvaliomimus watti |
| | Gaioxenus pilipalpis |
| | Holcaspis hispida |
| | Lecanomerus atriceps |
| | Lecanomerus sharpi |
| | Mecodema hauturu |
| | Molopsida sp. |
| | Neocicendela tuberculata |
| | Notagonum lawsoni |
| | Pelodiaetodes moorei |

| ORDER/Family | Species |
|---|---|
| Cerambycidae | Rhytisternus miser |
| | Scopodes multipunctatus |
| | Syllectus anomalus |
| | Ambeodontus tristis |
| | Coptomma lineatum |
| | Hybolasius lanipes |
| | Ochrocydus huttoni |
| | Oemona hirta |
| | Prionoplus reticularis (huhu beetle) |
| | Zorion sp. |
| Chrysomelidae | Alema paradoxa |
| | Eucolaspis brunnea |
| | Pleuraltica cyanea |
| | Trachytetra rugulosa |
| Clambidae | Clambus simsoni |
| Cleridae | Phymatophaea fuscitarsis |
| Coccinellidae | Stethorus bifidus |
| Corylophidae | Holopsis sp. |
| Curculionidae | Aesiotes notabilis |
| | Ancistropterus quadrispinosus |
| | Anagotus fairburnii (flax weevil) |
| | Arecocryptus bellus |
| | Didymus erroneus |
| | Eurynotia hochstetteri |
| | Exomesites optimus |
| | Mitrastethus baridioides |
| | Nyxetes bidens |
| | Pactola variabilis |
| | Paedaretus hispidus |
| | Phorostichus linearis |
| | Psepholax sp. |
| | Scolopterus penicillatus |
| | Stenotrupis debilis |
| | Stenotrupis wollastonianus |
| | Stephanorhynchus lawsoni |

| ORDER/Family | Species |
|---|---|
| | Synacalles sp. |
| | Tysius bicornis |
| Dasytidae | Halyles sp. |
| Elateridae | Metablax acutipennis |
| | Metablax cinctiger |
| | Panspoeus guttatus |
| | Thoramus sp. |
| | Sp.1 |
| | Sp.2 |
| Endomychidae | Holoparamecus sp. |
| Erotylidae | Hapalips prolixus |
| | Kushelengis politus |
| | Loberus nitens |
| Eucnemidae | Dromarolus nigellus |
| Hydraenidae | Podaena hauturu |
| | Podaena latipalpis |
| Hydrophilidae | Cyloma lawsonus |
| | Exydrus gibbosus |
| Latridiidae | Aridius nodifer |
| Leiodidae | Paracatops sp. |
| Lucanidae | Holloceratognathus cylindricus |
| | Mitophyllus arcuatus |
| | Paralissotes planus |
| Melyridae | Sp.1 |
| Mordellidae | Hoshihananomia antarctica |
| Nitidulidae | Hisparonia hystrix |
| Oedemeridae | Thelyphassa lineata |
| Scarabaeidae | Eucolaspis brunnea |
| | Saphobius sp. |
| | Sericospilus glabratus |
| | Stethaspis longicornis |
| Scirtidae | Contacyphon decussatus |
| | Sp.1 |
| Staphylinidae (rove beetles) | Agnosthaetus sp. |
| | Baeocera tekootii |
| | Brachynopus latus |
| | Brachynopus scutellaris |

| ORDER/Family | Species |
|---|---|
| | *Edalus pleuralis* |
| | *Gyrophaena* sp. |
| | *Nototrochus ferrugineus* |
| | *Omaliomimus litoreus* |
| | *Sepedophilus castaneui* |
| **Tenebrionidae** | *Chrysopeplus expolitus* |
| | *Mimopeus elongatus* |
| | *Mitua tuberculicostata* |
| | *Uloma tenebrionoides* |
| | *Xenocera* sp. |
| **Ulodidae** | *Exohadrus volutithorax* |
| **Zopheridae** | *Brouniphylax* sp. |
| | *Epistranus lawsoni* |
| | *Pristoderus bakewelli* |
| | *Pycnomerus* sp. |

# Reptiles

| Scientific name | Māori/common name | New Zealand Threat Classification* |
|---|---|---|
| **TUATARA** | | |
| Sphenodon punctatus | tuatara | Relict |
| **GECKOS** | | |
| Hoplodactylus duvaucelii | Duvaucel's gecko | Relict |
| Dactylocnemis pacificus | Pacific gecko | Relict |
| Mokopirirakau granulatus | forest gecko | Declining |
| Naultinus elegans | elegant gecko (previously Auckland green gecko) | Declining |
| Woodworthia maculata | Raukawa gecko (previously common gecko) | Not threatened |
| **SKINKS** | | |
| Oligosoma smithi | shore skink | Not threatened |
| Oligosoma aeneum | copper skink | Not threatened |
| Oligosoma moco | moko skink | Relict |
| Oligosoma ornatum | ornate skink | Declining |
| Oligosoma homalonotum | chervon skink | Nationally vulnerable |
| Oligosoma striatum | striped skink | Declining |
| Oligosoma townsi | Hauraki skink (previously Towns' or marbled skink) | Recovering |
| Oligosoma suteri | egg-laying skink | Relict |
| **Species in nearby locations possibly once present on Hauturu** | | |
| Toropuku sp. 'Coromandel' | northern striped gecko (previously known as the Coromandel striped skink) | Taxonomically indeterminate; nationally vulnerable |
| Oligosoma alani | robust skink | Recovering |
| Oligosoma macgregori | McGregor's skink | Recovering |

*from www.doc.govt.nz/our-work/reptiles-and-frogs-distribution/atlas/

# Birds

| Scientific name | Māori/common name | Conservation status |
|---|---|---|
| Apteryx mantelli | North Island brown kiwi | Declining |
| Coturnix ypsilophora | Australian brown quail | Introduced and naturalised |
| Tadorna variegata | putangitangi/paradise shelduck | Not threatened |
| Anas chlorotis | pāteke/brown teal | Recovering |
| Eudyptula minor | kororā/little penguin | Declining |
| Pterodroma gouldi | ōi/grey-faced petrel | Not threatened |
| Pterodroma cookii | tītī/Cook's petrel | Relict |
| Pachyptila turtur | tītīwainui/fairy prion | Relict |
| Procellaria parkinsoni | tāiko/black petrel | Nationally vulnerable |
| Puffinus bulleri | Buller's shearwater | Naturally uncommon |
| Puffinus griseus | tītī/sooty shearwater | Declining |
| Puffinus gavia | pakahā/fluttering shearwater | Relict |
| Puffinus assimilis | little shearwater | Recovering |
| Pelagodroma marina | takahikare-moana/white-faced storm petrel | Relict |
| Fregetta maoriana | New Zealand storm petrel | Nationally vulnerable |
| Pelecanoides urinatrix | kuaka/common diving petrel | Relict |
| Morus serrator | tākapu/Australasian gannet | Not threatened |
| Phalacrocorax melanoleucos | kawaupaka/little shag | Not threatened |
| Phalacrocorax carbo | kawau/black shag | Naturally uncommon |
| Phalacrocorax varius | kāruhiruhi/pied shag | Recovering |
| Ardea intermedia | plumed egret | Vagrant |
| Egretta sacra | matuku-moana/reef heron | Nationally endangered |
| Ardea ibis | cattle egret | Migrant |
| Egretta novaehollandiae | white-faced heron | Not threatened |
| Circus approximans | kāhu/swamp harrier | Not threatened |
| Gallirallus philippensis | moho-pererū/banded rail | Declining |
| Porzana pusilla | koitareke/marsh crake | Declining |
| Porzana tabuensis | pūweto/spotless crake | Declining |
| Porphyrio melanotus | pūkeko | Not threatened |
| Numenius phaeopus | Asiatic whimbrel | Migrant |
| Vanellus miles | spur-winged plover | Not threatened |
| Pluvialis fulva | Pacific golden plover | Migrant |
| Limosa lapponica | kuaka/bar-tailed godwit | Declining |
| Himantopus himantopus | poaka/pied stilt | Not threatened |
| Larus dominicanus | karoro/southern black-backed gull | Not threatened |
| Larus novaehollandiae | tarāpunga/red-billed gull | Declining |

| Scientific name | Māori/common name | Conservation status |
| --- | --- | --- |
| *Hydroprogne caspia* | taranui/Caspian tern | Nationally vulnerable |
| *Sterna striata* | tara/white-fronted tern | Declining |
| *Hemiphaga novaeseelandiae* | kererū/New Zealand pigeon | Not threatened |
| *Columba livia* | rock pigeon | Introduced and naturalised |
| *Strigops habroptilis* | kākāpō | Nationally critical |
| *Platycercus eximius* | eastern rosella | Introduced and naturalised |
| *Nestor meridionalis* | North Island kākā | Recovering |
| *Cyanoramphus novaezelandiae* | kākāriki/red-crowned parakeet | Relict |
| *Cyanoramphus auriceps* | kākāriki/yellow-crowned parakeet | Not threatened |
| *Cuculus optatus* | Oriental cuckoo | Vagrant |
| *Chrysococcyx lucidus* | pīpīwharauroa/shining cuckoo | Not threatened |
| *Eudynamys taitensis* | koekoeā/long-tailed cuckoo | Naturally uncommon |
| *Ninox novaeseelandiae* | ruru/morepork | Not threatened |
| *Tyto alba* | barn owl | Coloniser |
| *Dacelo novaeguineae* | laughing kookaburra | Introduced and naturalised |
| *Todiramphus sanctus* | kōtare/New Zealand kingfisher | Not threatened |
| *Acanthisitta chloris* | titipounamu/North Island rifleman | Declining |
| *Callaeas wilsoni* | North Island kōkako | Recovering |
| *Philesturnus rufusater* | tīeke/North Island saddleback | Recovering |
| *Notiomystis cincta* | hihi/stitchbird | Nationally vulnerable |
| *Gerygone igata* | riroriro/grey warbler | Not threatened |
| *Anthornis melanura* | korimako/bellbird | Not threatened |
| *Prosthemadera novaeseelandiae* | tūī | Not threatened |
| *Mohoua albicilla* | pōpokotea/whitehead | Declining |
| *Rhipidura fuliginosa* | pīwakawaka/North Island fantail | Not threatened |
| *Corvus frugilegus* | rook | Introduced and naturalised |
| *Petroica macrocephala* | miromiro/North Island tomtit | Not threatened |
| *Petroica longipes* | toutouwai/North Island robin | Declining |
| *Alauda arvensis* | Eurasian skylark | Introduced and naturalised |
| *Zosterops lateralis* | tauhou/silvereye | Not threatened |
| *Hirundo neoxena* | welcome swallow | Not threatened |
| *Turdus merula* | European blackbird | Introduced and naturalised |
| *Turdus philomelos* | song thrush | Introduced and naturalised |
| *Sturnus vulgaris* | common starling | Introduced and naturalised |
| *Acridotheres tristis* | common myna | Introduced and naturalised |
| *Anthus novaeseelandiae* | New Zealand pipit | Declining |
| *Passer domesticus* | house sparrow | Introduced and naturalised |
| *Prunella modularis* | dunnock | Introduced and naturalised |

| Scientific name | Māori/common name | Conservation status |
|---|---|---|
| *Fringilla coelebs* | chaffinch | Introduced and naturalised |
| *Carduelis chloris* | European greenfinch | Introduced and naturalised |
| *Carduelis carduelis* | European goldfinch | Introduced and naturalised |
| *Carduelis flammea* | common redpoll | Introduced and naturalised |
| *Emberiza citrinella* | yellowhammer | Introduced and naturalised |
| *Emberiza cirlus* | cirl bunting | Introduced and naturalised |

## Bats

| Scientific name | Māori/common name | Conservation status |
|---|---|---|
| *Mystacina tuberculata aupourica* | pekapeka/northern lesser short-tailed bat | Nationally vulnerable |
| *Chalinolobus tuberculatus* | pekapeka/long-tailed bat | Nationally critical |

# Vegetation and vascular flora

## Key

\* = naturalised
\*\* = exotic plants sparingly naturalised from rangers' garden
E = presumed eradicated
H = historic record (pre-1978)
M = being actively managed to eradicate (see Chapter 4)
[M] = present but no longer being actively managed (see Chapter 4)

## Vegetation zones

1 = Pasture-fernland/sedgeland
2 = Kānuka/kānuka–mānuka forest
3 = Coastal forest and scrub
4 = Kauri and kauri–hard beech forest
5 = Northern rātā/pūriri–tawaroa forest
6 = Tōwai–tawaroa forest
7 = Quintinia–tāwari–southern rātā forest and scrub

| Taxon name, family and plant group | Māori/common name | Vegetation zone | | | | | | |
|---|---|---|---|---|---|---|---|---|
| **LYCOPHYTES (5 indigenous + 1 naturalised)** | | | | | | | | |
| **Lycopodiaceae** | | | | | | | | |
| *Lycopodiella cernua* | | | | | | 5 | | |
| *Lycopodium deuterodensum* | | | 2 | | 4 | | | |
| *Lycopodium scariosum* | | | | | | | 6 | 7 |
| *Lycopodium volubile* | waewaekoukou | | 2 | | 4 | | | 7 |
| *Phlegmariurus varius (Huperzia varia)* | iwituna/tassel fern | | 2 | | 4 | 5 | 6 | 7 |
| **Selaginellaceae** | | | | | | | | |
| \*\**Selaginella kraussiana* M | selaginella | 1 | | | | | | |
| **FERNS (109 indigenous + 0 naturalised)** | | | | | | | | |
| **Aspleniaceae** | | | | | | | | |
| *Asplenium bulbiferum* | hen and chickens fern | | | | | 5 | 6 | 7 |
| *Asplenium bulbiferum × A. flaccidum* | | | | | | | 6 | 7 |
| *Asplenium decurrens (A. northlandicum)* | | | | 3 | | | | |
| *Asplenium flaccidum* | hanging spleenwort | | 2 | | | 5 | 6 | 7 |
| *Asplenium haurakiense* | coastal spleenwort | 1 | | 3 | | | | |
| *Asplenium haurakiense × A. oblongifolium* | | 1 | | 3 | | | | |
| *Asplenium hookerianum* | | | | | | 5 | | |
| *Asplenium lamprophyllum* | | | 2 | 3 | 4 | 5 | 6 | |
| *Asplenium ?lamprophyllum × A. oblongifolium* | | | | 3 | | | | |

| Taxon name, family and plant group | Māori/common name | 1 | 2 | 3 | 4 | 5 | 6 | 7 |
|---|---|---|---|---|---|---|---|---|
| Asplenium oblongifolium | shining spleenwort | 1 | 2 | 3 |  | 5 | 6 |  |
| Asplenium polyodon | sickle spleenwort | 1 | 2 | 3 | 4 | 5 | 6 | 7 |
| **Blechnaceae** | | | | | | | | |
| Blechnum blechnoides |  |  |  | 3 |  |  |  |  |
| Blechnum chambersii |  |  | 2 | 3 | 4 | 5 | 6 |  |
| Blechnum colensoi |  |  |  | 3 |  |  | 6 |  |
| Blechnum discolor | piupiu/crown fern |  | 2 |  | 4 | 5 | 6 |  |
| Blechnum filiforme | thread fern |  | 2 |  |  | 5 |  |  |
| Blechnum fluviatile |  |  |  |  |  | 5 | 6 | 7 |
| Blechnum fraseri |  |  |  |  |  | 5 |  | 7 |
| Blechnum membranaceum |  |  | 2 |  |  |  |  |  |
| Blechnum nigrum |  |  |  |  |  |  | 6 |  |
| Blechnum norfolkianum |  |  | 2 | 3 |  | 5 | 6 |  |
| Blechnum novae-zelandiae | kiokio |  |  | 3 | 4 | 5 | 6 | 7 |
| Blechnum procerum |  |  |  |  |  |  |  | 7 |
| Blechnum triangularifolium | Green Bay kiokio |  |  | 3 |  |  |  |  |
| Blechnum vulcanicum H |  |  |  |  |  |  |  | 7 |
| Doodia australis | rasp fern | 1 | 2 | 3 |  |  |  |  |
| **Cyathaceae** | | | | | | | | |
| Cyathea cunninghamii H | gully tree fern |  |  |  |  | 5 |  |  |
| Cyathea dealbata | ponga/silver fern |  | 2 |  | 4 | 5 |  |  |
| Cyathea medullaris | mamaku/black tree fern |  | 2 |  |  | 5 | 6 |  |
| Cyathea smithii | soft tree fern |  |  |  |  |  | 6 | 7 |
| **Dennstaedtiaceae** | | | | | | | | |
| Histiopteris incisa | mātātā/water fern |  | 2 |  | 4 | 5 | 6 | 7 |
| Hypolepis ambigua |  | 1 |  | 3 |  |  |  |  |
| Hypolepis dicksonioides |  |  |  | 3 |  |  |  |  |
| Hypolepis distans |  |  |  |  |  | 5 |  |  |
| Hypolepis lactea |  |  |  |  |  | 5 |  |  |
| Hypolepis rufobarbata |  |  |  |  |  |  |  | 7 |
| Leptolepia novae-zelandiae H |  |  |  |  |  |  |  |  |
| Paesia scaberula | ring fern |  | 2 | 3 |  |  |  |  |
| Pteridium esculentum | rārahu/bracken | 1 |  |  |  |  |  | 7 |
| **Dicksoniaceae** | | | | | | | | |
| Dicksonia squarrosa | wheki |  |  |  |  | 5 | 6 | 7 |
| **Dryopteridaceae** | | | | | | | | |
| Lastreopsis glabella |  |  | 2 |  |  |  |  |  |
| Lastreopsis hispida | hairy legs |  |  |  |  | 5 | 6 |  |

| Taxon name, family and plant group | Māori/common name | Vegetation zone | | | | | |
|---|---|---|---|---|---|---|---|
| *Lastreopsis microsora* ssp. *pentangularis* | | 2 | 3 | | | | |
| *Lastreopsis velutina* | velvet fern | 2 | | | | | |
| *Polystichum wawranum* | shield fern | | 3 | | | | |
| *Rumohra adiantiformis* | climbing shield fern | | | | | 6 | 7 |
| **Gleicheniaceae** | | | | | | | |
| *Gleichenia dicarpa* | tangle fern/umbrella fern | 2 | | | | | |
| *Gleichenia microphylla* H | waewae kākā | 2 | | | | | |
| *Sticherus cunninghamii* | tapuwae kōtuku/umbrella fern | | | 4 | | | 7 |
| *Sticherus flabellatus* | | 2 | | | 5 | 6 | |
| **Hymenophyllaceae** | | | | | | | |
| *Hymenophyllum nephrophyllum* | raurenga/kidney fern | 2 | | 4 | 5 | 6 | 7 |
| *Hymenophyllum armstrongii* | | | | | | | 7 |
| *Hymenophyllum demissum* | irirangi/filmy fern | 2 | | 4 | 5 | 6 | 7 |
| *Hymenophyllum dilatatum* | | | | 4 | 5 | 6 | 7 |
| *Hymenophyllum flabellatum* | | | | 4 | | 6 | |
| *Hymenophyllum flexuosum* | | | | | 5 | 6 | 7 |
| *Hymenophyllum frankliniae* | | | | | 5 | 6 | 7 |
| *Hymenophyllum lyallii* | | | | | | 6 | 7 |
| *Hymenophyllum multifidum* | | 2 | | 4 | | 6 | 7 |
| *Hymenophyllum rarum* | | | | | | 6 | 7 |
| *Hymenophyllum revolutum* | | | | 4 | 5 | 6 | 7 |
| *Hymenophyllum sanguinolentum* | piripiri | 2 | | 4 | 5 | 6 | 7 |
| *Hymenophyllum scabrum* | | | | | | 6 | 7 |
| *Trichomanes elongatum* | bristle fern | | | | 5 | 6 | |
| *Trichomanes endlicherianum* | | | | | 5 | 6 | |
| *Trichomanes strictum* | | | | | | 6 | 7 |
| *Trichomanes venosum* H | | | | | 5 | 6 | |
| **Lindsaeaceae** | | | | | | | |
| *Lindsaea linearis* | | 2 | | | | | |
| *Lindsaea trichomanoides* | | 2 | | 4 | 5 | 6 | 7 |
| *Lindsaea viridis* | | | | | 5 | | |
| **Loxsomataceae** | | | | | | | |
| *Loxsoma cunninghamii* H | | 2 | | | | | |
| **Lygodiaceae** | | | | | | | |
| *Lygodium articulatum* | mangemange/bushmen's mattress | 2 | | 4 | 5 | 6 | 7 |
| **Marattiaceae** | | | | | | | |
| *Ptisana salicina* | para/king fern | | | | 5 | | |

| Taxon name, family and plant group | Māori/common name | Vegetation zone | | | | | | |
|---|---|---|---|---|---|---|---|---|
| **Ophioglossaceae** | | | | | | | | |
| *Botrychium australe* H | parsley fern | | 2? | | | | | |
| *Ophioglossum coriaceum* | adder's tongue fern | | | 3 | | | | |
| **Osmundaceae** | | | | | | | | |
| *Leptopteris hymenophylloides* | heruheru | | | | | 5 | 6 | 7 |
| *Leptopteris superba* | heruheru/Prince of Wales feathers | | | | | | | |
| **Polypodiaceae** | | | | | | | | |
| *Loxogramme dictyopteris* | lance fern | | | | 4 | 5 | 6 | |
| *Microsorum pustulatum* | hound's tongue fern | | 2 | 3 | 4 | 5 | 6 | 7 |
| *Microsorum scandens* | mokimoki/fragrant fern | | | | | 5 | 6 | |
| *Notogrammitis billardierei (Grammitis billardierei)* | strap fern | | | | | | 6 | 7 |
| *Notogrammitis ciliata (Grammitis ciliata)* | | | 2 | | 4 | | 6 | |
| *Notogrammitis heterophylla (Ctenopteris heterophylla)* | | | | | | | 6 | 7 |
| *Notogrammitis pseudociliata (Grammitis pseudociliata)* | | | | | | | 6 | |
| *Notogrammitis rawlingsii (Grammitis rawlingsii)* | | | 2 | | 4 | | | |
| *Pyrrosia elaeagnifolia* | leather-leaf fern | 1 | 2 | 3 | 4 | 5 | 6 | |
| **Psilotaceae** | | | | | | | | |
| *Tmesipteris elongata* | chain fern | | 2 | | | | | 7 |
| *Tmesipteris lanceolata* | | | | | | 5 | | |
| *Tmesipteris sigmatifolia* | | | | | | 5 | 6 | |
| *Tmesipteris tannensis* | | | | | | 5 | 6 | 7 |
| **Pteridaceae** | | | | | | | | |
| *Adiantum aethiopicum* | true maidenhair | 1 | | 3 | | | | |
| *Adiantum cunninghamii* | common maidenhair | | 2 | | 4 | 5 | | |
| *Adiantum diaphanum* | small maidenhair | | 2 | 3 | | | | |
| *Adiantum fulvum* | | | 2 | | | | | |
| *Adiantum hispidulum* | rosy maidenhair | 1 | 2 | 3 | | | | |
| *Adiantum viridescens* | | | | | | 5 | | |
| *Cheilanthes distans* H | woolly cloak fern | | | 3 | | | | |
| *Cheilanthes sieberi* H | rock fern | | | 3 | | | | |
| *Pellaea rotundifolia* | tarawera/button fern | 1 | | | | | | |
| *Pteris comans* | | | | 3 | | | | |
| *Pteris comans × P. saxatilis* | | | | 3 | | | | |
| *Pteris macilenta* | | | | | | | 6 | |

| Taxon name, family and plant group | Māori/common name | Vegetation zone | | | | | |
|---|---|---|---|---|---|---|---|
| *Pteris tremula* | turawera | | 2 | 3 | | | |
| **Schizaeaceae** | | | | | | | |
| *Schizaea bifida* | bifurcated comb fern | | 2 | | 4 | | |
| *Schizaea dichotoma* | fan fern | | | | 4 | | |
| **Tectariaceae** | | | | | | | |
| *Arthropteris tenella* | jointed fern | | 2 | 3 | | | |
| **Thelypteridaceae** | | | | | | | |
| *Pneumatopteris pennigera* | pākau/gully fern | | | | | 5 | 6 |
| **Woodsiaceae** | | | | | | | |
| *Deparia petersenii* ssp. *congrua* | | 1 | 2 | 3 | | 5 | |
| *Diplazium australe* | | 1 | 2 | 3 | | 5 | |
| **GYMNOSPERMS (10 indigenous + 0 naturalised)** | | | | | | | |
| **Araucariaceae** | | | | | | | |
| *Agathis australis* | kauri | | 2 | | 4 | 5 | 7 |
| **Podocarpaceae** | | | | | | | |
| *Dacrycarpus dacrydioides* | kahikatea | | | | 4 | 5 | |
| *Dacrydium cupressinum* | rimu | | | | | 5? | |
| *Phyllocladus toatoa* | toatoa | | | | 4 | 5 | 7 |
| *Phyllocladus trichomanoides* | tānekaha | | 2 | 3 | | | |
| *Phyllocladus ?toatoa* × *P. trichomanoides* | | | | | | | |
| *Podocarpus laetus* (*P. cunninghamii*) | Hall's tōtara | | | | 4 | 5 | 6 | 7 |
| *Podocarpus totara* | tōtara | | | | 4 | | |
| *Prumnopitys ferruginea* | miro | | 2 | | 4 | 5 | 6 | 7 |
| *Prumnopitys taxifolia* | matai | | | | 4 | | |
| **ANGIOSPERMS – Dicotyledons (212 indigenous + 142 naturalised)** | | | | | | | |
| **Actinidiaceae** | | | | | | | |
| **Actinidia chinensis* var. *deliciosa* E | kiwifruit | | | | | | 7 |
| **Aizoaceae** | | | | | | | |
| *Disphyma australe* ssp. *australe* | horokaka/New Zealand ice plant | | | 3 | | | |
| *Tetragonia implexicoma* | native spinach | 1 | | 3 | | | |
| **Alseuosmiaceae** | | | | | | | |
| *Alseuosmia macrophylla* | toropapa | | 2 | | 4 | 5 | 6 | 7 |
| *Alseuosmia quercifolia* | oak-leaved toropapa | | 2 | | | | |
| **Amaranthaceae** | | | | | | | |
| **Amaranthus lividus* | purple amaranth | 1 | | | | | |
| **Atriplex prostrata* | orache | 1 | | | | | |
| **Chenopodiastrum murale* (*Chenopodium murale*) | nettle-leaved fathen | 1 | | | | | |

| Taxon name, family and plant group | Māori/common name | Vegetation zone | | | | | | |
|---|---|---|---|---|---|---|---|---|
| Chenopodium triandrum (Einadia triandra) | berry saltbush | 1 | | 3 | | | | |
| Chenopodium trigonon ssp. trigonon (Einadia trigonos) | | | | 3 | | | | |
| Salicornia quinqueflora (Sarcocornia quinqueflora) | glasswort | | | 3 | | | | |
| **Apiaceae** | | | | | | | | |
| Apium prostratum ssp. prostratum var. filiforme | shore celery | | | 3 | | | | |
| Centella uniflora | centella | 1 | | 3 | | | | |
| Daucus glochidiatus H | native carrot | | | | | | | |
| *Foeniculum vulgare E | fennel | 1 | | | | | | |
| Hydrocotyle moschata | hydrocotyle | 1 | | | | | | |
| *Oenanthe pimpinelloides | parsley dropwort | 1 | | | | | | |
| Scandia rosifolia | | | | | | | | |
| **Apocynaceae** | | | | | | | | |
| *Araujia hortorum M | moth plant | 1 | | | | | | |
| Parsonsia capsularis var. capsularis | New Zealand jasmine | | 2 | | | | | |
| **Araliaceae** | | | | | | | | |
| **Hedera helix E | ivy | 1 | 2 | | | | | |
| Pseudopanax arboreus | whauwhaupaku/five finger | | 2 | | 4 | 5 | 6 | 7 |
| Pseudopanax colensoi var. colensoi | mountain five finger | | | | | | 6 | 7 |
| Pseudopanax crassifolius | horoeka/lancewood | | 2 | | 4 | | | |
| Pseudopanax crassifolius × P. lessonii | | | 2 | 3 | | | | |
| Pseudopanax discolor | | | 2 | | 4 | 5 | 6 | 7 |
| Pseudopanax lessonii | houpara/coastal five finger | | 2 | 3 | | | | |
| Raukaua edgerleyi | raukawa | | | | | 5 | 6 | 7 |
| Raukaua simplex var. simplex | | | 2 | | 4 | | | |
| Schefflera digitata | patē | | | | | 5 | 6 | |
| **Argophyllaceae** | | | | | | | | |
| Corokia buddleioides | korokia | | 2 | | 4 | | | |
| **Asteraceae** | | | | | | | | |
| *Achillea millefolium E | yarrow | 1 | | | | | | |
| *Ageratina adenophora [M] | Mexican devil | 1 | | 3 | | 5 | | |
| *Ageratina riparia M | mist flower | 1 | 2 | 3 | | | | |
| Anaphalioides trinervis | everlasting daisy | | | | | 5 | | |
| *Bellis perennis | daisy | 1 | | | | | | |
| *Bidens pilosa H | cobbler's pegs | 1 | | | | | | |
| Brachyglottis kirkii s.l. | kohurangi/Kirk's daisy | | 2 | | 4 | | 6 | 7 |

| Taxon name, family and plant group | Māori/common name | Vegetation zone | | | | | | |
|---|---|---|---|---|---|---|---|---|
| Brachyglottis repanda | rangiora | | | | | 5 | | |
| *Carduus tenuiflorus | winged thistle | 1 | | | | | | |
| Centipeda minima ssp. minima | sneezeweed | 1 | | | | | | |
| *Cirsium arvense | Californian thistle | 1 | | | | | | |
| *Cirsium vulgare | Scotch thistle | 1 | | 3 | | | | |
| Cotula australis | soldier's buttons | 1 | | | | | | |
| Cotula coronopifolia H (reinstated) | bachelor's button | | | 3 | | | | |
| *Crepis capillaris | hawksbeard | 1 | | | | | | |
| *Erechtites valerianifolia M | Brazilian fireweed | 1 | | 3 | | | | |
| *Erigeron bonariensis (Conyza bonariensis) | wavy-leaved fleabane | 1 | | | | | | |
| *Erigeron karvinskianus [M] | Mexican daisy | | | | | 5 | | |
| *Erigeron sumatrensis (Conyza sumatrensis) | broad-leaved fleabane | 1 | | 3 | | | | |
| Euchiton delicatus | | | 2 | 3 | | | | |
| Euchiton involucratus H | creeping cudweed | | | 3 | | | | |
| Euchiton japonicus | creeping cudweed | 1 | | | | 5 | | |
| Euchiton sphaericus | Japanese cudweed | | | 3 | | | | |
| *Gamochaeta calviceps | silky cudweed | 1 | | | | | | |
| *Gamochaeta coarctata | purple cudweed | 1 | 2 | 3 | | | | |
| *Gamochaeta simplicicaulis | | 1 | | 3 | | | | |
| Helichrysum lanceolatum | niniao | | 2 | 3 | | | | |
| *Helminthotheca echioides | oxtongue | 1 | | | | | | |
| *Hypochaeris glabra | smooth catsear | 1 | | | | | | |
| *Hypochaeris radicata | catsear | 1 | | | | | | |
| *Jacobaea vulgaris | ragwort | | | 3 | | | | |
| Lagenophora pumila (Lagenifera pumila) | papatāniwhaniwha | | | 3 | | | | |
| *Lapsana communis | nipplewort | 1 | | | | | | |
| *Leontodon saxatilis (L. taraxacoides) | hawkbit | 1 | | | | | | |
| Leptinella tenella | | | | 3 | | | | |
| *Matricaria discoidea | rayless chamomile | 1 | | | | | | |
| Olearia albida | | | 2 | | | | | |
| Olearia furfuracea | akepiro | | 2 | 3 | | | | |
| Olearia rani | heketara | | 2 | | 4 | 5 | 6 | 7 |
| Ozothamnus leptophyllus H | tauhinu/cottonwood | | | 3 | | | | |
| Pseudognaphalium luteoalbum | Jersey cudweed | | | 3 | | | | |
| Senecio bipinnatisectus | Australian fireweed | 1 | | | | | | |
| *Senecio elegans M | purple groundsel | | | 3 | | | | |

| Taxon name, family and plant group | Māori/common name | Vegetation zone | | | | |
|---|---|---|---|---|---|---|
| *Senecio esleri* | Esler's fireweed | 1 | | | | |
| *Senecio glomeratus* | grey fireweed | 1 | | | | |
| *Senecio hispidulus* | fireweed | | | 3 | | |
| *Senecio lautus* | shore groundsel | 1 | | 3 | | |
| *Senecio minimus* | fireweed | | | 3 | 5 | |
| *Senecio quadridentatus* | cotton fireweed | | | 3 | | |
| *Senecio scaberulus* | fireweed | | | 3 | | |
| *\*Senecio sylvaticus* | wood groundsel | 1 | 2 | | | |
| *\*Sigesbeckia orientalis* H | punawaru | 1? | | | | |
| *\*Soliva sessilis* | Onehunga weed | 1 | | | | |
| *\*Sonchus asper* | prickly sow thistle | | | 3 | | |
| *Sonchus kirkii* | shore sow thistle | | | | 5 | |
| *\*Sonchus oleraceus* | sow thistle | 1 | | 3 | | |
| *\*Symphiotrichum subulatum (Aster subulatus)* | sea aster | | | 3 | | |
| *\*Taraxacum officinale* | dandelion | 1 | | | | |
| *\*Tragopogon porrifolius* H | salsify | 1 | | | | |
| **Atherospermataceae** | | | | | | |
| *Laurelia novae-zelandiae* | pukatea | | | | 5 | |
| **Balanophoraceae** | | | | | | |
| *Dactylanthus taylorii* | pua o Te Rēinga/flowers of Hades | | | | 5 | |
| **Bignoniaceae** | | | | | | |
| *\*\*Pandorea jasminoides* M | bower vine | 1 | 2 | | | |
| *\*\*Tecomaria capensis* E | Cape honeysuckle | 1 | | | | |
| **Boraginaceae** | | | | | | |
| *\*Myosotis arvensis* E | field forget-me-not | 1 | | | | |
| *\*\*Myosotis sylvatica* E | garden forget-me-not | 1 | | | | |
| **Brassicaceae** | | | | | | |
| *\*Barbarea verna* E | | 1 | | | | |
| *\*Brassica napus* | rape | | | 3 | | |
| *\*Brassica oleracea* H | wild cabbage | 1 | | | | |
| *\*Brassica oleracea var. italica* M | broccoli | | | 3 | | |
| *\*Brassica rapa var. oleifera (var. sylvestris)* | wild turnip | | | 3 | | |
| *\*Capsella bursa-pastoris* | shepherd's purse | 1 | | | | |
| *Cardamine chlorina* | New Zealand bitter cress | | | 3 | | |
| *\*Cardamine flexuosa* | wavy bitter cress | 1 | | | | |
| *\*Cardamine hirsuta* | bitter cress | 1 | | | | |

| Taxon name, family and plant group | Māori/common name | Vegetation zone | | | | | |
|---|---|---|---|---|---|---|---|
| Cardamine 'northern robusta' (C. debilis) | bitter cress | 1 | | 3 | | | |
| *Lepidium didymum | twin cress | 1 | | | | | |
| Lepidium oleraceum H | nau/Cook's scurvy grass | | | 3 | | | |
| Rorippa divaricata | matangaoa/New Zealand water cress | | | 3 | | | |
| *Sisymbrium officinale | hedge mustard | 1 | | | | | |
| **Campanulaceae** | | | | | | | |
| Lobelia anceps | shore lobelia | 1 | | | | | |
| Lobelia angulata | pānakenake | | | | | 5 | |
| Wahlenbergia vernicosa | coastal harebell | | | 3 | | | |
| Wahlenbergia violacea | violet harebell | | 2 | | | | |
| **Caprifoliaceae** | | | | | | | |
| *Lonicera japonica E | Japanese honeysuckle | 1 | | | | | |
| **Caryophyllaceae** | | | | | | | |
| *Cerastium fontanum ssp. vulgare | mouse-ear chickweed | 1 | | | | | |
| Cerastium glomeratum | annual mouse-ear chickweed | 1 | | | | | |
| *Polycarpon tetraphyllum | allseed | | | 3 | | | |
| *Sagina apetala | annual pearlwort | 1 | | | | | |
| *Sagina procumbens | pearlwort | 1 | | 3 | | | |
| *Silene gallica | catchfly | | | 3 | | | |
| Spergularia tasmanica | sea spurrey | | | 3 | | | |
| *Stellaria media ssp. media | chickweed | 1 | | 3 | | | |
| Stellaria parviflora | native chickweed | | | 3 | | | |
| **Chloranthaceae** | | | | | | | |
| Ascarina lucida | hutu | | | | | | 7 |
| **Convolvulaceae** | | | | | | | |
| Calystegia sepium ssp. roseata | bindweed | 1 | | 3 | | | |
| Calystegia sepium ssp. roseata × C. tuguriorum | | 1 | | 3 | | | |
| Calystegia soldanella | shore bindweed | 1 | | 3 | | | |
| Calystegia tuguriorum H | climbing convolvulus | 1 | | | | | |
| Dichondra brevifolia | | | | 3 | | | |
| Dichondra repens | Mercury Bay weed | 1 | 2 | 3 | | | |
| **Coriariaceae** | | | | | | | |
| Coriaria arborea | tutu | | | 3 | | | 7 |
| **Corynocarpaceae** | | | | | | | |
| Corynocarpus laevigatus | karaka | 1 | 2 | 3 | | 5 | |

| Taxon name, family and plant group | Māori/common name | Vegetation zone | | | | | | |
|---|---|---|---|---|---|---|---|---|
| **Crassulaceae** | | | | | | | | |
| **Crassula multicava* E | fairy crassula | 1 | | | | | | |
| *Crassula sieberiana* | | 1 | | 3 | | | | |
| **Sedum mexicanum* E | | 1 | | | | | | |
| **Cucurbitaceae** | | | | | | | | |
| *Sicyos mawhai* | māwhai/native cucumber | | | 3 | | | | |
| **Cunoniaceae** | | | | | | | | |
| *Weinmannia sylvicola* | tōwai/tawhero | | | | 4 | 5 | 6 | 7 |
| **Droseraceae** | | | | | | | | |
| *Drosera auriculata* | sundew | | 2 | | | | | |
| **Elaeocarpaceae** | | | | | | | | |
| *Aristotelia serrata* | makomako/wineberry | | | | | 5 | | |
| *Elaeocarpus dentatus* | hīnau | | | | 4 | 5 | | 7 |
| **Ericaceae** | | | | | | | | |
| *Archeria racemosa* | | | | | | | 6 | 7 |
| *Dracophyllum latifolium* | neinei/spiderwood | | | | 4 | | 6 | |
| *Dracophyllum sinclairii* | | | 2 | | 4 | | | |
| *Dracophyllum traversii* | neinei/spiderwood | | | | | | | 7 |
| *Gaultheria antipoda* | snowberry | | | | | | | 7 |
| *Leptecophylla juniperina* | prickly mingimingi | | 2 | | 4 | | | |
| *Leucopogon fasciculatus* | mingimingi | | 2 | | 4 | | | 7 |
| *Leucopogon fraseri* H | pātōtara | | | | | | | 7 |
| **Euphorbiaceae** | | | | | | | | |
| *Euphorbia glauca* | sand milkweed | 1 | | 3 | | | | |
| **Euphorbia peplus* | milkweed | 1 | | | | | | |
| ***Homalanthus populifolius* M | Queensland poplar | 1 | 2 | | | | | |
| **Fabaceae** | | | | | | | | |
| *Carmichaelia australis* | native broom | | 2 | 3 | | 5 | | |
| *Carmichaelia williamsii* | giant-flowered broom | | | 3 | | | | |
| ***Chamaecytisus palmensis* E | tree lucerne | 1 | | | | | | |
| **Lotus angustissimus* | slender birdsfoot trefoil | 1 | | | | | | |
| **Lotus pedunculatus* | lotus | 1 | | | | | | |
| **Lotus suaveolens* | hairy birdsfoot trefoil | 1 | | 3 | | | | |
| **Medicago arabica* | spotted bur medick | 1 | | | | | | |
| **Medicago lupulina* | black medick | 1 | | | | | | |
| **Medicago nigra* | bur medick | 1 | | | | | | |
| *Sophora chathamica* | kōwhai | | | 3 | | | | |
| **Trifolium dubium* | suckling clover | 1 | | | | | | |

| Taxon name, family and plant group | Māori/common name | Vegetation zone | | | | | | |
|---|---|---|---|---|---|---|---|---|
| *Trifolium glomeratum* | clustered clover | 1 | | | | | | |
| *Trifolium pratense* H | red clover | 1 | | | | | | |
| *Trifolium repens* | white clover | 1 | | | | | | |
| *Trifolium subterraneum* | subterranean clover | 1 | | | | | | |
| *Ulex europaeus* E | gorse | 1? | | | | | | |
| *Vicia hirsuta* | hairy vetch | 1 | | | | | | |
| *Vicia sativa* | narrow-leaved vetch | 1 | | | | | | |
| *Vicia tetrasperma* | four-seeded vetch | 1 | | | | | | |
| **Fumariaceae** | | | | | | | | |
| *Fumaria muralis* | scrambling fumitory | 1 | | | | | | |
| **Gentianaceae** | | | | | | | | |
| *Centaurium erythraea* | centaury | 1 | | 3 | | | | |
| **Geraniaceae** | | | | | | | | |
| *Geranium dissectum* | cut-leaved geranium | 1 | | | | | | |
| *Geranium gardneri* | turnip-rooted geranium | 1 | | | | | | |
| Geranium homeanum | | 1 | | | | | | |
| *Geranium molle* | dove's foot cranesbill | 1 | | | | | | |
| Pelargonium inodorum | kōpata | | 2 | 3 | | | | |
| **Gesneriaceae** | | | | | | | | |
| Rhabdothamnus solandri | taurepo | | | | | 5 | | |
| **Griseliniaceae** | | | | | | | | |
| Griselinia littoralis | pāpāuma/broadleaf | | | | | | | 7 |
| Griselinia lucida | puka | | | | 4 | 5 | | |
| **Haloragaceae** | | | | | | | | |
| Gonocarpus incanus | | | 2 | | | | | |
| Haloragis erecta | shrubby haloragis | 1 | | 3 | | | | |
| **Juglandaceae** | | | | | | | | |
| **Juglans ailantifolia* M | Japanese walnut | 1 | | | | | | |
| **Lamiaceae** | | | | | | | | |
| **Lamium galeobdolon* E | aluminium plant | 1 | | | | | | |
| *Lamium purpureum* | red dead-nettle | 1 | | | | | | |
| *Mentha ×piperata var. piperita* M | peppermint | 1 | | | | | | |
| *Mentha pulegium* H | penny royal | 1 | | | | | | |
| **Plectranthus ecklonii* E | | 1 | | | | | | |
| *Prunella vulgaris* | selfheal | 1 | | | | | | |
| *Stachys arvensis* | staggerweed | 1 | | | | | | |
| **Lauraceae** | | | | | | | | |
| Beilschmiedia tarairi | taraire | 1 | 2 | 3 | 4 | 5 | | |

| Taxon name, family and plant group | Māori/common name | Vegetation zone | | | | | | |
|---|---|---|---|---|---|---|---|---|
| *Beilschmiedia tawaroa* | tawaroa | | | | 4 | 5 | 6 | |
| **Laurus nobilis* E | bay laurel | 1 | | | | | | |
| *Litsea calicaris* | mangeao | | | | 4 | 5 | | |
| **Linaceae** | | | | | | | | |
| **Linum bienne* H | pale flax | 1 | | | | | | |
| *Linum monogynum* | rauhuia | | | 3 | | | | |
| **Loganiaceae** | | | | | | | | |
| *Geniostoma ligustrifolium* | hangehange | | 2 | 3 | | 5 | | |
| **Loranthaceae** | | | | | | | | |
| *Peraxilla tetrapetala* | mistletoe | | | | | | | 7 |
| **Lythraceae** | | | | | | | | |
| **Lythrum hyssopifolia* | loosestrife | 1 | | | | | | |
| **Malvaceae** | | | | | | | | |
| ***Abutilon darwinii* | Chinese lantern | 1 | | | | | | |
| *Entelea arborescens* | whau | | | | | 5 | | |
| *Hoheria populnea* | houhere/lacebark | | | 3 | | | | |
| **Malva neglecta* | dwarf mallow | 1 | | | | | | |
| **Malva parviflora* H | small-flowered mallow | 1 | | | | | | |
| ***Malvaviscus arboreus* E | Turk's cap | 1 | | | | | | |
| **Modiola caroliniana* | creeping mallow | 1 | | | | | | |
| *Plagianthus divaricatus* | salt-marsh ribbonwood | 1 | | | | | | |
| **Meliaceae** | | | | | | | | |
| *Dysoxylum spectabile* | kohekohe | | 2 | 3 | 4 | 5 | | |
| ***Melianthus major* E | Cape honey flower | 1 | | | | | | |
| **Monimiaceae** | | | | | | | | |
| *Hedycarya arborea* | porokaiwhiri/pigeonwood | 1 | 2 | 3 | | 5 | | |
| **Moraceae** | | | | | | | | |
| ***Ficus carica* | fig | 1 | | | | | | |
| **Myrtaceae** | | | | | | | | |
| *Kunzea amathicola (K. aff. ericoides [a])* E | rawiritoa, sand kānuka | | | 3 | | | | |
| *Kunzea robusta (K. aff. ericoides [b])* | kānuka | 1 | 2 | | 4 | | | |
| *Kunzea linearis × K. robusta* E | hybrid kānuka | | 2 | | | | | |
| *Kunzea robusta × K. sinclairii* E | hybrid kānuka | | 2 | | | | | |
| *Leptospermum scoparium* | mānuka | 1 | 2 | 3 | | | | 7 |
| *Lophomyrtus bullata* | ramarama | | | | 4 | 5 | | |
| *Metrosideros albiflora* | akatea/white rātā | | | | | | | 7 |
| *Metrosideros carminea* | carmine rātā | | | 3 | | | | |

| Taxon name, family and plant group | Māori/common name | Vegetation zone | | | | | | |
|---|---|---|---|---|---|---|---|---|
| *Metrosideros diffusa* | white rātā vine | | | | | 5 | | |
| *Metrosideros excelsa* | pōhutukawa | 1 | 2 | 3 | | | | |
| *Metrosideros excelsa* × *M. robusta* | | | | | 4 | | | |
| *Metrosideros excelsa* × *M. umbellata* | | | | | 4 | | | |
| *Metrosideros fulgens* | akakura | | 2 | | 4 | 5 | | 7 |
| *Metrosideros parkinsonii* | | | | | | | | 7 |
| *Metrosideros perforata* | akatea/small white rātā | | 2 | 3 | 4 | 5 | | 7 |
| *Metrosideros robusta* | northern tree rātā | | 2 | | 4 | 5 | 6 | |
| *Metrosideros umbellata* | southern rātā | | | | | | | 7 |
| **Nanodeaceae** | | | | | | | | |
| *Mida salicifolia* | maire taiki/mida | | 2 | | | | 6 | |
| **Nothofagaceae** | | | | | | | | |
| *Fuscospora solandri (Nothofagus solandri)* | black beech | | | | 4 | | | |
| *Fuscospora solandri* × *F. truncata* | | | | | 4 | | | |
| *Fuscospora truncata (Nothofagus truncata)* | tawhairaunui/hard beech | | 2 | | 4 | 5 | | |
| **Nyctaginaceae** | | | | | | | | |
| *Pisonia brunoniana* | parapara, bird-catching tree | | | 3 | | | | |
| **Oleaceae** | | | | | | | | |
| *Nestegis cunninghamii* | black maire | | | | | 5 | | |
| *Nestegis lanceolata* | white maire | | 2 | | 4 | 5 | | 7 |
| *Nestegis montana* | narrow-leaved maire | | 2 | | | 5 | | |
| **Onagraceae** | | | | | | | | |
| *Epilobium chionanthum* H | marsh willow herb | | | | | | | |
| *\*Epilobium ciliatum* H | tall willow herb | | | 3 | | | | |
| *Epilobium cinereum* H | | 1 | | | | | | |
| *Epilobium hirtigerum* H | hairy willow herb | | | | | | | |
| *Epilobium nummulariifolium* H | creeping willow herb | | | | | 5 | | |
| *Epilobium rotundifolium* | round-leaved willow herb | | | | | 5 | | |
| *Fuchsia excorticata* | kōtukutuku | | | | | 5 | | |
| **Orobanchaceae** | | | | | | | | |
| *\*Orobanche minor* H | broomrape | 1 | | | | | | |
| **Oxalidaceae** | | | | | | | | |
| *\*Oxalis debilis* | pink shamrock | 1 | | | | | | |
| *Oxalis exilis* | creeping oxalis | 1 | | | | | | |
| *\*Oxalis latifolia* | fish-tail oxalis | 1 | | | | | | |
| *\*Oxalis pes-caprae* | Bermuda buttercup | 1 | | | | | | |

| Taxon name, family and plant group | Māori/common name | Vegetation zone | | | | | | |
|---|---|---|---|---|---|---|---|---|
| Oxalis rubens | | | | 3 | | | | |
| **Paracryphiaceae** | | | | | | | | |
| Quintinia serrata | tāwheowheo/quintinia | | | | 4 | | 6 | 7 |
| **Passifloraceae** | | | | | | | | |
| Passiflora tetrandra | kōhia/native passion flower | | | 3 | | | | |
| **Phytolaccaceae** | | | | | | | | |
| *Phytolacca octandra | inkweed | 1 | | 3 | | | | |
| **Piperaceae** | | | | | | | | |
| Peperomia urvilleana | | 1 | | 3 | | | | |
| Piper excelsum ssp. excelsum (Macropiper excelsum) | kawakawa | 1 | 2 | 3 | | 5 | | |
| **Pittosporaceae** | | | | | | | | |
| Pittosporum cornifolium | tāwhiri karo | | 2 | 3 | 4 | 5 | | |
| Pittosporum crassifolium | karo | 1 | | 3 | | | | |
| Pittosporum eugenioides | tarata/lemonwood | | | | | 5 | | |
| Pittosporum kirkii | | | | | | | | 7 |
| Pittosporum tenuifolium | kōhūhū | | 2 | | 4 | | | |
| Pittosporum umbellatum | haekaro | | 2 | 3 | 4 | | | |
| **Plantaginaceae** | | | | | | | | |
| Callitriche muelleri | starwort | 1 | | | | 5 | | |
| *Callitriche stagnalis E | starwort | 1 | | | | | | |
| Hebe macrocarpa var. latisepala | koromiko | | 2 | 3 | 4 | | | 7 |
| Hebe pubescens ssp. sejuncta | koromiko | | 2 | 3 | | | | |
| Hebe ?pubescens ssp. sejuncta × H. stricta | koromiko | | 2 | | | | | |
| *Plantago australis | swamp plantain | 1 | | | | | | |
| *Plantago lanceolata | narrow-leaved plantain | 1 | | 3 | | | | |
| *Plantago major | broad-leaved plantain | 1 | | | | | | |
| Plantago raoulii | | | | 3 | | | | |
| *Veronica arvensis | field speedwell | 1 | | | | | | |
| *Veronica persica | scrambling speedwell | 1 | | | | | | |
| *Veronica serpyllifolia | turf speedwell | 1 | | | | | | |
| **Polygonaceae** | | | | | | | | |
| Muehlenbeckia complexa var. complexa | wire vine | 1 | | 3 | | | | |
| Muehlenbeckia complexa var. grandifolia | pōhuehue | | | | | 5? | | |
| Persicaria decipiens | swamp willow weed | 1 | | | | | | |
| *Persicaria maculosa H | willow weed | 1 | | | | | | |
| *Polygonum aviculare | wireweed | 1 | | | | | | |

| Taxon name, family and plant group | Māori/common name | Vegetation zone | | | | | | |
|---|---|---|---|---|---|---|---|---|
| *Rumex acetosella | sheep's sorrel | 1 | | | | | | |
| *Rumex brownii | hooked dock | 1 | | | | | | |
| *Rumex conglomeratus | clustered dock | 1 | | | | | | |
| *Rumex crispus H | curled dock | 1 | | | | | | |
| *Rumex obtusifolius | broad-leaved dock | 1 | | | | | | |
| *Rumex pulcher | fiddle dock | 1 | | | | | | |
| **Portulacaceae** | | | | | | | | |
| *Portulaca oleracea | wild portulaca | 1 | | | | | | |
| **Primulaceae** | | | | | | | | |
| *Lysimachia arvensis s.str. (Anagallis arvensis) | scarlet pimpernel | 1 | | 3 | | | | |
| Myrsine australis | māpou | | 2 | 3 | 4 | 5 | | 7 |
| Myrsine salicina | toro | | 2 | | 4 | 5 | 6 | |
| Samolus repens var. repens | sea primrose | 1 | | 3 | | | | |
| **Proteaceae** | | | | | | | | |
| *Hakea sericea M | prickly hakea | | 2 | | | | | |
| Knightia excelsa | rewarewa | | 2 | | 4 | 5 | 6 | 7 |
| Toronia toru | toru | | 2 | | 4 | | | |
| **Ranunculaceae** | | | | | | | | |
| Clematis cunninghamii | yellow clematis | | 2 | | | | | |
| Clematis cunninghamii × C. paniculata H | | | | | | | | |
| Clematis paniculata | clematis | | 2 | | | | | |
| Ranunculus amphitrichus H | waoriki | 1? | | | | | | |
| *Ranunculus bulbosus H | bulbous buttercup | 1 | | | | | | |
| *Ranunculus muricatus | spiny buttercup | 1 | | | | | | |
| *Ranunculus parviflorus | small-flowered buttercup | 1 | | | | | | |
| Ranunculus reflexus | mārūrū/hairy buttercup | | 2 | | | | | |
| *Ranunculus repens | creeping buttercup | 1 | | | | | | |
| *Ranunculus sardous | hairy buttercup | 1 | | | | | | |
| Ranunculus urvilleanus | | 1 | | | | | | |
| **Rhamnaceae** | | | | | | | | |
| Pomaderris amoena | tauhinu | | 2 | | | | | |
| Pomaderris kumeraho | kūmarahou | | 2 | | | | | |
| **Rosaceae** | | | | | | | | |
| Acaena anserinifolia H | piripiri/bidibid | 1? | 2? | | | | | |
| **Eriobotrya japonica M | loquat | 1 | | | | | | |
| **Prunus persica E | peach | 1 | | | | | | |
| Rubus australis | bush lawyer | | | | 4 | | | |

| Taxon name, family and plant group | Māori/common name | Vegetation zone | | | | | | |
|---|---|---|---|---|---|---|---|---|
| Rubus cissoides | bush lawyer | | 2 | | 4 | 5 | | |
| **Rubiaceae** | | | | | | | | |
| Coprosma arborea | māmāngi | | 2 | | 4 | 5 | | |
| Coprosma dodonaeifolia | | | | | | 5 | 6 | 7 |
| Coprosma autumnalis (C. grandifolia) | kanono | | 2 | | 4 | 5 | 6 | 7 |
| Coprosma lucida | shining karamū | | 2 | 3 | 4 | | | 7 |
| Coprosma macrocarpa ssp. minor | coastal karamū | 1 | 2 | 3 | | | | |
| Coprosma macrocarpa ssp. minor × C. propinqua | | | 2 | | | | | |
| Coprosma macrocarpa × C. robusta | | | 2 | | | | | |
| Coprosma neglecta | | | | 3? | | | | 7 |
| Coprosma repens | taupata | 1 | 2 | 3 | | 5 | | |
| Coprosma robusta | karamū | | 2 | 3 | | 5 | | 7 |
| Coprosma spathulata | | | 2 | | 4 | 5 | | |
| *Galium aparine | cleavers | 1 | | | | | | |
| *Galium divaricatum | slender bedstraw | 1 | | | | | | |
| Galium propinquum H | | | 2? | | | | | |
| Nertera depressa | | | | | | 5 | | |
| Nertera dichondrifolia | | | | | 4 | | 6 | 7 |
| Nertera villosa | hairy forest nertera | | | | | | | 7 |
| *Sherardia arvensis H | field madder | 1 | | | | | | |
| **Rutaceae** | | | | | | | | |
| Leionema nudum | mairehau | | 2 | | 4 | | | |
| Melicope ternata | wharangi | 1 | 2 | 3 | | | | |
| **Santalaceae** | | | | | | | | |
| Korthalsella salicornioides | dwarf mistletoe | 1 | 2 | | | | | |
| **Sapindaceae** | | | | | | | | |
| Alectryon excelsus ssp. excelsus | tītoki | 1 | | 3 | | | | |
| Dodonaea viscosa | akeake | | 2 | | | 5 | | |
| **Sapotaceae** | | | | | | | | |
| Planchonella costata | tawāpou | | | 3 | | | | |
| **Saxifragaceae** | | | | | | | | |
| **Saxifraga stolonifera E | creeping saxifrage | 1 | | | | | | |
| **Scrophulariaceae** | | | | | | | | |
| Myoporum laetum | ngaio | 1 | | 3 | | | | |
| **Solanaceae** | | | | | | | | |
| **Brugmansia candida E | angel's trumpet | 1 | | | | | | |
| *Physalis peruviana M | cape gooseberry | 1 | | | | | | |

| Taxon name, family and plant group | Māori/common name | Vegetation zone | | | | | |
|---|---|---|---|---|---|---|---|
| *Solanum americanum* (*S. nodiflorum*) | nightshade | 1 | 2 | 3 | | | |
| *Solanum aviculare* var. *aviculare* | poroporo | | | | | 5 | |
| *Solanum opacum* (*S. nigrum*) | nightshade | 1 | 2 | 3 | | | |
| **Strasburgeriaceae (Ixerbiaceae)** | | | | | | | |
| *Ixerba brexioides* | tāwari | | | | 4 | 5 | 6 | 7 |
| **Thymelaeaceae** | | | | | | | |
| *Pimelea acra* | | | | | | | | 7 |
| *Pimelea carnosa* | | | | 3 | | | |
| *Pimelea tomentosa* | | | 2 | | | | |
| *Pimelea urvilleana* s.l. | pinātoro | | | 3 | | | |
| **Tropaeolaceae** | | | | | | | |
| **Tropaeolum majus* E | garden nasturtium | 1 | | | | | |
| **Urticaceae** | | | | | | | |
| *Parietaria debilis* | native pellitory | | | 3 | | | |
| **Verbenaceae** | | | | | | | |
| *Vitex lucens* | pūriri | 1 | 2 | 3 | | 5 | |
| **Violaceae** | | | | | | | |
| *Melicytus lanceolatus* | narrow-leaved māhoe | | | | | | | 7 |
| *Melicytus micranthus* | small-leaved māhoe | | 2 | | | 5 | |
| *Melicytus novae-zelandiae* | coastal māhoe | 1 | | 3 | | | |
| *Melicytus ramiflorus* | māhoe | 1 | 2 | 3 | 4 | 5 | |
| ***Viola odorata* E | violet | 1 | | | | | |
| **Winteraceae** | | | | | | | |
| *Pseudowintera axillaris* | horopito | | | | | 5 | 6 | 7 |
| **ANGIOSPERMS – Monocotyledons (102 indigenous + 58 naturalised)** | | | | | | | |
| **Agapanthaceae** | | | | | | | |
| ***Agapanthus praecox* ssp. *orientalis* E | agapanthus | 1 | | | | | |
| **Alliaceae** | | | | | | | |
| ***Allium triquetrum* M | three-cornered garlic | 1 | | | | | |
| **Amaryllidaceae** | | | | | | | |
| *** Crinum* ×*powelli* E | Natal lily | 1 | | | | | |
| ***Leucojum aestivum* E | snowflake | 1 | | | | | |
| ***Narcissus pseudonarcissus* E | daffodil | 1 | | | | | |
| ***Narcissus tazetta* E | jonquil | 1 | | | | | |
| **Araceae** | | | | | | | |
| ***Alocasia brisbanensis* | elephant's ear | 1 | | | | | |
| **Arum italicum* M | Italian arum | 1 | | | | | |

| Taxon name, family and plant group | Māori/common name | Vegetation zone | | | | | | |
|---|---|---|---|---|---|---|---|---|
| **Colocasia esculenta* E | taro | 1 | | | | | | |
| **Monstera deliciosa* E | fruit salad plant | 1 | | | | | | |
| *Zantedeschia aethiopica* M | arum lily | 1 | | | | | | |
| **Arecaceae** | | | | | | | | |
| *Rhopalostylis sapida* | nīkau | 1 | 2 | 3 | 4 | 5 | | |
| **Trachycarpus fortunei* E | fan palm | | 2 | | | | | |
| **Asparagaceae** | | | | | | | | |
| *Arthropodium cirratum* | rengarenga lily | | | 3 | | | | |
| *Asparagus asparagoides* E | smilax | 1 | | 3 | | | | |
| *Asparagus scandens* M | climbing asparagus | 1 | 2 | | | | | |
| *Cordyline australis* H | tī/cabbage tree | 1? | | | | | | |
| *Cordyline banksii* | tī ngahere/bush cabbage tree | | | | | 5 | | |
| *Cordyline banksii × C. pumilio* | | | 2 | 3 | | | | |
| *Cordyline pumilio* | tī rauriki/dwarf cabbage tree | | 2 | | | | | |
| **Asteliaceae** | | | | | | | | |
| *Astelia banksii* | coastal astelia | 1 | 2 | 3 | 4 | 5 | | 7 |
| *Astelia fragrans* | kakaha/bush lily | | | | | | 6 | 7 |
| *Astelia hastata (Collospermum hastatum)* | kahakaha | | 2 | | 4 | 5 | | 7 |
| *Astelia microsperma (Collospermum microspermum)* | | | | | | | 6 | 7 |
| *Astelia solandri* | perching lily | | 2 | | 4 | 5 | 6 | 7 |
| *Astelia trinervia* | kauri grass | | | | 4 | | | 7 |
| **Cannaceae** | | | | | | | | |
| **Canna ×generalis* E | canna | 1 | | | | | | |
| **Cyperaceae** | | | | | | | | |
| *Carex banksiana (Uncinia banksii)* | fine hooked sedge | | 2 | | 4 | 5 | 6 | |
| *Carex dissita* | forest sedge | | 2 | | | | | |
| *Carex flagellifera* | trip-me-up | 1 | | 3 | | | | |
| *Carex forsteri* | | | 2 | 3 | | 5 | | |
| *Carex horizontalis (Uncinia rupestris)* | hooked sedge | | 2 | | | | | |
| *Carex inversa* | creeping sedge | 1 | | | | | | |
| *Carex lambertiana* | | | | | | 5? | | |
| *Carex pumila* H | sand sedge | | | 3 | | | | |
| *Carex solandri* | | | | 3 | | | | |
| *Carex spinirostris* | coastal sedge | | | 3 | | | | |
| *Carex testacea* | | 1 | 2 | 3 | | | | |
| *Carex uncinata (Uncinia uncinata)* | hooked sedge | | 2 | | 4 | | | |

| Taxon name, family and plant group | Māori/common name | Vegetation zone | | | | | | |
|---|---|---|---|---|---|---|---|---|
| Carex virgata | swamp sedge | 1 | | | | | | |
| *Cyperus albostriatus | | 1 | | | | | | |
| *Cyperus brevifolius | | 1 | | | | | | |
| *Cyperus eragrostis M | umbrella sedge | 1 | | | | | | |
| Cyperus ustulatus | giant umbrella sedge | 1 | | | | | | |
| Ficinia nodosa | knobby club-rush | 1 | | 3 | | | | |
| Gahnia lacera | cutty grass | | 2 | | | | | |
| Gahnia pauciflora | small cutty grass | | | | 4 | | | 7 |
| Gahnia setifolia | cutty grass | | 2 | | 4 | | | 7 |
| Gahnia xanthocarpa | cutty grass | | 2 | | | | | |
| Isolepis cernua | slender club-rush | | | 3 | | | | |
| Lepidosperma australe | square sedge | | 2 | | | | | |
| Lepidosperma laterale | sword sedge | | 2 | | | | | |
| Morelotia affinis | | | 2 | | | | | |
| Schoenus tendo | | | 2 | | | | | 7 |
| **Hyacinthaceae** | | | | | | | | |
| **Hyacinthoides non-scripta E | bluebell | 1 | | | | | | |
| **Iridaceae** | | | | | | | | |
| **Chasmanthe bicolor | | 1 | | | | | | |
| *Crocosmia ×crocosmiiflora M | montbretia | 1 | 2 | | | | | |
| **Iris foetidissima E | stinking iris | 1 | | | | | | |
| Libertia grandiflora | native iris | | 2 | | | | | |
| Libertia micrantha | | | | | 4 | 5 | 6 | 7 |
| **Watsonia meriana 'Bulbillifera' M | bulbil watsonia | 1 | | | | | | |
| **Juncaceae** | | | | | | | | |
| Juncus australis | leafless rush | 1 | | | | | | |
| *Juncus bufonius | toad rush | 1 | | | | | | |
| Juncus edgariae | leafless rush | 1 | | | | | | |
| *Juncus flavidus | | 1 | | | | | | |
| *Juncus subsecundus | | 1 | | | | | | |
| *Juncus tenuis var. tenuis | slender rush | 1 | | | | | | |
| Juncus usitatus | leafless rush | 1 | | | | | | |
| Luzula ?picta var. picta H | wood rush | 1 | | | | | | |
| **Orchidaceae** | | | | | | | | |
| Acianthus sinclairii | pixie cap orchid | 1 | 2 | | | | | |
| Bulbophyllum pygmaeum (Ichthyostomum pygmaeum) | pygmy orchid | | | | 4 | | | |

| Taxon name, family and plant group | Māori/common name | Vegetation zone | | | | | |
|---|---|---|---|---|---|---|---|
| Caladenia chlorostyla (Petalochilus chlorostylus) | lady's fingers | 2 | | | | | |
| Chiloglottis cornuta (Simpliglottis cornuta) | green bird orchid | | | | | 6 | |
| Corunastylis pumila | | 2 | | | | | |
| Corybas acuminatus (Nematoceras acuminatum) | spider orchid | | | | 5 | 6 | 7 |
| Corybas cheesemanii | helmet orchid | 2 | | | | | |
| Corybas oblongus (Singularybas oblongus) | spider orchid | | | 4 | 5 | | 7 |
| Corybas rivularis (Nematoceras rivulare) | spider orchid | | | | 5 | | |
| Corybas trilobus (Nematoceras trilobum) | spider orchid | 2 | | 4 | | | |
| Cyrtostylis oblonga | gnat orchid | | | 4 | | | |
| Danhatchia australis | | 2 | | | | | |
| Dendrobium cunninghamii (Winika cunninghamii) | winikā | 2 | | 4 | 5 | 6 | 7 |
| Drymoanthus adversus | | 2 | 3 | | | | |
| Earina aestivalis | | 2 | 3 | | | | |
| Earina autumnalis | autumn orchid | | | 4 | 5 | 6 | 7 |
| Earina mucronata | spring orchid | 2 | | | 5 | 6 | 7 |
| Gastrodia cunninghamii | black orchid | | | 4 | | 6 | |
| Gastrodia sesamoides | potato orchid | 2 | | | | | |
| Microtis parviflora | | 2 | | | | | |
| Microtis unifolia | onion orchid | 1 | 2 | | | | |
| Orthoceras novae-zeelandiae | horned orchid | 2 | | | | | |
| Pterostylis agathicola | kauri greenhood | | | 4 | | | |
| Pterostylis alobula | small greenhood orchid | 2 | | | | | |
| Pterostylis banksii | tutukiwi/greenhood orchid | 2 | | | 5 | | 7 |
| Pterostylis graminea | greenhood orchid | 2 | | | | | |
| Pterostylis trullifolia | small greenhood orchid | 2 | | | | | |
| Thelymitra cyanea | swamp sun orchid | 2 | | | | | |
| Thelymitra ixioides | spotted sun orchid | 2 | | | | | |
| Thelymitra aff. longifolia | scented sun orchid | 1 | | | | | |
| Thelymitra longifolia | sun orchid | 2 | 3 | | | | 7 |
| Thelymitra pauciflora | sun orchid | 2 | | | | | |
| **Pandanaceae** | | | | | | | |
| Freycinetia banksii | kiekie | 2 | 3 | 4 | 5 | 6 | 7 |
| **Poaceae** | | | | | | | |
| *Agrostis capillaris | browntop | 1 | | | | | |

| Taxon name, family and plant group | Māori/common name | Vegetation zone | | | | | |
|---|---|---|---|---|---|---|---|
| *Aira caryophyllea* H | silvery hair grass | 1 | | | | | |
| Anthosachne kingiana ssp. multiflora (A. multiflora) | blue wheat grass | | | 3 | | | |
| *Anthoxanthum odoratum | sweet vernal | 1 | | | | | |
| Austroderia splendens | toetoe | | | 3 | | | |
| *Avena barbata | slender oat | 1 | | 3 | | | |
| *Axonopus fissifolius | carpet grass | 1 | | | | | |
| *Briza minor H | shivery grass | 1 | | | | | |
| Bromus arenarius | | 1 | | 3 | | | |
| *Bromus catharticus (B. willdenowii) | prairie grass | 1 | | 3 | | | |
| *Bromus diandrus | ripgut brome | 1 | | | | | |
| *Bromus hordeaceus H | soft brome | 1 | | | | | |
| *Bromus lithobius | Chilean brome | 1 | | | | | |
| *Cenchrus clandestinus M | kikuyu | 1 | | | | | |
| *Cortaderia jubata M | purple pampas grass | | | 3 | | | |
| *Cortaderia selloana M | pampas grass | | | 3 | | | |
| *Cynodon dactylon | Indian doab | 1 | | | | | |
| *Dactylis glomerata | cocksfoot | 1 | | | | | |
| Dichelachne crinita | long-hair plume grass | | | 3 | | | |
| *Digitaria sanguinalis | summer grass | 1 | | | | | |
| Echinopogon ovatus | hedgehog grass | | 2 | | | 5 | |
| *Ehrharta erecta E | panic veldt grass | 1 | | | | | |
| *Festuca rubra ssp. commutata | Chewing's fescue | 1 | | | | | |
| *Holcus lanatus | Yorkshire fog | 1 | | | | | |
| Lachnagrostis billardierei | sand wind grass | | | 3 | | | |
| Lachnagrostis filiformis | New Zealand wind grass | 1 | | 3 | | | |
| Lachnagrostis littoralis ssp. littoralis | coastal wind grass | | | 3 | | | |
| *Lolium arundinaceum s.str. (Schedonorus arundinaceus) | tall fescue | 1 | | | | | |
| *Lolium perenne | perennial ryegrass | 1 | | | | | |
| Microlaena avenacea | bush rice grass | | | 3 | | 5 | |
| Microlaena polynoda | bamboo rice grass | | 2 | | | | |
| Microlaena stipoides | meadow rice grass | 1 | | | | | |
| Oplismenus hirtellus ssp. imbecillis | bush panic grass | 1 | 2 | | | | |
| *Paspalum dilatatum | paspalum | 1 | | 3 | | | |
| Paspalum orbiculare | scrobic | 1 | | | | | |
| *Phalaris minor | lesser canary grass | | | 3 | | | |
| Poa anceps | | 1 | | | | | |

| Taxon name, family and plant group | Māori/common name | Vegetation zone | | | | | | |
|---|---|---|---|---|---|---|---|---|
| *Poa annua | annual poa | 1 | | | | | | |
| Poa imbecilla H | thread grass | | | | | | | |
| *Poa pratensish | Kentucky bluegrass | 1 | | | | | | |
| *Poa trivialis | rough-stalked meadow grass | 1? | | | | | | |
| Rytidosperma biannulare | danthonia | | 2 | | | | | |
| Rytidosperma gracile | danthonia | | 2 | | | | | |
| *Rytidosperma racemosum | danthonia | 1 | | | | | | |
| Rytidosperma unarede | danthonia | 1 | 2 | 3 | | | | 7 |
| *Sporobolus africanus | ratstail | 1 | | | | | | |
| Trisetum arduanum | | | | 3 | | | | |
| *Vulpia bromoides | vulpia hair grass | | | 3 | | | | |
| *Vulpia myuros H | vulpia hair grass | 1 | | 3 | | | | |
| **Ripogonaceae** | | | | | | | | |
| Ripogonum scandens | kareao/supplejack | | 2 | | | 5 | 6 | 7 |
| **Thismiaceae** | | | | | | | | |
| Thismia rodwayi | thismia | | | | 4 | | | |
| **Typhaceae** | | | | | | | | |
| Typha orientalis H | raupō | 1 | | | | | | |
| **Xanthorrhoeaceae** | | | | | | | | |
| Dianella latissima | tūrutu/blueberry | | 2 | | | | | |
| Dianella nigra | tūrutu/blueberry | | 2 | | 4 | | 6 | 7 |
| Phormium cookianum ssp. hookeri | wharariki/mountain flax | | | | | | | 7 |
| Phormium tenax | harakeke/flax | 1 | | 3 | | | | |
| **Zingiberaceae** | | | | | | | | |
| **Hedychium gardnerianum E | wild ginger | 1 | | | | | | |
| **Zosteraceae** | | | | | | | | |
| Zostera muelleri ssp. novazelandica H | eel grass, sea grass | | | 3 | | | | |
| **UNACCEPTED RECORDS of Beever et al. 2012 (0 indigenous + 5 naturalised)** | | | | | | | | |
| **Aloe maculata (Asphodelaceae) | in garden not naturalised | | | | | | | |
| **Jasminum polyanthum (Oleaceae) | in garden not naturalised | | | | | | | |
| *Myosotis scorpioides (Boraginaceae) | 'garden weed' of Hamilton (1961), more likely based on M. sylvatica | | | | | | | |
| **Polygala myrtifolia (Polyalaceae) | in garden not naturalised (perhaps the sterile cultivar 'Grandiflora') | | | | | | | |
| *Solanum nigrum (Solanaceae) | redetermined to S. opacum | | | | | | | |

# Mosses

## Key

\* = endemic
\*\* = exotic
† = naturally uncommon (as classified by the Department of Conservation)

| Family | Species |
|---|---|
| **Amblystegiaceae** | Leptodictyum riparium |
| **Anomodontaceae** | Haplohymenium pseudotriste |
| **Bartramiaceae** | Breutelia pendula |
| | Philonotis tenuis |
| **Brachytheciaceae** | Brachythecium plumosum |
| | Eurhynchium praelongum\*\* |
| | Rhynchostegium muriculatum |
| | Rhynchostegium tenuifolium |
| **Bryaceae** | Bryum argenteum |
| | Bryum clavatum |
| | Bryum dichotomum |
| | Bryum duriusculum |
| | Bryum ?laevigatum |
| | Bryum sauteri |
| | Rosulabryum campylothecium |
| | Rosulabryum capillare |
| | Rosulabryum subtomentosum |
| **Catagoniaceae** | Catagonium nitens |
| **Cyrtopodaceae** | Cyrtopus setosus |
| **Daltoniaceae** | Achrophyllum dentatum |
| | Achrophyllum quadrifarium\* |
| | Calyptrochaeta apiculata |
| | Calyptrochaeta brownii |
| | Calyptrochaeta cristata\* |
| | Distichophyllum crispulum var. adnatum |

| Family | Species |
|---|---|
| | Distichophyllum crispulum var. crispulum\* |
| | Distichophyllum microcarpum |
| | Distichophyllum pulchellum |
| | Distichophyllum rotundifolium |
| **Dicnemonaceae** | Dicnemon calycinum |
| **Dicranaceae** | Campylopodium medium |
| | Campylopus clavatus |
| | Campylopus introflexus |
| | Campylopus pallidus |
| | Campylopus purpureocaulis |
| | Dicranella vaginata |
| | Dicranoloma billardierei |
| | Dicranoloma fasciatum\* |
| | Dicranoloma menziesii |
| | Dicranoloma robustum |
| | Holomitrium perichaetiale |
| | Holomitrium trichopodum |
| | Sclerodontium pallidum |
| **Ditrichaceae** | Ceratodon purpureus |
| | Ditrichum difficile |
| | Ditrichum punctulatum |
| | Pleuridium subulatum\*\* |
| | Wilsoniella blindioides |
| **Echinodiaceae** | Echinodium hispidum |
| | Echinodium umbrosum |
| **Fabroniaceae** | Ischyrodon lepturus † |

| Family | Species |
|---|---|
| **Fissidentaceae** | Fissidens asplenioides |
| | Fissidens blechnoides* |
| | Fissidens capitatus* |
| | Fissidens curvatus** |
| | Fissidens dealbatus |
| | Fissidens leptocladus |
| | Fissidens linearis var. angustifolius* |
| | Fissidens pallidus |
| | Fissidens rigidulus var. rigidulus |
| | Fissidens taxifolius** |
| | Fissidens tenellus var. tenellus |
| | aquatic form of F. tenellus var. tenellus |
| | Fissidens tenellus var. australiensis |
| **Funariaceae** | Entosthodon radians |
| | Funaria hygrometrica |
| | Physcomitrium pyriforme |
| **Grimmiaceae** | Racomitrium crispulum |
| | Racomitrium lanuginosum |
| **Hypnaceae** | Austrohondaella limata |
| | Ctenidium pubescens |
| | Hypnum chrysogaster |
| | Hypnum cupressiforme var. cupressiforme |
| | Hypnum cupressiforme var. filiforme |
| | Isopterygium albescens |
| | Pseudotaxiphyllum falcifolium* |
| **Hypnodendraceae** | Hypnodendron arcuatum* |
| | Hypnodendron marginatum* |
| | Hypnodendron spininervium* |
| | Mniodendron colensoi* |
| | Mniodendron comatum* |

| Family | Species |
|---|---|
| | Sciadocladus kerrii* |
| | Sciadocladus menziesii |
| **Hypopterygiaceae** | Canalohypopterygium tamariscinum* |
| | Catharomnion ciliatum* |
| | Cyathophorum bulbosum |
| | Dendrohypopterygium filiculiforme* |
| | Hypopterygium didictyon |
| | Hypopterygium tamarisci |
| | Lopidium concinnum |
| **Lembophyllaceae** | Camptochaete arbuscula var. arbuscula |
| | Camptochaete deflexa |
| | Camptochaete pulvinata |
| | Fallaciella gracilis |
| | Weymouthia cochlearifolia |
| | Weymouthia mollis |
| **Leptodontaceae** | Leptodon smithii |
| **Leptostomataceae** | Leptostomum inclinans |
| | Leptostomum macrocarpum |
| **Leucobryaceae** | Leucobryum javense |
| **Meteoriaceae** | Papillaria crocea |
| | Papillaria flavolimbata |
| **Mitteniaceae** | Mittenia plumula |
| **Neckeraceae** | Pendulothecium oblongifolium* |
| | Pendulothecium punctatum |
| | Thamnobryum pandum |
| **Orthodontiaceae** | Hymenodon pilifer |
| | Leptotheca gaudichaudii |
| **Orthorrhynchiaceae** | Orthorrhynchium elegans |
| **Orthotrichaceae** | Leratia obtusifolia |
| | Macromitrium brevicaule |
| | Macromitrium gracile* |

| Family | Species |
|---|---|
| | Macromitrium ligulaefolium |
| | Macromitrium ligulare |
| | Macromitrium longipes* |
| | Macromitrium prorepens* |
| | Zygodon intermedius |
| | Zygodon minutus |
| **Pylaisiadelphaceae** | Isopterygium minutirameum |
| **Polytrichaceae** | Dawsonia superba |
| | Oligotrichum tenuirostre* |
| | Pogonatum subulatum |
| | Polytrichadelphus magellanicus |
| | Polytrichum commune |
| | Polytrichum juniperinum |
| **Pottiaceae** | Barbula unguiculata** |
| | Chenia leptophylla** |
| | Didymodon australasiae |
| | Syntrichia laevipila |
| | Syntrichia papillosa |
| | Tortella cirrhata † |
| | Tortella flavovirens |
| | Tortella knightii |
| | Tortula muralis |
| | Tortula truncata |
| | Trichostomum sciophilum* |
| | Weissia controversa var. controversa |
| | Weissia controversa var. gymnostoma sensu Sainsbury |
| | Weissia austrocrispa |
| **Ptychomitriaceae** | Ptychomitrium australe |
| **Ptychomniaceae** | Cladomnion ericoides* |
| | Glyphothecium sciuroides |
| | Ptychomnion aciculare |

| Family | Species |
|---|---|
| **Racopilaceae** | Racopilum cuspidigerum var. convolutaceum |
| | Racopilum robustum* |
| | Racopilum strumiferum* |
| **Rhizogoniaceae** | Calomnion complanatum |
| | Cryptopodium bartramioides* |
| | Pyrrhobryum bifarium |
| | Rhizogonium distichum |
| | Rhizogonium novae-hollandiae |
| | Rhizogonium pennatum |
| **Sematophyllaceae** | Rhaphidorrhynchium amoenum |
| | Sematophyllum homomallum |
| | Sematophyllum jolliffii |
| | Sematophyllum subhumile var. contiguum |
| | Wijkia extenuata var. extenuata |
| **Splachnaceae** | Tayloria callophylla |
| **Thuidiaceae** | Thuidiopsis furfurosa |
| | Thuidiopsis sparsa |
| | Thuidium cymbifolium † |
| **Trachylomataceae** | Trachyloma diversinerve |
| | Trachyloma planifolium |

# Liverworts and hornworts

| Family | Species |
|---|---|
| **LIVERWORTS / MARCHANTIOPHYTA** | |
| **Acrobolbaceae** | Acrobolbus concinnus |
| | Acrobolbus epiphytus (Syn. Marsupidium epiphytum) |
| | Acrobolbus knightii (Syn. Marsupidium knightii) |
| | Acrobolbus lophocoleoides |
| | Acrobolbus papillosus (Syn. Marsupidium papillosum) |
| | Acrobolbus saccatus (Syn. Tylimanthus saccatus) |
| | Acrobolbus setulosus (Syn. Marsupidium setulosum) |
| | Acrobolbus tenellus (Syn. Tylimanthus tenellus) |
| | Saccogynidium australe |
| **Adelanthaceae** | Adelanthus falcatus |
| | Cuspidatula monodon |
| | Pseudomarsupidium piliferum |
| | Syzygiella colorata |
| | Syzygiella nigrescens |
| | Syzygiella tasmanica |
| **Anastrophyllaceae** | Chandonanthus squarrosus |
| **Aneuraceae** | Lobatiriccardia alterniloba |
| | Riccardia alba |
| | Riccardia bipinnatifida |
| | Riccardia colensoi |
| | Riccardia crassa |
| | Riccardia eriocaula |
| **Aytoniaceae** | Asterella tenera |
| **Balantiopsaceae** | Balantiopsis convexiuscula |
| | Balantiopsis diplophylla |

| Family | Species |
|---|---|
| | Balantiopsis diplophylla var. hockenii |
| | Isotachis lyallii |
| | Isotachis montana |
| **Cephaloziellaceae** | Anastrophyllopsis subcomplicata (Syn. Anastrophyllum schismoides var. schismoides) |
| **Dumortieriaaceae** | Dumortiera hirsuta |
| **Frullaniaceae** | Frullania aterrima |
| | Frullania chevalieri |
| | Frullania deplanata |
| | Frullania fugax |
| | Frullania hodgsoniae |
| | Frullania ptychantha |
| | Frullania rostellata |
| | Frullania rostrata |
| | Frullania solanderiana |
| | Frullania squarrosula |
| | Frullania toropuku |
| **Goebeliellaceae** | Goebeliella cornigera |
| **Hymenophytaceae** | Hymenophyton leptopodum |
| **Lejeuneaceae** | Acrolejeunea mollis |
| | Cheilolejeunea sp. (50 records at a range of altitudes. Cheilolejeunea species are poorly understood at this stage in New Zealand and often do not have accurate names. There are probably several species represented in the collections.) |
| | Cololejeunea floccosa |
| | Cololejeunea hodgsoniae |
| | Cololejeunea laevigata |
| | Cololejeunea pulchella |
| | Colura pulcherrima var. bartlettii |

| Family | Species |
|---|---|
| | *Cumulolejeunea ocellata* (Syn. *Rectolejeunea ocellata*) |
| | *Diplasiolejeunea plicatiloba* |
| | *Drepanolejeunea aucklandica* |
| | *Echinolejeunea papillata* |
| | *Lejeunea* species |
| | *Lejeunea epiphylla* |
| | *Lejeunea flava* |
| | *Lejeunea gracilipes* |
| | *Lejeunea helmsiana* |
| | *Lejeunea hodgsoniana* |
| | *Lejeunea oracola* |
| | *Lejeunea* sp. (Sometimes incorrectly called *L. subelobata*) |
| | *Lejeunea tumida* |
| | *Lopholejeunea* species |
| | *Metalejeunea cucullata* |
| | *Microlejeunea filicupsis* (Syn. *Harpalejeunea filicupsis*) |
| | *Microlejeunea latitans* (Syn. *Harpalejeunea latitans*) |
| | *Myriocoleopsis minutissima* (Syn. *Cololejeunea minutissima*) |
| | *Nephelolejeunea hamata* (Syn. *Austrolejeunea hamata*) |
| | *Nephelolejeunea* sp. aff. *nudipes* (was *Siphonolejeunea*) |
| | *Siphonolejeunea nudipes* |
| | *Spruceanthus olivaceus* (Syn. *Archilejeunea olivacea*) |
| | *Thysananthus anguiformis* (Syn. *Mastigolejeunea anguiformis*) |

| Family | Species |
|---|---|
| **Lepicoleaceae** | *Lepicolea attenuata* |
| **Lepidolaenaceae** | *Lepidolaena clavigera* |
| | *Lepidolaena palpebrifolia* |
| | *Lepidolaena taylorii* |
| **Lepidoziaceae** | *Acromastigum anisostomum* |
| | *Acromastigum colensoanum* |
| | *Acromastigum marginatum* (Only record is on ground in forest at mid-high altitude) |
| | *Bazzania adnexa* |
| | *Bazzania hochstetteri* |
| | *Bazzania involuta* var. *submutica* |
| | *Bazzania nitida* |
| | *Bazzania novae-zelandiae* |
| | *Bazzania tayloriana* |
| | *Ceremanus perfragilis* (Syn. *Telaranea perfragilis*) |
| | *Kurzia fragilifolia* (Syn. *Telaranea fragilifolia*) |
| | *Kurzia hippuroides* |
| | *Lembidium longifolium* |
| | *Lembidium nutans* |
| | *Lepidozia bidens* |
| | *Lepidozia concinna* |
| | *Lepidozia elobata* |
| | *Lepidozia laevifolia* |
| | *Lepidozia microphylla* |
| | *Lepidozia spinosissima* |
| | *Neolepidozia patentissima* (Syn. *Telaranea patentissima* var. *patentissima*) |
| | *Neolepidozia praenitens* (Syn. *Telaranea praenitens* var. *praenitens*) |

| Family | Species |
|---|---|
| | Neolepidozia tetrapila (Syn. Telaranea tetrapila var. tetrapila) |
| | Paracromastigum macrostipum |
| | Psiloclada clandestina |
| | Telaranea herzogii |
| | Tricholepidozia lindenbergii (Syn. Telaranea lindenbergii var. lindenbergii) |
| | Tricholepidozia remotifolia (Syn. Telaranea remotifolia) |
| | Zoopsidella caledonica |
| | Zoopsis argentea |
| | Zoopsis leitgebiana |
| | Zoopsis nitida |
| | Zoopsis setulosa |
| **Lophocoleaceae** | Chiloscyphus bispinosus |
| | Chiloscyphus caniculatus (Syn. C. semiteres var. canaliculatus) |
| | Chiloscyphus cuspidatus |
| | Chiloscyphus erosus |
| | Chiloscyphus lentus |
| | Chiloscyphus muricatus |
| | Chiloscyphus novaezeelandiae var. grandistipulus |
| | Chiloscyphus rupicola |
| | Chiloscyphus semiteres |
| | Chiloscyphus subporosus |
| | Clasmatocolea strongylophylla |
| | Heteroscyphus allodontus |
| | Heteroscyphus ammophilus |
| | Heteroscyphus ciliatus) |
| | Heteroscyphus coalitus |
| | Heteroscyphus cunestipulus |
| | Heteroscyphus deceptifrons |

| Family | Species |
|---|---|
| | Heteroscyphus echinellus |
| | Heteroscyphus fissistipus |
| | Heteroscyphus fissistipus var. multispinus |
| | Heteroscyphus fissistipus var. repandus |
| | Heteroscyphus lingulatus |
| | Heteroscyphus menziesii (Syn. Chiloscyphus menziesii) |
| | Heteroscyphus triacanthus |
| | Heteroscyphus allodontus |
| | Heteroscyphus parallelifolius |
| | Lamellocolea granditexta |
| | Leptoscyphus compactus (Syn. Heteroscyphus compactus) |
| | Leptoscyphus idiodontus |
| | Tetracymbaliella decipiens |
| **Lophoziaceae** | Anastrophyllopsis subcomplicata (Syn. Anastrophyllum schismoides var. schismoides) |
| **Lunulariaceae** | Lunularia cruciata |
| **Marchantiaceae** | Marchantia pileata |
| **Mastigophoraceae** | Dendromastigophora flagellifera |
| **Metzgeriaceae** | Metzgeria flavovirens |
| | Metzgeria furcata |
| | Metzgeria leptoneura |
| | Metzgeria sp. |
| **Monocleaceae** | Monoclea forsteri |
| **Pallaviciniaceae** | Pallavicinia innovans |
| | Pallavicinia lyellii |
| | Pallavicinia tenuinervis (Syn. Symphyogyna tenuinervis) |
| | Pallavicinia xiphoides |
| | Podomitrium phyllanthus |

| Family | Species |
|---|---|
| | *Symphyogyna hymenophyllum* |
| **Plagiochilaceae** | *Chiastocaulon biserialis* (Syn. *Acrochila biserialis*) |
| | *Chiastocaulon conjugatum* (Syn. *Plagiochilion conjugatus*) |
| | *Dinckleria fruticella* |
| | *Dinckleria pleurata* |
| | *Plagiochila annotina* |
| | *Plagiochila circinalis* |
| | *Plagiochila colensoi* |
| | *Plagiochila colensoi* var. *quinquespina* |
| | *Plagiochila deltoidea* Lindenb. var. *deltoidea* |
| | *Plagiochila fasciculata* |
| | *Plagiochila gigantea* |
| | *Plagiochila incurvicolla* |
| | *Plagiochila intertexta* (Syn. *P. sinclairii*) |
| | *Plagiochila microdictyon* |
| | *Plagiochila trispicata* (was known in NZ as *P. arbuscula* var. *arbuscula*) |
| | *Plagiochila rutlandii* |
| | *Plagiochila stephensoniana* |
| **Porellaceae** | *Porella elegantula* |
| **Pseudolepicoleaceae** | *Temnoma pulchellum* |
| **Radulaceae** | *Radula allisonii* |
| | *Radula aneurismalis* |
| | *Radula demissa* (was known in NZ as *R buccinifera*) |
| | *Radula cuspidata* (Syn. *R. dentifolia*) |
| | *Radula grandis* |
| | *Radula marginata* |
| | *Radula physoloba* |
| | *Radula plicata* |

| Family | Species |
|---|---|
| | *Radula pseudoscripta* |
| | *Radula strangulata* |
| | *Radula tasmanica* |
| **Schistochilaceae** | *Gottschea pinnatifolia* |
| | *Gottschea tuloides* |
| | *Schistochila appendiculata* |
| | *Schistochila balfouriana* |
| | *Schistochila glaucescens* |
| | *Schistochila nobilis* |
| | *Schistochila repleta* |
| **Solenostomataceae** | *Solenostoma inundatum* |
| | *Solenostoma novaezelandiae* |
| **Trichocoleaceae** | *Leiomitra lanata* |
| | *Trichocolea hatcherii* |
| | *Trichocolea mollissima* |
| | *Trichocolea rigida* |
| **HORNWORTS/ANTHOCEROTOPHYTA** | |
| **Dendrocerotaceae** | *Dendroceros granulatus* |
| | *Phaeomegaceros hirticalyx* (Syn. *Phaeoceros hirticalyx*) |
| | *Megaceros pellucidus* (Syn. *M. flagellaris*) |
| **Notothyladaceae** | *Phaeoceros carolinianus* |

# Lichens

This species list originated from the records published by the author and colleagues in 1991. The list has been updated using David Galloway's monumental revised second edition of *Flora of New Zealand — Lichens* published in 2007. A number of previously unidentified lichen species collected for the 1991 paper and lodged in Auckland Museum herbarium have subsequently been identified by specialist lichenologists and are here included in the species list. The family classification used here is from the global revision of 2016. Over 1000 voucher specimens comprising virtually all species listed here are held in the Auckland Museum herbarium.

## Key

* = northernmost record in New Zealand

| Family | Species |
|---|---|
| **Arthoniaceae** | Arthonia cinnabarina |
| | Arthonia sp. |
| | Arthothelium fusconigrum |
| **Baeomycetaceae** | Baeomyces heteromorphus |
| **Brigantiaceae** | Brigantiaea chrysosticta |
| | Brigantiaea phaeomma |
| **Caliciaceae** | Amandinea decedens |
| | Amandinea otagensis |
| | Amandinea punctata |
| | Buellia stellulata |
| | Calicium hyperelloides |
| | Calicium robustellum |
| | Calicium tricolor |
| **Candelariaceae** | Candelariella vitellina |
| | Stirtoniella kelica |
| **Chryostrichaceae** | Chrysothrix candelaris |
| | Chrysothrix granulosa* |
| **Cladoniaceae** | Cladia aggregata |
| | Cladia retipora |
| | Cladia sullivanii |
| | Cladonia capitellata |
| | Cladonia cervicornis verticillata |
| | Cladonia chlorophaea |
| | Cladonia confusa |

| Family | Species |
|---|---|
| | Cladonia darwinii |
| | Cladonia fimbriata |
| | Cladonia floerkeana |
| | Cladonia furcata |
| | Cladonia gallowayi |
| | Cladonia glebosa* |
| | Cladonia imbricate* |
| | Cladonia incerta |
| | Cladonia macilenta |
| | Cladonia nudicaulis* |
| | Cladonia ochrochlora |
| | Cladonia polycarpoides |
| | Cladonia pyxidata |
| | Cladonia rigida |
| | Cladonia scabriuscula |
| | Cladonia strangulata |
| | Cladonia sulcata |
| | Cladonia tenerrima |
| | Cladonia wilsonii |
| | Cladonia spp. |
| | Metus conglomeratus |
| | Neophyllis melacarpa |
| | Thysanothecium scutellatum |
| **Coccocarpiaceae** | Coccocarpia erythroxyli |
| | Coccocarpia palmicola |

| Family | Species |
|---|---|
| | *Coccocarpia pellita* |
| **Coccotremaceae** | *Coccotrema cucurbitula* * |
| **Coenogoniaceae** | *Coenogonium implexum* |
| | *Coenogonium luteum* |
| | *Coenogonium zonatum* |
| **Collemataceae** | *Collema fasciculare* |
| | *Collema laeve* |
| | *Collema subconveniens* |
| | *Collema subflaccidum* * |
| | *Collema* sp. |
| | *Leptogium aucklandicum* |
| | *Leptogium coralloideum* |
| | *Leptogium crispatellum* |
| | *Leptogium cyanescens* |
| | *Leptogium denticulatum* |
| | *Leptogium propaguliferum* |
| | *Leptogium* spp. |
| **Gomphillaceae** | *Gyalidea lecanorina* * |
| **Graphidaceae** | *Fissurina monospora* |
| | *Fissurina novae-zelandiae* |
| | *Graphis librata* |
| | *Leiorreuma exaltatum* * |
| | *Phaeographis inusta* |
| | *Phaeographis mucronata* |
| | *Ramonia* sp. |
| | *Thelotrema lepadinum* |
| **Icmadophilaceae** | *Dibaeis absoluta* |
| | *Dibaeis arcuata* |
| | *Siphula decumbens* * |
| | *Siphula gracilis* * |
| **Lecanoraceae** | *Lecanora argentata* * |
| | *Lecanora melacarpella* |
| | *Lecanora subcoarctica* |
| | *Lecanora symmicta* |
| **Lecideaceae** | *Lecidea conisalea* |
| | *Pyrrhospora laeta* * |
| **Lichinaceae** | *Lichina confinis* |
| | *Polychidium contortum* |

| Family | Species |
|---|---|
| **Lobariaceae** | *Crocodia aurata* |
| | *Crocodia poculifera* |
| | *Lobarina scrobiculata* * |
| | *Lopadium monosporum.* |
| | *Pseudocyphellaria carpoloma* |
| | *Pseudocyphellaria chloroleuca* |
| | *Pseudocyphellaria cinnamomea* |
| | *Pseudocyphellaria coriacea* |
| | *Pseudocyphellaria coronata* |
| | *Pseudocyphellaria crocata* |
| | *Pseudocyphellaria dissimilis* |
| | *Pseudocyphellaria episticta* |
| | *Pseudocyphellaria faveolata* |
| | *Pseudocyphellaria fimbriatoides* * |
| | *Pseudocyphellaria glabra* |
| | *Pseudocyphellaria haywardiorum* |
| | *Pseudocyphellaria intricata* |
| | *Pseudocyphellaria lividofusca* |
| | *Pseudocyphellaria montagnei* |
| | *Pseudocyphellaria multifida* |
| | *Pseudocyphellaria pickeringii* |
| | *Pseudocyphellaria rubella* * |
| | *Pseudocyphellaria rufovirescens* |
| | *Pseudocyphellaria wilkinsii* |
| | *Sticta cineroglauca* |

| Family | Species |
|---|---|
| | Sticta filix |
| | Sticta fuliginosa |
| | Sticta lacera |
| | Sticta latifrons |
| | Sticta livida* |
| | Sticta squamata |
| | Sticta subcaperata |
| **Megalosporaceae** | Megaloblastenia flavidoatra* |
| | Megaloblastenia marginiflexa |
| | Megalospora bartlettii |
| | Megalospora campylospora |
| | Megalospora gompholoma |
| | Megalospora knightii |
| **Melaspileaceae** | Melaspilea subeffigurans |
| **Miltideaceae** | Miltidea ceroplasta |
| **Nephromataceae** | Nephroma plumbeum* |
| **Opegraphaceae** | Chiodecton sp. |
| | Lecanactis subfarinosa |
| | Opegrapha agelaeoides |
| | Opegrapha diaphoriza |
| | Opegrapha intertexta |
| **Pannariaceae** | Erioderma sorediatum |
| | Fuscopannaria subimmixta |
| | Leioderma duplicatum |
| | Leioderma sorediatum |
| | Pannaria allorhiza |
| | Pannaria araneosa |
| | Pannaria athroophylla* |
| | Pannaria crenulata |
| | Pannaria elixii |
| | Pannaria fulvescens |
| | Pannaria immixta |
| | Pannaria leproloma |
| | Pannaria minutiphylla |

| Family | Species |
|---|---|
| | Pannaria pyxinoides |
| | Parmeliella nigrata |
| | Parmeliella nigrocincta |
| | Parmeliella thysanota* |
| | Psoroma asperellum* |
| | Psoroma implexum* |
| | Psoroma spp. |
| | Psoromaria rosulata |
| | Psoromidium aleuroides* |
| **Parmeliaceae** | Canomaculina subtinctoria |
| | Flavoparmelia haysomii |
| | Flavoparmelia soredians |
| | Flavoparmelia sp. |
| | Gowardia nigricans* |
| | Hypogymnia subphysodes* |
| | Hypotrachyna immaculata |
| | Hypotrachyna osseoalba |
| | Hypotrachyna revoluta* |
| | Menegazzia aucklandica |
| | Menegazzia eperforata* |
| | Menegazzia neozelandica |
| | Menegazzia nothofagi |
| | Parmelia erumpens |
| | Parmelia testacea |
| | Parmelinopsis jamesii* |
| | Parmelinopsis spumosa |
| | Parmelinopsis subfatiscens* |
| | Parmotrema arnoldii |
| | Parmotrema cetratum |
| | Parmotrema crinitum |
| | Parmotrema mellissii |
| | Parmotrema perlatum |
| | Parmotrema reticulatum |
| | Parmotrema tinctorum |
| | Punctelia borreri |
| | Usnea ciliifera |

| Family | Species |
|---|---|
| | Usnea cornuta |
| | Usnea inermis |
| | Usnea molliuscula* |
| | Usnea rubicunda |
| | Usnea spp. |
| | Xanthoparmelia australasica |
| | Xanthoparmelia flavescentireagens* |
| | Xanthoparmelia furcata |
| | Xanthoparmelia isidiigera |
| | Xanthoparmelia mexicana* |
| | Xanthoparmelia mougeotina |
| | Xanthoparmelia neotinctina |
| | Xanthoparmelia pulla |
| | Xanthoparmelia scabrosa |
| | Xanthoparmelia thamnoides |
| | Xanthoparmelia verrucella |
| | Xanthoparmelia sp. |
| **Peltigeraceae** | Peltigera dolichorhiza |
| | Peltigera nana |
| | Peltigera polydactylon |
| **Pertusariaceae** | Ochrolechia parella |
| | Pertusaria hypoxantha |
| | Pertusaria lavata |
| | Pertusaria melanospora |
| | Pertusaria novaezelandiae |
| | Pertusaria psoromica* |
| | Pertusaria sorodes |
| | Pertusaria xanthoplaca |
| | Pertusaria spp. |
| **Phlyctidaceae** | Phlyctis sp. |
| **Physciaceae** | Diplotomma canescens canescens |
| | Dirinaria applanata |

| Family | Species |
|---|---|
| | Heterodermia isidiophora |
| | Heterodermia japonica |
| | Heterodermia leucomela |
| | Heterodermia microphylla |
| | Heterodermia obscurata |
| | Heterodermia speciosa |
| | Heterodermia spp. |
| | Physcia caesia |
| | Physcia erumpens |
| | Physcia poncinsii |
| | Pyxine subcinerea |
| **Pilocarpaceae** | Badimiella pteridophylla |
| | Bapalmuia buchanani |
| | Byssoloma subdiscordans |
| | Roccellinastrum neglectum |
| **Porpidiaceae** | Paraporpidia leptocarpa* |
| | Poeltiaria turgescens |
| | Porpidia albocaerulescens |
| | Porpidia sp. |
| **Pyrenulaceae** | Pyrenula deliquescens* |
| | Pyrenula sp. |
| **Ramalinaceae** | Megalaria grossa |
| | Phyllopsora corallina* |
| | Phyllopsora microdactyla |
| | Ramalina australiensis |
| | Ramalina celastri celastri |
| **Sarrameanaceae** | Sarrameana albidoplumbea |
| **Scoliciosporaceae** | Scoliciosporum sp. |
| **Sphaerophoraceae** | Bunodophoron australe |
| | Bunodophoron flaccidum |
| | Bunodophoron murrayi |
| | Bunodophoron patagonicum |
| | Leifidium tenerum |
| **Stereocaulaceae** | Lepraria incana |
| | Stereocaulon corticatulum |
| | Stereocaulon ramulosum |

| Family | Species |
|---|---|
| | *Stereocaulon vesuvianum**  |
| **Strigulaceae** | *Strigula* sp. |
| **Teloschistaceae** | *Caloplaca flavorubescens* |
| | *Caloplaca holocarpa* |
| | *Caloplaca rosei* |
| | *Teloschistes chrysophthalmus* |
| | *Teloschistes flavicans* |
| | *Teloschistes xanthorioides* |
| | *Xanthoria ligulata* |
| | *Xanthoria parietina* |
| **Tephromelataceae** | *Tephromela atra* |
| **Trapeliaceae** | *Placopsis argillacea* |
| | *Trapelia coarctata* |
| | *Trapeliopsis congregans** |
| | *Trapeliopsis granulosa* |
| **Trichotheliaceae** | *Porina exocha* |
| **Trypetheliaceae** | *Aptrootia elatior** |
| **Verrucariaceae** | *Catapyrenium cinereum* |
| | *Verrucaria maura* |

# Fungi

## Key

(type) = has type specimen from Hauturu
incertae sedis = of uncertain placement

| CLASS/Family | Species | Classification |
|---|---|---|
| **DIVISION ASCOMYCOTA** | | |
| **incertae sedis** | *Chaetopsis probosciophora* | Endemic |
| | *Chalarodes bisetis* | Endemic |
| | *Circinotrichum* sp. | Unknown |
| | *Cryptophiale pusilla* | Endemic |
| | *Gyrothrix* sp. | Unknown |
| | *Hansfordia* sp. | Unknown |
| | *Spiropes* sp. | Unknown |
| | *Staphylotrichum* sp. | Unknown |
| | *Sympodiella fragilis* | Endemic |
| | *Sympodiella nodosa* | Endemic |
| | *Ulocoryphus mastigophorus* | Endemic |
| | *Waydora typica* | Indigenous |
| | *Xylohypha novae-zelandiae* | Endemic |
| | *Zebrospora bicolor* | Indigenous |
| | *Zygosporium gibbum* | Indigenous |
| **CLASS DOTHIDEOMYCETES** | | |
| **incertae sedis** | *Colensoniella torulispora* | Endemic |
| | *Coronospora novae-zelandiae* | Endemic |
| **Anteagloniaceae** | *Anteaglonium abbreviatum* | Unknown |
| **Cladosporiaceae** | *Cladosporium colocasiae* | Exotic naturalised |
| **Corynesporascaceae** | *Corynespora citricola* | Exotic naturalised |
| **Hysteriaceae** | *Hysterium* sp. | Unknown |
| **Kirschsteiniotheliaceae** | *Kirschsteiniothelia aethiops* | Unknown |
| **Leptosphaeriaceae** | *Leptosphaeria* sp. | Unknown |
| | *Sphaerellopsis filum* | Indigenous |
| **Massarinaceae** | *Helminthosporium* sp. | Unknown |
| **Meliolinaceae** | *Meliolina novae-zealandiae* | Endemic |
| **Mycosphaerellaceae** | *Cercospora chenopodii* | Exotic naturalised |
| | *Cymadothea trifolii* | Exotic naturalised |
| | *Mycosphaerella geniostomatis* | Indigenous |
| | *Mycosphaerella pittospori* | Indigenous |

| CLASS/Family | Species | Classification |
|---|---|---|
| | *Passalora graminis* | Exotic naturalised |
| | *Ramularia sphaeroidea* | Exotic naturalised |
| | *Septoria apiicola* | Exotic naturalised |
| | *Septoria chrysanthemella* | Exotic naturalised |
| | *Septoria convolvuli* | Exotic naturalised |
| | *Zasmidium dianellae* | Indigenous |
| | *Zasmidium gahniae* | Endemic |
| **Ohleriaceae** | *Ohleria brasiliensis* | Indigenous |
| **Parmulariaceae** | *Placosoma nothopanacis* | Endemic |
| | *Rhagadolobium bakerianum* | Indigenous |
| **Patellariaceae** | *Patellaria* sp. | Unknown |
| **Pleosporaceae** | *Bipolaris sorokiniana* | Exotic naturalised |
| **Pseudoperisporiaceae** | *Episphaerella dodonaeae* | Indigenous |
| **Tubeufiaceae** | *Acanthostigma affine* | Indigenous |
| | *Acanthostigma scopulum* | Indigenous |
| | *Helicosporium* sp. | Unknown |
| | *Podonectria novae-zealandiae* | Endemic |
| | *Thaxteriella helicoma* | Indigenous |
| **CLASS EUROTIOMYCETES** | | |
| **Aspergillaceae** | *Aspergillus fumigatus* | Unknown; recorded from literature |
| **Chaetothyriaceae** | *Chaetothyrium strigosum* | Indigenous |
| **Elaphomycetaceae** | *Elaphomyces bollardii* (type) | Endemic |
| | *Elaphomyces* sp. | Indigenous |
| **Herpotrichiellaceae** | *Capronia* sp. | Unknown |
| **CLASS GEOGLOSSOMYCETES** | | |
| **Geoglossaceae** | *Geoglossum umbratile* | Indigenous |
| **CLASS LECANOROMYCETES** | | |
| **Ectolechiaceae** | *Lopadium coralloideum* | Indigenous |
| **Stictidaceae** | *Delpontia* sp. | Unknown |
| | *Stictis asteliae* | Endemic |
| | *Stictis clavata* | Endemic |
| | *Stictis collospermi* | Endemic |
| | *Stictis dealbata* | Endemic |
| | *Stictis gigantea* | Unknown |
| | *Stictis lata* | Endemic |
| | *Stictis radiata* | Indigenous |
| | *Stictis ramuligera* var. *minor* | Endemic |

| CLASS/Family | Species | Classification |
|---|---|---|
| | *Stictis ramuligera* | Unknown |
| | *Stictis subiculata* | Indigenous |
| | *Stictis trinervia* | Endemic |
| **CLASS LEOTIOMYCETES** | | |
| **incertae sedis** | *Chalara acuaria* | Indigenous |
| | *Chalara constricta* (type) | Indigenous |
| | *Chalara scabrida* (type) | Endemic |
| | *Chalara stipitata* | Endemic |
| | *Micropeziza* sp. | Indigenous |
| | *Sorokina* sp. | Unknown |
| **Bulgariaceae** | *Bovista aestivalis* | Indigenous |
| **Dermateaceae** | *Mollisia* sp. | Unknown |
| **Erysiphaceae** | *Golovinomyces cichoracearum* | Exotic naturalised |
| | *Podosphaera* sp. | Unknown |
| **Helotiaceae** | *Hymenoscyphus metrosideri* | Endemic |
| | *Hymenoscyphus quintiniae* | Endemic |
| | *Hymenotorrendiella* sp. | Exotic naturalised |
| | *Strossmayeria basitricha* | Indigenous |
| **Hyaloscyphaceae** | *Arachnopeziza* sp. | Unknown |
| | *Calycellina* sp. | Unknown |
| | *Hispidula tokerau* (type) | Endemic |
| | *Lachnum abnorme* | Indigenous |
| | *Lachnum varians* | Indigenous |
| | *Proliferodiscus dingleyae* | Indigenous |
| **Leotiaceae** | *Microglossum rufum* | Indigenous |
| **Rhytismataceae** | *Coccomyces globosus* | Indigenous |
| | *Coccomyces lauraceus* | Indigenous |
| | *Coccomyces limitatus* | Indigenous |
| | *Coccomyces phyllocladi* | Indigenous |
| | *Coccomyces radiatus* | Indigenous |
| | *Hypoderma campanulatum* | Endemic |
| | *Hypoderma carinatum* | Endemic |
| | *Hypoderma cookianum* (type) | Endemic |
| | *Hypoderma cordylines* | Endemic |
| | *Hypoderma liliense* (type) | Endemic |
| | *Hypoderma obtectum* (type) | Endemic |
| | *Hypoderma rubi* | Indigenous |
| | *Lophodermium agathidis* | Indigenous |
| | *Lophodermium atrum* (type) | Endemic |

| CLASS/Family | Species | Classification |
|---|---|---|
| | *Lophodermium brunneolum* | Endemic |
| | *Lophodermium hauturuanum* (type) | Endemic |
| | *Lophodermium inclusum* | Endemic |
| | *Lophodermium irregulare* (type) | Endemic |
| | *Lophodermium nigrofactum* | Endemic |
| | *Lophodermium richeae* | Indigenous |
| | *Lophodermium tindalii* | Endemic |
| | *Lophodermium unciniae* | Endemic |
| | *Marthamyces emarginatus* | Indigenous |
| | *Marthamyces quadrifidus* | Indigenous |
| | *Propolis* sp. | Unknown |
| | *Terriera brevis* | Endemic |
| **Rutstroemiaceae** | *Dicephalospora chrysotricha* | Endemic |
| | *Lanzia allantospora* | Endemic |
| **Vibrisseaceae** | *Phialocephala* sp. | Unknown |
| **CLASS ORBILIOMYCETES** | | |
| **Orbiliaceae** | *Hyalorbilia inflatula* | Indigenous |
| | *Orbilia juruensis* | Indigenous |
| | *Orbilia xanthostigma* | Exotic naturalised |
| **CLASS PEZIZOMYCETES** | | |
| **incertae sedis** | *Verticicladium* sp. | Unknown |
| **Pyronemataceae** | *Paurocotylis pila* | Endemic |
| | *Scutellinia scutellata* | Indigenous |
| **Sarcosomataceae** | *Conoplea novae-zelandiae* | Endemic |
| | *Plectania rhytidia* | Indigenous |
| | *Pseudoplectania affinis* | Endemic |
| | *Urnula campylospora* | Indigenous |
| **CLASS SORDARIOMYCETES** | | |
| **incertae sedis** | *Acremonium luzulae* | Unknown |
| **Amphisphaeriaceae** | *Amphisphaeria* sp. | Unknown |
| **Bertiaceae** | *Bertia moriformis* | Indigenous |
| **Bionectriaceae** | *Clonostachys* cf. *byssicola* | Indigenous |
| | *Clonostachys ralfsii* (type) | Indigenous |
| | *Clonostachys rosea* | Unknown |
| | *Hydropisphaera erubescens* | Indigenous |
| | *Hydropisphaera macrarenula* | Indigenous |
| | *Hydropisphaera multiloculata* | Endemic |
| | *Ijuhya peristomialis* | Indigenous |
| | *Stilbocrea macrostoma* | Indigenous |

| CLASS/Family | Species | Classification |
|---|---|---|
| Calosphaeriaceae | *Pachytrype princeps* | Exotic naturalised |
| Catabotrydaceae | *Catabotrys decidua* | Indigenous |
| Chaetosphaeriaceae | *Chaetosphaeria raciborskii* | Indigenous |
| | *Dictyochaeta fertilis* | Indigenous |
| | *Menisporopsis novae-zelandiae* | Indigenous |
| Clavicipitaceae | *Claviceps purpurea* | Exotic naturalised; ergot |
| | *Harposporium* sp. | Unknown |
| Cordycipitaceae | *Cordyceps hauturu* (type) | Endemic; āwheto, vegetable caterpillar |
| | *Cordyceps kirkii* | Endemic; recorded from photograph only |
| | *Isaria tenuipes* | Unknown |
| Cryphonectriaceae | *Endothia fluens* | Indigenous |
| Cytosporaceae | *Xenotypa* sp. | Unknown |
| Diatrypaceae | *Diatrypella* sp. | Unknown |
| | *Eutypella* sp. | Indigenous |
| Glomerellaceae | *Colletotrichum graminicola* | Exotic naturalised |
| Hypocreaceae | *Hypocrea tawa* | Indigenous |
| | *Hypomyces chrysospermus* | Indigenous |
| | *Trichoderma catoptron* | Indigenous |
| | *Trichoderma gelatinosum* | Unknown |
| | *Trichoderma semiorbis* | Indigenous |
| | *Trichoderma sulphureum* | Unknown |
| | *Trichoderma vinosum* | Indigenous |
| Hyponectriaceae | *Physalospora* sp. | Unknown |
| Lasiosphaeriaceae | *Lasiosphaeria* sp. | Unknown |
| Nectriaceae | *Calonectria* sp. | Unknown |
| | *Cosmospora episphaeria* | Indigenous |
| | *Fusarium haematococcum* | Unknown |
| | *Fusarium heterosporum* | Exotic naturalised |
| | *Fusarium illudens* | Indigenous |
| | *Ilyonectria destructans* | Indigenous |
| | *Microcera coccophila* | Unknown |
| | *Nectria* cf. *inventa* | Unknown |
| | *Nectria pseudotrichia* | Indigenous |
| | *Nectria ruapehu* | Endemic |
| | *Neonectria punicea* | Indigenous |
| | *Neonectria* sp. | Unknown |
| | *Pseudocosmospora vilior* | Unknown |

| CLASS/Family | Species | Classification |
|---|---|---|
| | *Stylonectria wegeliniana* | Unknown |
| | *Volutella ciliata* | Indigenous |
| **Niessliaceae** | *Niesslia artocarpi* | Indigenous |
| | *Acremonium luzulae* | Unknown |
| **Ophiocordycipitaceae** | *Ophiocordyceps robertsii* | Indigenous; āwheto, vegetable caterpillar |
| **Pestalotiopsidaceae** | *Pestalotiopsis* sp. | Unknown |
| **Phyllachoraceae** | *Phyllachora cyperi* | Unknown |
| | *Phyllachora hauturu* subsp. *hauturu* (type) | Endemic |
| | *Phyllachora setariicola* | Exotic naturalised |
| **Stachybotriaceae** | *Stachybotrys freycinetiae* | Endemic |
| **Valsaceae** | *Cytospora* sp. | Unknown |
| **Xylariaceae** | *Annulohypoxylon truncatum* | Unknown |
| | *Anthostomella* sp. | Unknown |
| | *Biscogniauxia capnodes* var. *rumpens* | Indigenous |
| | *Biscogniauxia capnodes* | Indigenous |
| | *Hypoxylon hypomiltum* | Unknown |
| | *Hypoxylon perforatum* | Indigenous |
| | *Hypoxylon subcorticeum* | Endemic |
| | *Hypoxylon subcrocopeplum* | Endemic |
| | *Hypoxylon torrendii* | Indigenous |
| | *Kretzschmaria clavus* | Indigenous |
| | *Kretzschmaria pavimentosa* | Indigenous |
| | *Nemania caries* | Indigenous |
| | *Nemania* cf. *diffusa* | Indigenous |
| | *Nemania serpens* | Indigenous |
| | *Penzigia discolor* | Indigenous |
| | *Rosellinia communis* | Endemic |
| | *Rosellinia johnstonii* | Endemic |
| | *Xylaria apiculata* | Endemic |
| | *Xylaria schreuderiana* | Indigenous |
| **DIVISION BASIDIOMYCOTA** | | |
| **CLASS AGARICOMYCETES** | | |
| **incertae sedis** | *Gymnopilus leptospermi* | Indigenous |
| | *Phlyctibasidium polyporoideum* | Indigenous |
| | cf. *Ripartites* sp. | Unknown |
| | *Sidera lenis* | Indigenous |
| | *Sidera lowei* | Indigenous |

| CLASS/Family | Species | Classification |
| --- | --- | --- |
| Agaricaceae | Leucocoprinus fragilissimus | Unknown; recorded from photograph only |
| | Lycoperdon glabrescens | Indigenous |
| | Lycoperdon lividum | Unknown |
| | Lycoperdon cf. perlatum | Indigenous |
| | Lycoperdon sp. | Unknown |
| | Lycoperdon pratense | Exotic naturalised |
| | Morganella compacta | Indigenous |
| Amanitaceae | Amanita nehuta | Endemic |
| Amylocorticiaceae | Amyloathelia amylacea | Indigenous |
| | Podoserpula pusio var. pusio | Indigenous |
| Auriculariaceae | Auricularia cornea | Indigenous; hakeke, wood ear |
| | Exidia glandulosa | Unknown |
| | Exidiopsis mucedinea | Indigenous |
| Auriscalpiaceae | Artomyces turgidus | Indigenous |
| Bankeraceae | Phellodon nothofagi | Endemic |
| | Phellodon sinclairii | Endemic |
| | Sarcodon thwaitesii | Indigenous |
| Bolbitiaceae | Descomyces cf. albellus | Indigenous |
| | Descomyces albus | Exotic naturalised |
| Boletaceae (boletes) | Austroboletus novae-zelandiae | Endemic |
| | cf. Boletus sp. | Endemic |
| | Boletus semigastroideus | Endemic |
| | Porphyrellus formosus | Endemic |
| | Rossbeevera pachydermis | Endemic |
| Bondarzewiaceae | Bondarzewia kirkii | Endemic |
| | Heterobasidion araucariae | Indigenous |
| Calostomataceae | Calostoma fuscum | Indigenous |
| | Calostoma rodwayi | Indigenous |
| Cantharellaceae | Cantharellus sp. | Unknown |
| | Cantharellus wellingtonensis | Endemic |
| Clathraceae | Ileodictyon cibarium | Indigenous; matakupenga, kōpurawhetū, basket fungus |
| | Pseudocolus fusiformis | Indigenous |
| Clavariaceae (coral fungi) | cf. Clavaria sp. | Unknown |
| | Clavulinopsis sulcata | Indigenous; recorded from photograph only |
| Clitocybaceae | Lepista antipoda | Endemic |
| Coniophoraceae | Coniophora dimitica | Indigenous |

| CLASS/Family | Species | Classification |
|---|---|---|
| Cortinariaceae | *Cortinarius castaneiceps* (type) | Endemic |
| | *Cortinarius gemmeus* (type) | Endemic |
| | *Cortinarius ignotus* (type) | Endemic |
| | *Cortinarius porphyroideus* | Endemic |
| | *Cortinarius porphyrophaeus* (type) | Endemic |
| | *Cortinarius subviolaceus* | Unknown |
| | *Dermocybe aurata* | Indigenous |
| | *Phlegmacium connatum* | Indigenous |
| Crepidotaceae | *Crepidotus mollis* | Unknown |
| Cyphellaceae | *Chondrostereum vesiculosum* | Endemic |
| Cyphellopsidaceae | *Dendrothele aucklandica* | Endemic |
| | *Dendrothele corniculata* | Endemic |
| Entolomataceae | *Alboleptonia sericella* | Exotic naturalised |
| | *Entoloma aromaticum* | Indigenous |
| | *Entoloma blandiodorum* (type) | Endemic |
| | *Entoloma consanguineum* (type) | Endemic |
| | *Entoloma hochstetteri* | Indigenous; werewere-kōkako, sky blue mushroom |
| | *Entoloma inventum* | Endemic |
| | *Entoloma persimile* | Endemic |
| | *Entoloma squamiferum* | Endemic |
| | *Entoloma translucidum* | Indigenous |
| | *Entoloma virescens* | Indigenous |
| Fomitopsidaceae | *Antrodia* cf. *albida* | Indigenous |
| | *Laetiporus portentosus* | Indigenous; pūtawa, puku tawai |
| | *Postia brunnea* | Indigenous |
| Ganodermataceae | *Ganoderma* cf. *applanatum* | Endemic |
| Geastraceae | *Geastrum velutinum* | Indigenous |
| Gomphaceae | *Ramaria junquilleovertex* (type) | Endemic |
| | *Ramaria rotundispora* (type) | Endemic |
| Hericiaceae | *Hericium* aff. *coralloides* | Recorded from photography only; pekepekekiore, icicle fungus |
| Hydnaceae | *Hydnum crocidens* | Indigenous |
| Hydnangiaceae | *Hydnangium kanuka* | Endemic |
| Hygrophoraceae | *Cuphophyllus griseorufescens* (type) | Endemic |
| | *Gliophorus chromolimoneus* | Indigenous |
| | *Gliophorus ostrinus* (type) | Endemic |

| CLASS/Family | Species | Classification |
|---|---|---|
| | *Gliophorus viridis* | Indigenous |
| | *Hygrocybe blanda* (type) | Endemic |
| | *Hygrocybe rubrocarnosa* | Endemic |
| **Hymenochaetaceae** | *Coltricia cinnamomea* | Indigenous |
| | *Coltricia salpincta* | Indigenous |
| | *Coltricia strigosa* (type) | Endemic |
| | *Coltriciella dependens* | Indigenous |
| | *Cyclomyces tabacinus* | Indigenous |
| | *Fomitiporia robusta* | Indigenous |
| | *Fuscoporia contigua* | Unknown |
| | *Fuscoporia ferrea* | Indigenous |
| | *Hymenochaete cruenta* | Indigenous |
| | *Hymenochaete magnahypha* (type) | Indigenous |
| | *Hymenochaete patelliformis* | Endemic |
| | *Hymenochaete plurimaesetae* | Endemic |
| | *Hymenochaete rhabarbarina* | Indigenous |
| | *Hymenochaete semistupposa* | Indigenous |
| | *Hymenochaete tasmanica* | Indigenous |
| | *Hymenochaete vallata* | Indigenous |
| | *Hymenochaete villosa* | Indigenous |
| | *Inonotus cf. glomeratus* | Indigenous |
| | *Phellinus dingleyae* | Endemic |
| | *Phellinus kamahi* | Endemic |
| | *Phellinus tawhai* | Endemic |
| | *Phylloporia pectinata* | Indigenous |
| | *Pseudochaete tabacina* | Indigenous |
| **Hymenogasteraceae** | *Hebeloma* sp. | Unknown |
| | *Hebeloma victoriense* | Indigenous; recorded from photograph only |
| | *Hymenogaster* sp. | Unknown |
| | *Psilocybe weraroa* | Endemic |
| **Hysterangiaceae** | *Aroramyces gelatinosporus* | Indigenous |
| | *Galerina patagonica* | Indigenous; recorded from photograph only |
| | *Hysterangium neotunicatum* | Indigenous |
| | *Hysterangium rugisporum* (type) | Endemic |
| **Inocybaceae** | *Inocybe albovestita* | Endemic |
| | *Inocybe amygdalina* | Endemic |
| | *Inocybe luteobulbosa* | Unknown |

| CLASS/Family | Species | Classification |
|---|---|---|
| **Lachnocladiaceae** | *Asterostroma persimile* | Indigenous |
| | *Dichostereum rhodosporum* | Indigenous |
| | *Vararia investiens* | Indigenous |
| **Lyophyllaceae** | *Calocybe* sp. | Unknown |
| **Marasmiaceae** | *Chaetocalathus cocciformis* | Endemic |
| | *Marasmius kanukaneus* | Indigenous |
| **Meripilaceae** | *Rigidoporus vinctus* | Indigenous |
| **Meruliaceae** | *Hyphoderma litschaueri* | Indigenous |
| | *Hyphoderma utriculosum* | Endemic |
| | *Hypochnicium lyndoniae* | Indigenous |
| **Mycenaceae** | *Favolaschia calocera* | Exotic naturalised; orange pore fungus |
| | *Favolaschia pustulosa* | Indigenous |
| | *Mycena interrupta* | Indigenous; recorded from photograph only |
| | *Mycena ura* | Indigenous; recorded from photograph only |
| **Nidulariaceae (bird's nest fungi)** | *Crucibulum laeve* | Indigenous |
| | *Nidula candida* | Indigenous; recorded from photograph only |
| | *Nidula niveotomentosa* | Indigenous |
| **Omphalotaceae** | *Mycetinis curraniae* | Endemic |
| **Peniophoraceae** | *Peniophora cinerea* | Indigenous |
| | *Peniophora coprosmae* | Indigenous |
| | *Peniophora crustosa* | Indigenous |
| **Phallaceae (stinkhorns)** | *Aseroe rubra* | Indigenous; puapuatai, flower fungus |
| | *Mutinus* sp. | Unknown |
| **Phanerochaetaceae** | *Antrodiella citrea* | Indigenous |
| | *Antrodiella hunua* | Endemic |
| | *Bulgaria inquinans* | Indigenous |
| | *Byssomerulius corium* | Indigenous |
| | *Candelabrochaete eruciformis* | Endemic |
| | *Gelatoporia dichroa* | Unknown |
| | *Phanerochaete sordida* | Indigenous |
| | *Phlebiopsis crassa* | Indigenous |
| | *Pseudolagarobasidium pronum* | Unknown |
| **Physalacriaceae** | *Armillaria limonea* | Indigenous; recorded from photograph only |

| CLASS/Family | Species | Classification |
|---|---|---|
| | Armillaria novae-zelandiae | Indigenous; harore, honey mushroom |
| | Cyptotrama sp. | Indigenous; recorded from photograph only |
| **Pleurotaceae** | Pleurotus australis | Indigenous; recorded from photograph only; oyster mushroom |
| | Resupinatus cf. huia | Endemic |
| **Pluteaceae** | Pluteus perroseus | Indigenous |
| | Pluteus readiarum | Endemic |
| **Polyporaceae** | Australoporus tasmanicus | Indigenous |
| | Cerrena zonata | Indigenous |
| | Coriolopsis sanguinaria | Unknown |
| | Datroniella scutellata | Indigenous |
| | Dichomitus newhookii (type) | Endemic |
| | Echinochaete russiceps | Indigenous |
| | Fomes hemitephrus | Indigenous |
| | Lopharia cinerascens | Unknown |
| | Polyporus arcularius | Indigenous |
| | Polyporus melanopus | Exotic naturalised |
| | Poria cf. byssina | Unknown |
| | Pycnoporus coccineus | Indigenous |
| | Trametes hirsuta | Indigenous |
| **Porotheleaceae** | Rectipilus fasciculatus | Indigenous |
| **Russulaceae** | Lactarius clarkeae var. aurantioruber | Endemic |
| | Lactifluus clarkeae | Unknown |
| | Russula acrolamellata | Endemic |
| | Russula cremeoochracea | Endemic |
| | Russula griseoviridis | Endemic |
| | Russula novae-zelandiae (type) | Endemic |
| | Russula osphranticarpa | Endemic |
| | Russula spinispora | Endemic |
| | Russula subvinosa | Endemic |
| **Schizoporaceae** | Hyphodontia arguta | Indigenous |
| | Hyphodontia novozelandica | Indigenous |
| | Hyphodontia radula | Unknown |
| **Sclerodermataceae (earthballs)** | Scleroderma sp. | Unknown |
| **Sclerogastraceae** | Sclerogaster sp. | Unknown |
| **Sphaerobolaceae** | Sphaerobolus stellatus | Indigenous |

| CLASS/Family | Species | Classification |
|---|---|---|
| Steccherinaceae | Flaviporus brownii | Unknown |
| | Steccherinum ochraceum | Unknown |
| Stephanosporaceae | Stephanospora kanuka | Endemic |
| Stereaceae | Aleurobotrys botryosus | Unknown |
| | Aleurodiscus berggrenii | Endemic |
| | Aleurodiscus coralloides | Endemic |
| | Aleurodiscus mirabilis | Indigenous |
| | Aleurodiscus ochraceoflavus | Endemic |
| | Aleurodiscus zealandicus | Indigenous |
| | Stereum aotearoa | Indigenous |
| | Stereum illudens | Indigenous; recorded from photography only |
| | Stereum scutellatum | Endemic |
| | Stereum vellereum | Indigenous |
| | Stereum versicolor | Indigenous |
| Strophariaceae | Clavogaster virescens | Endemic |
| | Hypholoma australianum | Indigenous; recorded from photograph only |
| | Leratiomyces erythrocephalus | Indigenous |
| | Pholiota multicingulata (type) | Indigenous |
| Tricholomataceae | Tricholoma viridiolivaceum | Endemic |
| Truncocolumellaceae | Truncocolumella sp. | Unknown |
| Tubariaceae | Cyclocybe parasitica | Indigenous; recorded from photograph only; tawaka, poplar mushroom |
| | Tubaria perplexa (type) | Endemic |
| Xenasmataceae | Xenasma rimicola | Indigenous |
| **CLASS ATRACTIELLOMYCETES** | | |
| | Infundibura adhaerens | Indigenous |
| **CLASS DACRYMYCETES** | | |
| | Calocera guepinioides | Indigenous |
| **CLASS EXOBASIDIOMYCETES** | | |
| | Entyloma parietariae | Exotic naturalised |
| | Jamesdicksonia dactylidis | Exotic naturalised |
| | Rhamphospora nymphaeae | Exotic naturalised |
| **CLASS MICROBOTRYOMYCETES** | | |
| | Bauerago gardneri | Indigenous |
| **CLASS PUCCINIOMYCETES (RUST FUNGI)** | | |
| | Aecidium otagense | Endemic |

| CLASS/Family | Species | Classification |
|---|---|---|
| | *Melampsora euphorbiae* | Exotic naturalised |
| | *Melampsora larici-populina* | Exotic naturalised |
| | *Melampsora lini* | Exotic naturalised |
| | *Melampsora novae-zelandiae* | Endemic |
| | *Puccinia atkinsonii* | Endemic |
| | *Puccinia clavata* | Endemic |
| | *Puccinia coprosmae* | Indigenous |
| | *Puccinia coronata* | Exotic naturalised |
| | *Puccinia cynodontis* | Exotic naturalised |
| | *Puccinia graminis* | Exotic naturalised |
| | *Puccinia hieracii* | Exotic naturalised |
| | *Puccinia lagenophorae* | Indigenous |
| | *Puccinia menthae* | Exotic naturalised |
| | *Puccinia pelargonii-zonalis* | Exotic naturalised |
| | *Puccinia tetragoniae* var. *novae-zelandiae* | Endemic |
| | *Tranzschelia discolor* | Exotic naturalised |
| | *Uredo puawhananga* | Endemic |
| | *Uromyces dactylidis* | Exotic naturalised |
| | *Uromyces ehrhartae* | Indigenous |
| | *Uromyces rumicis* | Exotic naturalised |
| | *Uromyces tenuicutis* | Exotic naturalised |
| | *Uromyces trifolii-repenti* | Exotic naturalised |
| **CLASS TREMELLOMYCETES** | | |
| | *Tremella fuciformis* | Indigenous |
| **CLASS USTILAGINOMYCETES (SMUT FUNGI)** | | |
| | *Farysporium endotrichum* | Indigenous |
| | *Ustilago striiformis* | Exotic naturalised |
| **DIVISION GLOMEROMYCOTA** | | |
| | *Glomus macrocarpum* | Unknown |
| **DIVISION MUCOROMYCOTA** | | |
| | *Endogone flammicorona* | Exotic naturalised |
| **DIVISION MYXOMYCOTA (SLIME MOULDS — KINGDOM PROTOZOA)** | | |
| | *Diderma donkii* | Indigenous |
| | *Didymium bahiense* | Indigenous |
| **DIVISION OOMYCOTA (WATER MOULDS — KINGDOM CHROMISTA)** | | |
| | *Albugo candida* | Exotic naturalised |
| | *Phytophthora cinnamomi* | Exotic naturalised; recorded from literature |

## Stream vertebrates

| Scientific name | Māori/common name | Conservation status |
| --- | --- | --- |
| Anguilla dieffenbachii | kūwharuwharu/longfin eel | Declining |
| Galaxias fasciatus | banded kōkopu | Not threatened |
| Gobiomorphus huttoni | redfin bully | Declining |

## Seaweeds

| |
| --- |
| **CYANOBACTERIA** |
| Calothrix scopulorum |
| Dermocarpella prasina |
| Entophysalis deusta |
| Isactis plana |
| Lyngbya confervoides |
| Lyngbya majuscula |
| Phormidium nigroviride |
| **BROWN ALGAE** |
| Bachelotia antillarum |
| Carpophyllum flexuosum |
| Carpophyllum maschalocarpum |
| Carpophyllum plumosum |
| Colpomenia sinuosa |
| Colpomenia peregrina |
| Cystophora retroflexa |
| Cystophora torulosa |
| Dictyota kunthii |
| Ecklonia radiata |
| Elachista australis |
| Feldmannia irregularis |
| Feldmannia mitchelliae |
| Halopteris congesta |
| Halopteris virgata |
| Hapalospongidion saxigenum |
| Hormosira banksii |
| Hydroclathrus clathrata |
| Perithalia capillaris |
| Petalonia binghamiae |
| Ralfsia verrucosa |
| Sargassum sinclairii |
| Scytosiphon lomentaria |
| Scytothamnus australis |
| Spatoglossum sp. |
| Sphacelaria pulvinata |
| Splachnidium rugosum |
| Tinocladia novae-zelandiae |
| Xiphophora chondrophylla |
| Zonaria aureomarginata |
| Zonaria turneriana |
| **DIATOMS** |
| Melosira nummuloides |
| **RED ALGAE** |
| Abroteia suborbiculare |
| Acrosorium ciliolatum |
| Adamsiella chauvinii |
| Antithamnion pectinatum |
| Antithamnionella adnata |
| Aphanocladia delicatula |
| Apophlaea sinclairii |
| Arthrocardia corymbosa |
| Asparagopsis armata |
| 'Bangia' sp. |
| Bostrychia gracilis |
| Bostrychia intricata |
| Bostrychia vaga |
| Capreolia implexa |
| Catenella fusiformis |

| |
|---|
| Catenellopsis oligarthra |
| Caulacanthus ustulatus |
| Centroceras clavulatum |
| Ceramium uncinatum |
| Champia laingii |
| Champia novae-zelandiae |
| Chondria lanceolata |
| Cladhymenia coronata |
| Corallina ferreyrae |
| Dasya subtilis |
| Dasyclonium incisum |
| Dasyclonium ovalifolium |
| Delisea compressa |
| Dipterosiphonia dendritica |
| Erythrotrichia carnea |
| Gayliella flaccida |
| Gelidium caulacantheum |
| Gigartina laingii |
| Gigartina minuta |
| Gigartina sp. A |
| Gymnogongrus furcatus |
| Haraldiophyllum crispatum (Dellow 1955, as Myriogramme denticulata) |
| Helminthocladia australis |
| Hildenbrandia |
| Hymenena variolosa |
| Jania rosea |
| Jania sagittata |
| Jania crassa |
| Laurencia distichophylla |
| Laurencia thyrsifera |
| Liagora harveyana |
| Lithophyllum carpophylli |
| Lithophyllum pustulatum |
| Lomentaria caespitosa |
| Lomentaria umbellata |
| Lophothamnion hirtum |
| Lophurella caespitosa |

| |
|---|
| Melanthalia abscissa |
| Melobesia |
| Nesophila hoggardii |
| Nancythalia humilis |
| Nothogenia pulvinata |
| Pachymenia lusoria |
| Peyssonnelia rugosa |
| Phacelocarpus labillardierei |
| Pleurostichidium falkenbergii |
| Plocamium angustum |
| Plocamium cartilagineum |
| Plocamium cirrhosum |
| Polysiphonia pernacola |
| Polysiphonia scopulorum |
| Pterocladia lucida |
| Pterocladiella capillacea |
| Pyropia plicata |
| Rhizopogonia asperata |
| Rhodochorton purpureum |
| Rhodophyllis membranacea |
| Spyridia filamentosa |
| Symphyocladia marchantioides |
| Tsengia feredayae |
| Vertebrata aterrima |
| Vidalia colensoi |
| **GREEN ALGAE** |
| Blidingia minima var. minima (as Enteromorpha nana, Dellow 1955, Dromgoole 1964) |
| Bryopsis plumosa |
| Caulerpa flexilis |
| Caulerpa geminata |
| Cladophora crinalis |
| Codium convolutum |
| Derbesia novae-zelandiae |
| Lychaete herpestica |
| Microdictyon mutabile |
| Ulva australis |
| Ulva procera |

# Appendices

## Appendix 1: List of weeds present on Hauturu

| Common name | Scientific name | First found | Control | Distribution | Status in 2019 |
|---|---|---|---|---|---|
| **Bryophytes** | | | | | |
| Rogue fissidens moss | Fissidens taxifolius | 1980 | Digging out proposed | Te Maraeroa and Thumb Track | Common, but potential to contain |
| **Lycophytes** | | | | | |
| Selaginella | Selaginella kraussiana | 1978 | Weeding & spraying | Cultivated rangers' garden | Keeps popping up |
| **Dicotyledons** | | | | | |
| Bower vine | Pandorea jasminoides | 1981 | Pulled out | Forest behind bunkhouse | Two shoots controlled 2018 |
| Brazilian fireweed | Erechtites valerianifolia | 1981 | 2011 pulled out | Te Maraeroa, Shag Track and Hingaia | Still present Te Maraeroa |
| Broccoli | Brassica oleracea var. italica | 2015 | Pulled out | North coast | None since 2017 |
| Cape gooseberry | Physalis peruviana | 1981 | Dug out | Te Maraeroa | One seen in 2011 |
| Fig | Ficus carica | 1980 | 2009 | Orchard and rangers' garden. Spreading slightly by layering | Four cultivated historic trees to be kept |
| Forget-me-not | Myosotis – 2 spp. | 1937 | Dug out | By fig tree | None found recently |
| Japanese walnut | Juglans ailantifolia | 1980 | Cut down. Seedlings pulled/dug out | One tree near cowshed | Occasional seedlings still appearing |
| Loquat | Eriobotrya japonica | 1978 | Cut down | One tree in orchard producing seedlings. Found 2007, Valley Extension | None since 2010 |
| Mexican daisy | Erigeron karvinskianus | 2004 | 2004 | Te Wairere and Kiriraukawa streams | Not controlled |
| Mexican devil | Ageratina adenophora | 1940 | Hand pulled | Initially Pōhutukawa Flat, now widespread, mainly coastal | No longer under active control |
| Mint | Mentha ×piperata | 1921 | Dug out | Local west end Te Maraeroa | Still present 2019 |
| Mist flower | Ageratina riparia | 1963 | 1978 | Mostly Te Maraeroa | Now scarce. Apart from Hingaia (?) |

| Common name | Scientific name | First found | Control | Distribution | Status in 2019 |
|---|---|---|---|---|---|
| Moth plant | *Araujia hortorum* | 1989 | Pulled out | West Landing, Ōrau Gorge, Thumb Track | Occasional new plants |
| Pitted crassula | *Crassula multicava* | 1980 | Pulled out | Rangers' garden and east of quarantine shed | Not found recently |
| Prickly hakea | *Hakea sericea* | 1978 | 1980s. Cut down. Seeds collected. Seedlings pulled | Shrubland near East Cape | Under two-yearly visits, last visited 2016 with two juveniles removed |
| Purple groundsel/ purple ragwort | *Senecio elegans* | 2005 | Hand pulled | Te Tītoki Point 2005, home plot 2010, near mouth of Ōrau River 2017 | Occasionally found |
| Queensland poplar | *Homalanthus populifolius* | 1977 | Pulled out | Valley plot | A few seedlings each year |
| Sedum | *Sedum mexicanum* | 1982 | Removed by hand | Naturalised on boulder bank in front of rangers' house | Eradicated (?) |
| **Monocotyledons** | | | | | |
| Agapanthus | *Agapanthus praecox* | 1978 | Dug out | Te Waikohare Stream mouth, rangers' garden | Eradicated |
| Arum lily | *Zantedeschia aethiopica* | 1978 | Dug out | Widespread around Te Maraeroa and Tirikakawa Stream mouth | Under active control |
| Chinese windmill palm | *Trachycarpus fortunei* | 1973 | Cut down. Seedlings pulled | Behind bunkhouse | Nothing found since 2012 |
| Climbing asparagus | *Asparagus scandens* | 1971 | Hand pulled, Ohakiri 2018 | Widespread in southwestern side of island from Te Tītoki Point to 450 m ASL. Ohakiri Stream and Ōrau Gorge | Under active control |
| Dwarf umbrella sedge | *Cyperus eragrostis* | 2011 | Hand pulled, dug out | Home plot. Bunkhouse plot. By old tuatara/ aviary area in microlaena pasture | Still present by aviary 2018 |
| Elephant's ear | *Alocasia brisbanensis* | 1988 | Dug out | Bunkhouse and Te Waikohare Stream mouth | Under active control 2018 |

| Common name | Scientific name | First found | Control | Distribution | Status in 2019 |
|---|---|---|---|---|---|
| Italian arum | *Arum italicum* | 1978 | Dug out | Widespread around Te Maraeroa | Possibly eradicated |
| Jonquil | *Narcissus tazetta* | 1988 | Dug out | Sparingly naturalised, Te Waikohare Stream mouth | None found recently |
| Kikuyu grass | *Cenchrus clandestinus* | 2010 | Sprayed and hand pulled | House and bunkhouse | By helicopter lawn and by sheds still being controlled 2018 |
| Montbretia | *Crocosmia ×crocosmiiflora* | 1977 | Dug out | Bunkhouse and home plots, Te Waikohare Stream mouth | None seen since 2015 |
| Onion weed | *Allium triquetrum* | 1978 | Dug out | Te Waikohare Stream mouth | Still present |
| Pampas grass | *Cortaderia selloana* | 1972? Conf. 1980 | Sprayed, Velpar granules and hand pulled | All open slip areas inland and most creek beds, and all coastal cliff faces | Under island-wide control 2018 |
| Purple pampas grass | *Cortaderia jubata* | 1972? Conf. 1977 | Combined with *C. selloana* | If present is mixed with *C. selloana* | Not listed as separate from *C. selloana* for years |
| Smilax | *Asparagus asparagoides* | 1966 | Dug out | Single plant, boulder bank east of bunkhouse | Not seen since 2007 |
| Veldt grass | *Ehrharta erecta* | 2008 | Sprayed then covered in polythene and planted over | Bunkhouse | No new incursions found |
| Watsonia | *Watsonia meriana* 'Bulbillifera' | 1997 | Dug out | Opposite workshop in Montbretia | Still re-occurring from cormils. One plant a season. Low weed score |

# Appendix 2: Translocations to Hauturu, 1903–2013

The motivation behind most translocations to Hauturu was species recovery. Miskelly and Powlesland's (2013) comprehensive review of bird translocations in Aotearoa New Zealand is referenced here for most records, but in some cases this is a secondary reference. Miskelly and Powlesland (2013) should be consulted for primary references.

| Year | No. | Species | Source population | Outcome | References |
|---|---|---|---|---|---|
| 1898 | 2 | North Island brown kiwi | Unknown | Reinforcement | |
| 1903 | 4 | Kākāpō | Resolution Is, Fiordland | Failed | Miskelly & Powlesland 2013, Oliver 1955 |
| 1903 | 3 | North Island brown kiwi | Unknown | Reinforcement | Miskelly & Powlesland 2013, Oliver 1955, Oliver 1922 |
| 1913 | 1 | North Island brown kiwi (albino) | Taupō | Reinforcement | Oliver 1955 |
| 1915 | 19 | Great spotted kiwi | Gouland Downs | Failed | Miskelly & Powlesland 2013, Oliver 1955 |
| 1919 | 16 | North Island brown kiwi | Unknown | Reinforcement | Miskelly & Powlesland 2013, Oliver 1955 |
| 1925 | 12 | Tīeke/North Island saddleback | Taranga/Hen Is | Failed | Miskelly & Powlesland 2013, Oliver 1955, Lovegrove 1996 |
| 1980–1988 | 32 | North Island kōkako | Rotorua & Waikato | Established | Innes et al. 2013 |
| 1982 | 4 | Kākāpō | Te Pākeka/Maud Is | Ongoing | Miskelly & Powlesland 2013 |
| 1982 | 18 | Kākāpō | Rakiura/Stewart Is | Ongoing | Miskelly & Powlesland 2013 |
| 1984 | 50 | Tīeke/North Island saddleback | Repanga/Cuvier Is | Established | Lovegrove 1996b |
| 1986 | 46 | Tāiko/black petrel | Aotea/Great Barrier Is | Reinforcement | Miskelly & Powlesland 2013, Imber et al. 2003a |
| 1986 | 42 | Tīeke/North Island saddleback | Marotere/Chicken Is | Established | Lovegrove 1996b |
| 1987 | 60 | Tāiko/black petrel | Aotea/Great Barrier Is | Reinforcement | Miskelly & Powlesland 2013, Imber et al. 2003a |
| 1987 | 47 | Tīeke/North Island saddleback | Repanga/Cuvier Is | Established | Lovegrove 1996b |
| 1988 | 40 | Tāiko/black petrel | Aotea/Great Barrier Is | Reinforcement | |
| 1988 | 49 | Tīeke/North Island saddleback | Repanga/Cuvier Is | Established | Lovegrove 1996b |

| Year | No. | Species | Source population | Outcome | References |
|---|---|---|---|---|---|
| 1989 | 49 | Tāiko/black petrel | Aotea/Great Barrier Is | Reinforcement | Miskelly & Powlesland 2013, Imber et al. 2003a |
| 1990 | 54 | Tāiko/black petrel | Aotea/Great Barrier Is | Reinforcement | Miskelly & Powlesland 2013, Imber et al. 2003a |
| 1994 | 2 | North Island kōkako | Aotea/Great Barrier Is | Unknown | Innes et al. 2013 |
| 2012 | 1 | Kākāpō | Whenua Hou/Codfish Is | Ongoing | Miskelly & Powlesland 2013 |
| 2012 | 1 | Kākāpō | Pukenui/Anchor Is | Ongoing | Miskelly & Powlesland 2013 |
| 2012 | 7 | Kākāpō | Whenua Hou/Codfish & Pukenui/Anchor Is | Ongoing | Miskelly & Powlesland 2013 |
| 2013 | 2 | Kākāpō | Te Kākahu Tamatea/Chalky Is | Ongoing | Miskelly & Powlesland 2013 |
| 2014 | 1 | Kākāpō | Whenua Hou/Codfish Is | Ongoing | |
| 2017 | 4 | Kākāpō | Whenua Hou/Codfish Is | Ongoing | |
| 2018 | 1 | Kākāpō | Whenua Hou/Codfish Is | Ongoing | |

# Appendix 3: Translocations from Hauturu, 1932–2018

Most of the translocations of species from Hauturu through to the late 1990s were undertaken for species recovery, whereas most undertaken from the early 2000s onward were motivated by ecosystem restoration. Kākāpō are moved on and off the island as adults, chicks and eggs on a fairly frequent basis as part of national recovery efforts.

| Year | No. | Species | Release site | Purpose | Outcome | References |
|---|---|---|---|---|---|---|
| 1932 | 15 | Korimako/bellbird | Waitākere Ranges | Species recovery | Failed | Miskelly & Powlesland 2013, Turbott 1961 |
| 1964 | 6 | North Island brown kiwi | Pōnui Is | Species recovery | Established | Miskelly & Powlesland 2013 |
| 1980–81 | 46 | Hihi/stitchbird | Taranga/Hen Is | Species recovery | Failed | Miskelly & Powlesland 2013, Taylor et al. 2005 |
| 1982–85 | 66 | Hihi/stitchbird | Repanga/Cuvier Is | Species recovery | Failed | Miskelly & Powlesland 2013, Taylor et al. 2005 |
| 1983 | 30 | Hihi/stitchbird | Kāpiti Is | Species recovery | Failed | Miskelly & Powlesland 2013, Taylor et al. 2005 |
| 1984 | 30 | Hihi/stitchbird | Kāpiti Is | Species recovery | Failed | Miskelly & Powlesland 2013, Taylor et al. 2005 |
| 1989 | 40 | Pōpokotea/whitehead | Tiritiri Matangi | Ecological restoration | Established | Miskelly & Powlesland 2013 |
| 1990 | 12 | Hihi/stitchbird | Kāpiti Is | Species recovery | Established | Miskelly & Powlesland 2013, Taylor et al. 2005 |
| 1990 | 40 | Pōpokotea/whitehead | Tiritiri Matangi | Ecological restoration | Established | Miskelly & Powlesland 2013 |
| 1991 | 48 | Hihi/stitchbird | Kāpiti Is | Species recovery | Established | Miskelly & Powlesland 2013, Taylor et al. 2005 |
| 1992 | 3 | Kākāpō | Auckland Zoo | Species recovery | Ongoing | Miskelly & Powlesland 2013 |
| 1992 | 47 | Hihi/stitchbird | Kāpiti Is | Species recovery | Established | Miskelly & Powlesland 2013, Taylor et al. 2005 |
| 1994 | 40 | Hihi/stitchbird | Mokoia Is | Species recovery | Failed | Miskelly & Powlesland 2013, Taylor et al. 2005 |
| 1995 | 37 | Hihi/stitchbird | Tiritiri Matangi | Species recovery | Established | Miskelly & Powlesland 2013, Taylor et al. 2005 |
| 1995–2002 | 16 | Hihi/stitchbird | Pūkaha Mt Bruce | Species recovery | Captive | Miskelly & Powlesland 2013, Taylor et al. 2005 |
| 1995–96 | 7 | North Island kōkako | Kāpiti Is | Species recovery | Established | Innes et al. 2013 |
| 1996 | 4 | Kākāpō | Te Pākeka/Maud Is & Pearl Is | Species recovery | Ongoing | Miskelly & Powlesland 2013 |
| 1996 | 12 | Hihi/stitchbird | Tiritiri Matangi | Species recovery | Established | Miskelly & Powlesland 2013, Taylor et al. 2005 |

| Year | No. | Species | Release site | Purpose | Outcome | References |
|---|---|---|---|---|---|---|
| 1997 | 2 | Kākāpō | Whenua Hou/ Codfish Is | Species recovery | Ongoing | Miskelly & Powlesland 2013 |
| 1998 | 6 | Kākāpō | Te Pākeka/ Maud Is & Nukuwaiata Is | Species recovery | Ongoing | Miskelly & Powlesland 2013 |
| 1998 | | Pua o Te Rēinga/ dactylanthus | Tiritiri Matangi | Ecological restoration | Failed | Miskelly & Powlesland 2013, R. Walle pers. comm. |
| 1999 | 1 | Kākāpō | Whenua Hou/ Codfish Is | Species recovery | Ongoing | Miskelly & Powlesland 2013 |
| 1999 | 6 | Kākāpō | Whenua Hou/ Codfish Is | Species recovery | Ongoing | Miskelly & Powlesland 2013 |
| 2002 | 12 | Hihi/stitchbird | Kāpiti Is | Species recovery | Established | Miskelly & Powlesland 2013, Taylor et al. 2005 |
| 2008 | 6 | Wētāpunga/ giant wētā | Butterfly Creek, Auckland | Species & ecological restoration | Captive | Miskelly & Powlesland 2013, Oliver 1955 |
| 2008 | 31 | Red-crowned kākāriki | Motuihe Is | Ecological restoration | Established | Miskelly & Powlesland 2013, Ortiz-Catedral 2010 |
| 2009 | 12 | Wētāpunga/ giant wētā | Butterfly Creek, Auckland | Species & ecological restoration | Captive | C. Green, pers. comm. 2018 |
| 2009 | 20 | Hihi/stitchbird | Maungatautari, Waikato | Species recovery | Established | Miskelly & Powlesland 2013, Smuts-Kennedy & Parker 2013 |
| 2009 | 60 | Pōpokotea/ whitehead | Maungatautari, Waikato | Ecological restoration | Established | Miskelly & Powlesland 2013, Smuts-Kennedy & Parker 2013 |
| 2009 | 16 | Red-crowned kākāriki | Motuihe Is | Ecological restoration | Established | Miskelly & Powlesland 2013, Ortiz-Catedral 2010 |
| 2009 | 20 | North Island brown kiwi | Remutaka Ranges | Ecological restoration | Established | Miskelly & Powlesland 2013 |
| 2009 | 10 | Red-crowned kākāriki | Tāwharanui Open Sanctuary | Ecological restoration | Established | Miskelly & Powlesland 2013, Ortiz-Catedral 2010 |
| 2009 | 14 | Red-crowned kākāriki | Tāwharanui Open Sanctuary | Ecological restoration | Established | Miskelly & Powlesland 2013, Ortiz-Catedral 2010 |
| 2009 | 31 | Titipounamu/ North Island rifleman | Tiritiri Matangi | Ecological restoration | Established | Miskelly & Powlesland 2013 |
| 2010 | 12 | Wētāpunga/ giant wētā | Butterfly Creek, Auckland | Species & ecological restoration | Captive | C. Green, pers. comm. 2018 |

| Year | No. | Species | Release site | Purpose | Outcome | References |
|---|---|---|---|---|---|---|
| 2010 | 50 | Tītī/Cook's petrel | Cape Kidnappers, Hawke's Bay | Ecological restoration | In progress | Miskelly & Powlesland 2013 |
| 2010 | 30 | North Island brown kiwi | Pūkaha Mt Bruce | Ecological restoration | In progress | Miskelly & Powlesland 2013 |
| 2010 | 50 | Red-crowned kākāriki | Tāwharanui Open Sanctuary | Ecological restoration | Established | Miskelly & Powlesland 2013, Ortiz-Catedral 2010 |
| 2010 | 20 | Hihi/stitchbird | Tiritiri Matangi | Species recovery | Established | |
| 2010 | 14 | Titipounamu/ North Island rifleman | Tiritiri Matangi | Ecological restoration | Established | Miskelly & Powlesland 2013 |
| 2011 | 100 | Tītī/Cook's petrel | Cape Kidnappers, Hawke's Bay | Ecological restoration | In progress | Miskelly & Powlesland 2013 |
| 2011 | 15 | Titipounamu/ North Island rifleman | Tiritiri Matangi | Ecological restoration | Established | Miskelly & Powlesland 2013 |
| 2012 | 1 | Kākāpō | Pukenui/ Anchor Is, Fiordland | Species recovery | Ongoing | Miskelly & Powlesland 2013 |
| 2012 | 12 | Wētāpunga/ giant wētā | Auckland Zoo | Species & ecological restoration | Captive | C. Green, pers. comm. 2018 |
| 2012 | 12 | Wētāpunga/ giant wētā | Butterfly Creek, Auckland | Species & ecological restoration | Captive | C. Green, pers. comm. 2018 |
| 2012 | 82 | Tītī/Cook's petrel | Cape Kidnappers, Hawke's Bay | Ecological restoration | In progress | Miskelly & Powlesland 2013 |
| 2012 | 40 | Pōpokotea/ whitehead | Motuihe Is | Ecological restoration | Established | Miskelly & Powlesland 2013 |
| 2012 | 30 | Pōpokotea/ whitehead | Motutapu Is | Ecological restoration | Established | Miskelly & Powlesland 2013 |
| 2012 | 20 | Tīeke/ North Island saddleback | Motutapu Is | Ecological restoration | Established | Miskelly & Powlesland 2013 |
| 2012 | 30 | Pōpokotea/ whitehead | Rangitoto Is | Ecological restoration | Established | Miskelly & Powlesland 2013 |
| 2012 | 20 | Tīeke/ North Island saddleback | Rangitoto Is | Ecological restoration | Established | Miskelly & Powlesland 2013 |
| 2013 | 50 | Tītī/Cook's petrel | Boundary Stream, Hawke's Bay | Ecological restoration | In progress | D. Fastier, pers. comm. 2018 |

| Year | No. | Species | Release site | Purpose | Outcome | References |
|---|---|---|---|---|---|---|
| 2013 | 68 | Tītī/Cook's petrel | Boundary Stream, Hawke's Bay | Ecological restoration | In progress | D. Fastier, pers. comm. 2018 |
| 2013 | 50 | Tītī/Cook's petrel | Cape Kidnappers, Hawke's Bay | Ecological restoration | In progress | D. Fastier, pers. comm. 2018 |
| 2014 | 1 | Kākāpō | Auckland Zoo | Species recovery | Ongoing | R. Walle, pers. comm. 2018 |
| 2014 | 86 | Tītī/Cook's petrel | Boundary Stream, Hawke's Bay | Ecological restoration | In progress | D. Fastier, pers. comm. 2018 |
| 2014 | 40 | Pacific gecko | Motuora Is | Ecological restoration | In progress | S. Sinclair, pers. comm. 2018 |
| 2014 | 31 | Pacific gecko | Papakohatu/ Crusoe Is | Ecological restoration | In progress | S. Sinclair, pers. comm. 2018 |
| 2014 | 60 | Pōpokotea/ whitehead | Rotokare, Taranaki | Ecological restoration | Established | K. A. Parker, unpub. data 2018 |
| 2014 | 21 | Tīeke/ North Island saddleback | Rotokare, Taranaki | Ecological restoration | Established | K. A. Parker, unpub. data 2018 |
| 2014 | 40 | Pōpokotea/ whitehead | Rotoroa Is | Ecological restoration | Established | K. A. Parker, unpub. data 2018 |
| 2014 | 40 | Tīeke/ North Island saddleback | Rotoroa Is | Ecological restoration | Established | K. A. Parker, unpub. data 2018 |
| 2014 | 1 | Kākāpō | Whenua Hou/ Codfish Is | Species recovery | Ongoing | R. Walle, pers. comm. 2018 |
| 2015 | 82 | Tītī/Cook's petrel | Boundary Stream, Hawke's Bay | Ecological restoration | In progress | |
| 2016 | 106 | Tītī/Cook's petrel | Boundary Stream, Hawke's Bay | Ecological restoration | In progress | |
| 2016 | 1 | Kākāpō | Auckland Zoo | Species recovery | Ongoing | |
| 2016 | 12 | Wētāpunga/ giant wētā | Auckland Zoo | Species recovery | Ongoing | |
| 2017 | 40 | Red-crowned kākāriki | Moturua Is | Ecological restoration | In progress | K. A. Parker, unpub. data 2018 |
| 2018 | 20 | North Island kōkako | Pouiatoa, Taranaki | Ecological restoration | In progress | Kokako Specialist Group, unpub. data 2018 |

# Appendix 4: Summary of lizard species with confirmed records on Hauturu

Compiled from unpublished reports to the Department of Conservation by Tony Whitaker; habits and habitats as observed on northern offshore islands.

| Species | Habits and habitat | Notes |
|---|---|---|
| **Geckos: Family Diplodactylidae** | | |
| Raukawa gecko<br>*Woodworthia maculata* | Nocturnal, coastal rocky beaches to forest | May be more abundant in coastal areas when Pacific geckos are present |
| Duvaucel's gecko<br>*Hoplodactylus duvaucelii* | Nocturnal, coastal rocky beaches, cliffs, to forest | Foraging can vary from arboreal to terrestrial |
| Elegant gecko<br>*Naultinus elegans* | Diurnal, forest edge often in mānuka–kānuka | May extend into tall forest |
| Forest gecko<br>*Mokopirirakau granulatus* | Nocturnal, usually in mature forest, often associated with tall kānuka | Commonly reported sun basking |
| Pacific gecko<br>*Dactylocnemis pacificus* | Nocturnal, coastal rocky beaches to forest | May be more abundant in forested areas when Raukawa geckos are present |
| **Skinks: Family Scincidae** | | |
| Chevron skink<br>*Oligosoma homalonotum* | Diurnal, often associated with debris near streams but will inhabit hollow tree trunks | Extends to tall forest and can be arboreal including crowns of tree ferns |
| Copper skink<br>*O. aeneum* | Diurnal–nocturnal inhabitant of diverse habitats from coastal to tall forest | Varies in time active and often crepuscular; often associated with damp sites |
| Egg-laying (Suter's) skink<br>*O. suteri* | Nocturnal and strongly coastal; occupies a wide range of rocky habitats | Becomes particularly abundant around rotting seaweed; will forage in the open at night |
| Moko skink<br>*O. moco* | Strictly diurnal in open areas from coast to forest | Often particularly abundant in grassland and flax areas |
| Ornate skink<br>*O. ornatum* | Crepuscular, from vegetation around rocky coastal areas to forest | Often associated with damp sites |
| Shore skink<br>*O. smithi* | Strictly diurnal and largely confined to rocky coastal habitats | Will forage near rotting seaweed and on carrion such as fish remains |
| Striped skink<br>*O. striatum* | Diurnal, forest inhabiting | Some evidence that at least partly arboreal |
| Hauraki skink<br>*O. townsi* | Crepuscular–nocturnal in well-vegetated coastal sites, including coastal seabird burrows | Not known to extend into tall forest, but seems most common in sunny coastal sites |

# Appendix 5: Checklist of the birds of Hauturu

This list includes species known to breed, or to have bred on Hauturu, regular non-breeding visitors, rare vagrants and species extinct in historic period.
**Status**: B – Breeding, V – Regular visitor, Va – Vagrant, E – Extinct
**Abundance**: A – Abundant, C – Common, U – Uncommon, R – Rare

| Common name | Status | Notes | Key references |
|---|---|---|---|
| North Island brown kiwi | B, C | Widespread on flat and all forested areas | Turbott 1961, Colbourne 2005 |
| Australian brown quail | B, R | Former breeding resident, no recent sightings | Reischek 1887, Turbott 1961, ranger diaries |
| Canada goose | Va, R | Singles on flat 1932, 1933, 1934 | Turbott 1961, ranger diaries |
| greylag goose | Va, R | One on flat early 1960s | L. Wagener, pers. comm. |
| paradise shelduck | B, R | Bred on flat 1980s, also visiting | Ranger diaries, T. C. Greene pers. comm. |
| brown teal | B, U | A few pairs on flat and south coast streams | Turbott 1961 |
| grey duck | Va, R | Rare visitor to flat | Turbott 1961 |
| mallard | Va, R | Rare visitor to flat | T. Lovegrove unpubl. data |
| little penguin | B, C | Breeds around coastline | Turbott 1961 |
| grey-faced petrel | B, C | Recolonising coastal clifftops, c. 200–300 pairs | Turbott 1961, Rayner et al. 2009 |
| Cook's petrel | B, A | Breeds Aug–Apr, c. 400,000 pairs | Turbott 1961, Imber et al. 2003, Rayner et al. 2007 |
| fairy prion | V, U | Attracted to lights during recent petrel research | Turbott 1961, Bishop 1963, C. P. Gaskin pers. comm. |
| black petrel | B, C | Breeds Oct–Jul, c. 600 pairs | Turbott 1961, Imber 1987, Bell et al. 2018 |
| Buller's shearwater | V, U | Attracted to lights during recent petrel research | Turbott 1961, C. P. Gaskin pers. comm. |
| sooty shearwater | B, R | Listed by Reischek | Reischek 1887 |
| fluttering shearwater | B, C | A few pairs on Lots Wife, recolonising Hauturu | Turbott 1961, Bishop 1963, M. J. Rayner unpubl. data |
| little shearwater | V, U | Listed by Reischek | Reischek 1887, Turbott 1961 |
| white-faced storm petrel | V, U | Attracted to lights during recent petrel research | Turbott 1961, C. P. Gaskin pers. comm. |
| New Zealand storm petrel | B, U | Breeds summer–autumn, c. 300–400 pairs | Gaskin et al. 2011, Rayner et al. 2013, Rayner et al. 2015 |
| common diving petrel | B, U | Breeds on Lots Wife, recolonising Hauturu | Reischek 1887, Turbott 1961, M. J. Rayner unpub. data |
| Australasian gannet | V, C | Feeds along coast and offshore throughout year | Reischek 1887, Turbott 1961 |

| Common name | Status | Notes | Key references |
|---|---|---|---|
| little shag | Va, R | Listed by Reischek, one dead in Awaroa 19/12/72 | Reischek 1887, T. G. Lovegrove unpub. data |
| black shag | Va, R | Listed by Oliver 1922 | Oliver 1922 |
| pied shag | B, C | c. 10–15 pairs at Tirikakawa colony | Reischek 1887, Turbott 1961 |
| lesser frigate bird | Va, R | One ashore, died near homestead 4/3/51 | Turbott 1961 |
| plumed egret | Va, R | One on flat 1988 | K. A. Parker unpub. data |
| reef heron | Va, R | One at Te Hue 11/9/91 | Ranger diaries |
| cattle egret | Va, R | One on 8/4/86, one on 14/4/91 | Ranger diaries |
| white-faced heron | V, R | Visitor to flat when livestock present | T. G. Lovegrove unpub. data |
| New Zealand falcon | E | Assumed locally extinct, pair at Whēkau 9/1/67 | Reischek 1887, Turbott 1961, ranger diaries |
| swamp harrier | B, C | Assumed breeding, ranges over flat and forest areas | Reischek 1887, Turbott 1961 |
| banded rail | B, U | Recolonising sedgelands and forest edge on flat | Turbott 1961, C. R. Veitch unpub. data |
| marsh crake | Va, R | 1904 record by R. H. Shakespear | Oliver 1922 |
| spotless crake | B, U | Assumed breeding, damp sedgelands on flat | C. R. Veitch unpub. data |
| pūkeko | B, C | Small groups resident on flat | Turbott 1961 |
| variable oystercatcher | V, R | Listed by Hutton and Reischek, not seen recently | Hutton 1869, Reischek 1887 |
| banded dotterel | Va, R | One in 1937–1938 | Turbott 1961 |
| eastern curlew | Va, R | Two near homestead 19/9/45 | Turbott 1961 |
| Asiatic whimbrel | Va, R | One on 9/9/81 | Ranger diaries |
| spur-winged plover | V, R | Two on 12/8/89 | Ranger diaries |
| golden plover | Va, R | One on 20/10/82 | Ranger diaries |
| bar-tailed godwit | Va, R | Occasional visitors | Turbott 1961 |
| North Island snipe | E | Globally extinct, one specimen Auckland Museum | Hutton 1871, Turbott 1961, Miskelly 1988 |
| pied stilt | V, R | 1/2/1947, 12 south coast Feb 2002 | Turbott 1961, T. G. Lovegrove unpub. data |
| southern black-backed gull | B, C | c. 20–30 pairs on coastal stacks and headlands | Reischek 1887, Turbott 1961 |
| red-billed gull | B, C | Regular visitor, feeds along coast and offshore | Reischek 1887, Turbott 1961 |
| Caspian tern | V, U | Feeds in shallow waters along south coast | Turbott 1961 |
| white-fronted tern | B, C | c. 50–100 pairs Lots Wife and stacks around coast | Reischek 1887, Turbott 1961 |

| Common name | Status | Notes | Key references |
|---|---|---|---|
| New Zealand pigeon | B, A | Widespread, in most forest types | Turbott 1961 |
| Barbary dove | Va, R | One 'ring-necked dove' in garden 22/4/79 | Ranger diaries |
| rock pigeon | Va, R | Visitor to flat | C. Smuts-Kennedy pers. comm. |
| kākāpō | B, R | Managed under Kakapo Recovery Programme | Eason et al. 2006, Miskelly & Powlesland 2013 |
| eastern rosella | Va, R | One at 20/9/62, one 'exotic parrot' shot 19/8/64 | Ranger diaries |
| kākā | B, C | Widespread, in most forest types | Turbott 1961 |
| red-crowned kākāriki | B, A | Widespread, in most forest types and on flat | Turbott 1961, Greene 1998, 2003, Girardet et al. 2001 |
| yellow-crowned kākāriki | B, U | Sparse, often in kānuka and kauri/beech forest | Turbott 1961, Greene 1998, 2003, Girardet et al. 2001 |
| orange-fronted kākāriki | E? | Listed by Reischek 1887, not recorded since | Reischek 1887 |
| budgerigar | Va, R | Stray birds seen 21/10/81 and 17/5/87 | Ranger diaries |
| oriental cuckoo | Va, R | One on flat Oct–Nov 1971 | Reed 1972 |
| shining cuckoo | B, C | Migrant, present in spring–summer | Turbott 1961 |
| long-tailed cuckoo | B, C | Migrant, present in spring–summer | Turbott 1961, McLean 1988 |
| morepork | B, C | Widespread | Turbott 1961 |
| barn owl | Va, R | One on flat July 1993 | Smuts-Kennedy & Lovegrove 1996 |
| laughing kookaburra | Va, R | One on 9/9/45, one in Feb 1941 | Ranger diaries, Turbott 1961 |
| New Zealand kingfisher | B, C | Common resident, especially near the coast | Turbott 1961 |
| North Island rifleman | B, U | Formerly common, now sparse on lower slopes | Turbott 1961, McCallum 1982, Girardet et al. 2001 |
| North Island kōkako | B, C | Releases from 1980, c. 400 pairs by 2013 | Innes et al. 2012, Flux et al. 2013 |
| North Island saddleback | B, C | Releases 1984–1988, c. 6800 by 2013 | Turbott 1961, Lovegrove 1996b, Toy et al. 2018 |
| hihi/stitchbird | B, C | Widespread most forest types, c. 3100 in 2013 | Reischek 1887, Turbott 1961, Angehr 1986, Toy et al. 2018 |
| satin flycatcher (?) | Va, R | Bird fitting male satin flycatcher seen 8/6/59 | Ranger diaries |
| grey warbler | B, C | Widespread, especially in seral forest | Turbott 1961, Girardet et al. 2001 |

| Common name | Status | Notes | Key references |
|---|---|---|---|
| bellbird | B, A | Probably most common forest bird | Turbott 1961, Girardet et al. 2001 |
| tūī | B, A | Widespread, in most forest types, c. 4600 birds | Turbott 1961, Girardet et al. 2001, Toy et al. 2018 |
| whitehead | B, A | Widespread, especially in seral forest | Turbott 1961, Girardet et al. 2001 |
| Australian magpie | V, R | Visitor to flat | Turbott 1961, ranger diaries |
| North Island fantail | B, C | Widespread, especially in seral forest | Turbott 1961, Girardet et al. 2001 |
| rook | Va, R | One in Oct 1945, one on 5/9/78, one on 4/10/91 | Turbott 1961, ranger diaries |
| North Island tomtit | B, C | Widespread, especially in seral forest | Turbott 1961, Girardet et al. 2001 |
| North Island robin | B, A | Widespread most forest types | Turbott 1961, Girardet et al. 2001 |
| Eurasian skylark | B, R | On flat before stock removed | Turbott 1961 |
| silvereye | B, U | Scarce, mostly in seral forest | Turbott 1961, Girardet et al. 2001 |
| welcome swallow | B, C | Around buildings, a few elsewhere around coast | Ranger diaries (for early records) |
| Eurasian blackbird | B, C | A few on flat, sparse in forest | Turbott 1961, McCallum 1982, Girardet et al. 2001 |
| song thrush | B, U | A few on flat | Turbott 1961, McCallum 1982, Girardet et al. 2001 |
| common starling | B, U | On flat, declined since stock removed | Turbott 1961, McCallum 1982 |
| common myna | B, U | On flat, declined since stock removed | Turbott 1961, McCallum 1982 |
| New Zealand pipit | Va, R | Listed by Reischek, occasional visitors | Reischek 1887, Turbott 1961 |
| house sparrow | B, C | A few around buildings and on flat | Turbott 1961, McCallum 1982 |
| dunnock | B, U | Sparse, forest and scrub edges | Turbott 1961, McCallum 1982, Girardet et al. 2001 |
| chaffinch | B, U | A few pairs on flat, sparse in seral forest | Reischek 1887, Turbott 1961, McCallum 1982 |
| European greenfinch | B, U | A few on flat in rank pasture and sedge areas | Turbott 1961, McCallum 1982 |
| European goldfinch | B, U | A few on flat in rank pasture and sedge areas | Turbott 1961 |
| common redpoll | B, R | Scarce, possibly breeds in scrubland areas | Turbott 1961 |
| yellowhammer | B, U | A few on flat in rank pasture and sedge areas | Turbott 1961 |
| cirl bunting | Va, R | One likely on 26/6/69 | Ranger diaries |

# Appendix 6: Nationally and regionally threatened and uncommon vascular plant species recorded on Hauturu

| Nationally threatened (de Lange et al. 2017) | Conservation status | Present status on Hauturu (dates for most recent record) |
|---|---|---|
| Korthalsella salicornioides | Nationally critical | occasional, kānuka fringe of flats |
| Lophomyrtus bullata | Nationally critical | sparse, in forest |
| Scandia rosifolia | Nationally critical | occasional, coastal cliffs |
| Senecio scaberulus | Nationally critical | scarce, coastal fringe |
| Centipeda minima ssp. minima | Nationally endangered | seasonal, muddy coastal tracks |
| Lepidium oleraceum | Nationally endangered | presumed extinct (1945) |
| Agathis australis | Nationally vulnerable | plentiful, mostly middle ridges |
| Brachyglottis kirkii | Nationally vulnerable | locally common, lower slopes |
| Dactylanthus taylorii | Nationally vulnerable | sparse, rātā/tawaroa forest |
| Kunzea robusta | Nationally vulnerable | plentiful, cleared ridges |
| Leptinella tenella | Nationally vulnerable | sparse, coastal gullies |
| Metrosideros albiflora | Nationally vulnerable | abundant, high ridges |
| Metrosideros carminea | Nationally vulnerable | occasional, rātā/tawaroa forest |
| Metrosideros diffusa | Nationally vulnerable | occasional, valley forest |
| Metrosideros excelsa | Nationally vulnerable | common, coastal & higher |
| Metrosideros fulgens | Nationally vulnerable | common, mixed forest |
| Metrosideros parkinsonii | Nationally vulnerable | sparse, two summit areas |
| Metrosideros perforata | Nationally vulnerable | common, mixed forest |
| Metrosideros robusta | Nationally vulnerable | common, ridges & valleys |
| Metrosideros umbellata | Nationally vulnerable | common, higher ridges |
| Paspalum orbiculare | Nationally vulnerable | presumed extinct, coastal flats |
| Pimelea tomentosa | Nationally vulnerable | sparse, mānuka scrub |
| Solanum aviculare var. aviculare | Nationally vulnerable | occasional, river beds |
| Anthosachne kingiana ssp. multiflora | At risk – declining | sparse, coastal |
| Daucus glochidiatus | At risk – declining | presumed extinct (1897–1910) |
| Euphorbia glauca | At risk – declining | sparse, coastal flats |
| Kunzea amathicola | At risk – declining | presumed extinct, coastal |
| Leptospermum scoparium | At risk – declining | abundant, cleared areas, poor soils |
| Linum monogynum var. monogynum | At risk – declining | sparse, coastal banks |
| Mida salicifolia | At risk – declining | common, poor soils, lower ridges |
| Peraxilla tetrapetala | At risk – declining | sparse, cloud forest |
| Pittosporum kirkii | At risk – declining | sparse, high ridges |

| Nationally threatened (de Lange et al. 2017) | Conservation status | Present status on Hauturu (dates for most recent record) |
|---|---|---|
| Ptisana salicina | At risk – declining | presumed extinct, stream beds |
| Ranunculus urvilleanus | At risk – declining | locally common, wet flats |
| Rorippa divaricata | At risk – declining | sparse, coastal |
| Sonchus kirkii | At risk – declining | sparse, coastal |
| Trisetum arduanum | At risk – declining | presumed extinct (1949) |
| Zostera muelleri ssp. novazelandica | At risk – declining | presumed extinct (1897–1910) |
| Epilobium hirtigerum | At risk – recovering | presumed extinct (1896) |
| Carmichaelia williamsii | At risk – relict | locally common, cliffs/talus |
| Pisonia brunoniana | At risk – relict | abundant, coastal forest |
| Planchonella costata | At risk – relict | occasional, coastal forest |
| Sicyos mawhai | At risk – relict | sparse, coastal fringe |
| Blechnum norfolkianum | Naturally uncommon | common, kānuka/kauri |
| Botrychium australe | Naturally uncommon | presumed extinct (1897–1910) |
| Bromus arenarius | Naturally uncommon | sparse, coastal flats |
| Coprosma dodonaeifolia | Naturally uncommon | locally common, higher altitude |
| Coprosma neglecta | Naturally uncommon | sparse, Herekohu & coastal forest |
| Corunastylis pumila | Naturally uncommon | occasional, mānuka scrub |
| Corybas rivularis | Naturally uncommon | sparse, cloud forest |
| Danhatchia australis | Naturally uncommon | occasional, lower stream slopes |
| Hebe pubescens ssp. sejuncta | Naturally uncommon | common, coastal shrublands |
| Hypolepis dicksonioides | Naturally uncommon | sparse, coastal rockfalls |
| Lindsaea viridis | Naturally uncommon | scarce, Awaroa streamside |
| Notogrammitis rawlingsii | Naturally uncommon | sparse, kauri/hard beech |
| Pimelea acra | Naturally uncommon | single record, rock face |
| Schizaea dichotoma | Naturally uncommon | sparse, kauri forest |
| Thelymitra ixioides | Naturally uncommon | single record (1980) |
| Thismia rodwayi | Naturally uncommon | single record (1987) |
| Muehlenbeckia complexa var. grandifolia | Data deficient | single record, tawaroa forest |

| Regionally threatened (Stanley et al. 2005) | Conservation status | Present status on Hauturu (dates for most recent record) |
|---|---|---|
| Ascarina lucida var. lucida | Regionally critical | sparse, cloud forest |
| Blechnum colensoi | Regionally critical | single collection |
| Blechnum vulcanicum | Regionally critical | presumed extinct, old lit. record |
| Gastrodia cunninghamii | Regionally critical | occasional, no recent records |
| Nertera villosa | Regionally critical | single record |
| Nestegis cunninghamii | Regionally critical | sparse, lower to mid altitudes |
| Fuscospora solandri | Regionally critical | sparse, present on one ridge |
| Plantago raoulii | Regionally critical | sparse, no recent records |
| Luzula picta var. picta | Regionally endangered | presumed extinct, old lit. record |
| Pimelea tomentosa | Regionally endangered | local, mānuka scrub |
| Raukaua edgerleyi | Regionally vulnerable | common, higher altitude |
| Myoporum laetum | Gradual decline | sparse, coastal |
| Blechnum blechnoides | Sparse | sparse, stream mouths |
| Blechnum triangularifolium | Sparse | single record, coastal clay bank |
| Carex forsteri | Sparse | occasional, streamsides |
| Earina aestivalis | Sparse | locally common, epiphytic |
| Chenopodium triandrum | Sparse | occasional, coastal cliffs |
| Hypolepis lactea | Sparse | sparse, disturbed sites |
| Hypolepis rufobarbata | Sparse | sparse, cloud forest |
| Leptinella tenella | Sparse | sparse, stream mouths, gully sides |
| Microlaena polynoda | Sparse | sparse, recently found, Ōrau Gorge |
| Nestegis montana | Sparse | sparse, mid altitude |
| Olearia albida | Sparse | local, one population near coast |
| Ophioglossum coriaceum | Sparse | sparse, stream mouth |
| Pelargonium inodorum | Sparse | no recent records (1981) |
| Pteris comans | Sparse | common, coastal forest |
| Senecio quadridentatus | Sparse | no recent records (1981) |
| Tmesipteris sigmatifolia | Sparse | occasional, on tree fern trunks |
| Wahlenbergia vernicosa | Sparse | occasional, coastal banks |
| Archeria racemosa | Range restricted | common, higher altitudes |
| Blechnum nigrum | Range restricted | sparse, higher altitudes |
| Blechnum procerum | Range restricted | common, summit ridge |
| Astelia microsperma | Range restricted | common, higher altitudes |
| Dracophyllum traversii | Range restricted | common, near summit |
| Notogrammitis billardierei | Range restricted | occasional, cloud forest |
| Notogrammitis pseudociliata | Range restricted | sparse, mid altitude |
| Griselinia littoralis | Range restricted | common, cloud forest |

| Regionally threatened (Stanley et al. 2005) | Conservation status | Present status on Hauturu (dates for most recent record) |
|---|---|---|
| Hebe macrocarpa var. latisepala | Range restricted | common, kauri/hard beech forest |
| Hymenophyllum armstrongii | Range restricted | sparse, epiphytic in cloud forest |
| Hymenophyllum lyallii | Range restricted | abundant, epiphytic, cloud forest |
| Libertia micrantha | Range restricted | common, at higher altitudes |
| Loxsoma cunninghamii | Range restricted | presumed extinct (1960) |
| Melicytus lanceolatus | Range restricted | single record (1993) |
| Poa imbecilla | Range restricted | presumed extinct (1901) |
| Pseudopanax colensoi | Range restricted | occasional, cloud forest |
| Pseudopanax discolor | Range restricted | common, mid to upper altitudes |
| Sticherus flabellatus | Range restricted | sparse, lower stream sides |
| Trichomanes strictum | Range restricted | occasional, higher altitudes |
| Astelia fragrans | Data deficient | common, understorey on ridges |
| Epilobium chionanthum | Data deficient | presumed extinct (1901) |
| Euchiton delicatus | Data deficient | single record (1978) |
| Galium propinquum | Data deficient | presumed extinct (1897) |
| Leptolepia novae-zelandiae | Data deficient | presumed extinct (1897–1910) |
| Raukaua simplex | Data deficient | sparse, mid altitude |
| Thelymitra cyanea | Data deficient | single record (1980) |

# Notes

## Chapter 2: Papakāinga

1. Paula Morris, *Rangatira* (Auckland: Penguin, 2011), 134–35.
2. Andreas Reischek, *Yesterdays in Maoriland: New Zealand in the 80s* (London: Jonathan Cape, 1930), 91–92.
3. Gerhard Mueller to Surveyor General, 29 June 1895, L & S 4411/29, p. 1, BAAZ 1109/151B.
4. James Cowan, *Pictures of Old New Zealand: The Partridge Collection of Maori Paintings by Gottfried Lindauer* (Auckland: Whitcombe & Tombs, 1930), 78.
5. Two days earlier, the 'They Say' column in the *Observer* was mocking the planned eviction: 'the Commissioner of Crown Lands and the Crown Solicitor each wants the other to evict Little Barrier Tenetahi, but each is scared of Mrs [Rahui Te Kiri] Tenetahi', *Observer*, vol. XV, issue 890, 18 January 1896.
6. Charles John Alexander, 'Notes of a Holiday Trip around the North Island of New Zealand in the Government Steamer *Hinemoa* from 9 March to 10 April 1896', *Marine News* 48, nos 2–4 (1999–2000): 119.
7. Ibid.
8. *New Zealand Herald*, vol. XXXIII, issue 10034, 22 January 1896.
9. Robert Shakespear, *Diaries*, entry from 2 January 1897, unpublished manuscript, MS-274, Auckland Museum.
10. Report to the Crown Acquisition of Hauturu (Little Barrier Island), commissioned by the Waitangi Tribunal, by Ralph Johnson, February 1999, p. 57.

## Chapter 3: Island life

1. W. M. Hamilton & I. A. Atkinson, 'Vegetation; List of plants observed on Little Barrier', in *Little Barrier Island (Hauturu)*, Bulletin of New Zealand Department of Scientific and Industrial Research no. 137, ed. W. M. Hamilton (Wellington: Government Printer, 1961), 87–121.
2. Herbert Guthrie-Smith, *Bird Life on Island and Shore* (Edinburgh and London: W. Blackwood and Sons, 1925).

## Chapter 4: Restoration of Hauturu

1. J. Campbell, 'Seedling recovery on Hauturu/Little Barrier Island, after eradication of Pacific rats *Rattus exulans*', DOC Research & Development Series 325, 2011.
2. J. A. Moore, N. J. Nelson, S. N. Keall & C. H. Daugherty, 'Implications of social dominance and multiple paternity for the genetic diversity of a captive-bred reptile population (tuatara)', *Conservation Genetics* 9, no. 5 (2008): 1243–51.
3. S. A. B. Girardet, C. R. Veitch & J. L. Craig, 'Bird and rat numbers on Little Barrier Island, New Zealand, over the period of cat eradication 1976–80', *New Zealand Journal of Zoology* 28 (2001): 13–29.
4. C. Miskelly, 'The Little Barrier Island snipe', *Notornis* 35 (1988): 273–81.)
5. D. Campbell & I. Atkinson, 'Depression of tree recruitment by the Pacific rat (*Rattus exulans Peale*) on New Zealand's northern offshore islands', *Biological Conservation* 107 (September 2002): 19–35
6. R. Johnson, 'Report on the Crown Acquisition of Hauturu (Little Barrier Island)', Report commissioned by the Waitangi Tribunal, WAI 567 A1, 1999.
7. C. Gillies, 'Advances in New Zealand mammalogy 1990–2000: House cat', *Journal of the Royal Society of New Zealand* 31 (March 2001): 205–18.
8. W. M. Hamilton, *Little Barrier Island (Hauturu)*, Bulletin of the New Zealand Department of Scientific and Industrial Research no. 137 (Wellington: Government Printer, 1961), 1–198.
9. C. R. Veitch, 'The eradication of feral cats (*Felis catus*) from Little Barrier Island, New Zealand', *New Zealand Journal of Zoology* 28 (March 2001): 1–12.
10. David Towns & Keith Broome, 'From small Maria to massive Campbell: Forty years of rat eradications from New Zealand islands', *New Zealand Journal of Zoology* 30 (December 2003): 377–98.
11. P. L. Cromarty et al., 'Eradication planning for invasive alien animal species on islands — the approach developed by the New Zealand Department of Conservation', in *Turning the Tide: The eradication of invasive species*, eds C. R. Veitch & M. N. Clout (IUCN SSC Invasive Species Specialist Group, 2002), 85–91.
12. Towns & Broome, 'From small Maria to massive Campbell'.
13. Richard Griffiths, 'Little Barrier Island Kiore Eradication — Post Operational Report' (Warkworth: Department of Conservation, 2004).
14. Ibid.
15. Richard Griffiths, 'Hauturu Kiore Eradication — Post Eradication Target Species Monitoring and Assessment of Operational Success'

(Warkworth: Department of Conservation, 2006).
16. G. Wilson, 'Little Barrier Island Weed Management Report 2006' (Auckland: Department of Conservation, 2007).
17. Hamilton, *Little Barrier Island*.
18. J. Innes et al., 'Successful recovery of North Island kokako *Callaeas cinerea wilsoni* populations, by adaptive management', *Biological Conservation* 87 (March 1999): 201–14.
19. C. Daugherty et al., 'Conservation of tuatara on Hauturu (Little Barrier Island)', *New Zealand Journal of Zoology* 28 (2001): 362–63.
20. P. Gaze, *Tuatara Recovery Plan, 2001–2011* (Wellington: Biodiversity Recovery Unit, Department of Conservation, 2001).
21. M. J. Rayner et al., 'Using miniaturized radiotelemetry to discover the breeding grounds of the endangered New Zealand Storm Petrel *Fregetta maoriana*', *Ibis* 157 (2015): 754–66.
22. M. J. Imber, 'Breeding ecology and conservation of the black petrel (*Procellaria parkinsoni*)', *Notornis* 34 (1987): 19–39.
23. Richard Griffiths et al., 'Costs and benefits for biodiversity following rat and cat eradication on Te Hauturu-o-Toi/Little Barrier Island', in *Island Invasives: Scaling up to meet the challenge*, eds C. R. Veitch, M. N. Clout, A. R. Martin, J. C. Russell & C. J. West (Gland, Switzerland: IUCN, 2019), 558–67.
24. M. J. Rayner, M. E. Hauber, M. J. Imber, R. K. Stamp & M. N. Clout, 'Spatial heterogeneity of mesopredator release within an oceanic island system', *Proceedings of the National Academy of Sciences of the United States of America* 104 (2007): 20862–65. DOI:10.1073/pnas.0707414105.
25. Griffiths et al., 'Costs and benefits'.
26. M. J. Rayner, B. J. Dunphy & T. J. Landers, 'Grey-faced petrel (*Pterodroma macroptera gouldi*) breeding on Little Barrier Island, New Zealand', *Notornis* 56 (2009): 222–23.
27. R. Sibson, 'A visit to Little Barrier Island', *New Zealand Birds Notes* 2 (1947): 134–44.
28. D. H. Brunton, B. A. Evans & W. Ji, 'Assessing natural dispersal of New Zealand bellbirds using song type and song playbacks', *New Zealand Journal of Ecology* 32, no. 2 (2008): 147–54.
29. Griffiths et al., 'Costs and benefits'.
30. Lyn Wade, *Hauturu (Little Barrier Island) Kiwi Survey July 2009* (Warkworth: Department of Conservation, 2009).
31. N. Brown, 'Little Barrier Island Annual Reptile Monitoring Programme Jan/Feb 2013 Report' (Warkworth: Department of Conservation, 2013).
32. Ibid.
33. J. M. Hoare, 'Hauturu (Little Barrier Island) Reptile Survey — 23 February to 3 March 2009' (Warkworth: Department of Conservation, 2009).
34. Griffiths et al., 'Costs and benefits'.
35. C. J. Green, G. W. Gibbs, P. A. Barrett & M. City, 'Wetapunga (*Deinacrida heterocantha*) population changes following Pacific Rat (*Rattus exulans*) eradication on Little Barrier Island', in *Island Invasives: Eradication and management*, eds C. R. Veitch, M. N. Clout & D. R. Towns (Gland, Switzerland: IUCN, 2011), 305–08.
36. Griffiths et al., 'Costs and benefits'.
37. G. W. Gibbs, *New Zealand Butterflies: Identification and natural history* (Auckland: Collins, 1980).
38. Lyn Wade, *A Survey of Aquatic Invertebrates and Fish in a Selection of Intermittent Streams on Hauturu/Little Barrier Island* (Whangarei: NorthTec, Applied and Environmental Sciences, 2014).
39. Campbell 'Seedling recovery on Hauturu/Little Barrier Island, after eradication of Pacific rats *Rattus exulans*', DOC Research & Development Series 325.
40. Tadashi Fukami et al., 'Above- and below-ground impacts of introduced predators in seabird-dominated island ecosystems', *Ecology Letters* 9, no. 2 (January 2007): 1299–1307.

## Chapter 5: Geology of Hauturu

1. W. M. Hamilton, *The Little Barrier Island, Hauturu*, Bulletin of the New Zealand Department of Scientific and Industrial Research no. 54 (Wellington: Government Printer, 1937); *New Zealand Journal of Science and Technology* 17: 465–95; A. M. Hopgood & R. H. Barron, 'Notes on the Geology of Little Barrier Island', *Tane* 6 (1954): 7–19; D. Kear, 'Geology', in *Little Barrier Island (Hauturu)*, Bulletin of the New Zealand Department of Scientific and Industrial Research no. 137, ed. W. M. Hamilton (Wellington: Government Printer, 1961).
2. J. M. Lindsay, T. J. Worthington, I. E. M. Smith & P. M. Black, 'Geology, petrology, and petrogenesis of Little Barrier Island, Hauraki Gulf, New Zealand', *New Zealand Journal of Geology and Geophysics* 42, no. 2 (1999): 155–68.
3. Ibid.
4. Kear, 'Geology'.
5. After J. Lindsay & P. Moore, 'Geological

features of Little Barrier Island, Hauraki Gulf', *Tane* 35 (1995): 25–38.
6. Kear, 'Geology'.
7. W. M. Hamilton, *Little Barrier Island (Hauturu)*, Bulletin of the New Zealand Department of Scientific and Industrial Research no. 137 (Wellington: Government Printer, 1961).

## Chapter 6: Biota of Hauturu: Flora, fauna and fungi

1. A. J. Townsend, P. J. de Lange, C. A. J. Duffy, C. M. Miskelly, J. Molloy & D. A. Norton, *New Zealand Threat Classification System Manual* (Wellington: Department of Conservation, 2008).
2. A list of the 2008 to 2011 publications is available at https://www.doc.govt.nz/about-us/science-publications/conservation-publications/nz-threat-classification-system/nz-threat-classification-system-lists-2008-2011/. Publications since 2011 are listed at https://www.doc.govt.nz/about-us/science-publications/series/new-zealand-threat-classification-series.

### 6.1: Aquatic and terrestrial invertebrates

1. D. J. Campbell, H. Moller, G. W. Ramsay & J. C. Watt, 'Observations on foods of kiore (*Rattus exulans*) found in husking stations on northern offshore islands of New Zealand', *New Zealand Journal of Ecology* 7 (1984): 131–38.
2. D. F. Ward & A. Ramón-Laca, 'Molecular identification of the prey range of the invasive Asian paper wasp', *Ecology and Evolution* 3, no. 13 (2013): 4408–14.
3. T. R. Buckley & S. Bartlam, 'Revising the Threat Classification status of Data Deficient earthworms from the Auckland and Northland regions', Landcare Research Contract Report: LC0910/140, Lincoln, New Zealand, 2010.
4. M. I. Stevens, D. J. Winter, J. McCartney & P. Greenslade, 'New Zealand's giant Collembolla: New information on distribution and morphology for *Holacanthella* Börner, 1906 (Neanuridae: Uchidanurinae)', *New Zealand Journal of Zoology* 34, no. 1 (2007): 63–78.
5. J. Allwood, D. M. Gleeson, G. Mayer, S. Daniels, J. R. Beggs & T. R. Buckley, 'Support for vicariant origins of the New Zealand Onychophora', *Journal of Biogeography* 37 (2010): 669–81.
6. J. Monge-Nájera & B. Morera-Brenes, 'Velvet worms (Onychophora) in folklore and art: Geographic pattern, types of cultural reference and public perception', *British Journal of Education, Society & Behavioural Science* 10, no. 3 (2015): 1–9.
7. W. M. Hamilton, *Little Barrier Island (Hauturu)*, Bulletin of the New Zealand Department of Scientific and Industrial Research no. 137 (Wellington: Government Printer, 1961).
8. K. A. J. Wise, 'Aquatic insects of Little Barrier Island', *Records of the Auckland Institute and Museum* 4, no. 6 (1956): 321–27.
9. M. J. Winterbourn, 'A survey of the stream fauna of Little Barrier Island', *Tane* 10 (1964): 59–69; Lyn Wade, 'A survey of aquatic insects and fish in a selection of intermittent streams on Hauturu/Little Barrier Island', unpublished report, 2014.
10. See Winterbourn, 'A survey of the stream fauna of Little Barrier Island'.
11. J. A. Delago & R. L. Palmer, 'A revision of the *Podaena* Ordesh (Insecta: Coleoptera: Hydraenidae)', *Zootaxa* 2678 (2010): 1–47.
12. Wise, 'Aquatic insects of Little Barrier Island'; Winterbourn, 'A survey of the stream fauna of Little Barrier Island'; Wade, 'A survey of aquatic insects and fish'.
13. D. A. Craig, R. E. G. Craig & T. K. Crosby, 'Simuliidae (Insecta: Diptera)', *Fauna of New Zealand* 68 (2012).
14. B. J. Sinclair, 'Ceratomirinae (Diptera: Empidoidea: Brachystomatidae)', *Fauna of New Zealand* 74 (2017).
15. E. N. Milligan & J. J. Sumich, 'Mollusc species of Little Barrier Island', in 'Studies of three offshore islands', eds T. C. Chambers & R. L. Bieleski, *Tane* 6 (1953–1954): 119–26.
16. C. J. Green, G. W. Gibbs & P. A. Barratt, 'Wetapunga (*Deinacrida heterocantha*) population changes following Pacific rat (*Rattus exulans*) eradication on Little Barrier Island', in *Island Invasives: Eradication and management*, eds C. R. Veitch, M. N. Clout & D. R. Towns (Gland, Switzerland: ICUN, 2011).
17. C. Watts & D. Thornburrow, 'Habitat use, behaviour and movement patterns of a threatened New Zealand giant weta, *Deinacrida heterocantha* (Anostostomatidae: Orthoptera)', *Journal of Orthoptera Research* 20, no. 1 (2011): 127–35.
18. C. Watts, I. Stringer, G. Sherley, G. Gibbs & C. Green, 'History of weta (Orthoptera: Anostostomatidae) translocation in New Zealand: Lessons learned, islands as sanctuaries and the future', *Journal of Insect Conservation* 12 (2008): 359–70. DOI 10.1007/s10841-008-9154-5.
19. B. A. Holloway, 'A new bat-fly family from New Zealand (Diptera: Mystacinobiidae)', *New Zealand Journal of Zoology* 3, no. 4 (1976): 279–301.
20. C. Amiot & W. Ji, 'New host record

for *Ornithomya variegata* (Diptera: Hippoboscidae) in New Zealand with a review of previous records in Australasia', *Notornis* 62 (2015): 47-50.

21 C. W. Dick & B. D. Patterson, 'Bat flies: Obligate ectoparasites of bats', in *Micromammals and Macroparasites*, eds S. Morand, B. R. Krasnov & R. Poulin (Tokyo: Springer, 2006).

22 D. E. Hartnett, F. H. MacDonald, N. A. Martin, G. P. Walker & D. F. Ward, 'A survey of the adventive parasitoid *Meteorus pulchricornis* (Hymenoptera: Braconidae) and larval parasitoids of native Lepidoptera, New Zealand', *New Zealand Journal of Zoology* (March 2018): 326-40. DOI:10.1080/03014223.2018.1426021

23 D. E. Pattemore & D. S. Wilcove, 'Invasive rats and recent colonist birds partially compensate for the loss of endemic New Zealand pollinators', *Proceedings of the Royal Society B: Biological Sciences* 279 (2011): 1597-605.

24 G. W. Gibbs, 'Micropterigidae (Lepidoptera) of the Southwestern Pacific: A revision with the establishment of five new genera from Australia, New Caledonia and New Zealand', *Zootaxa* 2520 (2010): 1-48.

25 C. J. Painting & G. I. Holwell, 'Observations on the ecology and behaviour of the New Zealand giraffe weevil (*Lasiorhynchus barbicornis*)', *New Zealand Journal of Zoology* 41, no. 2 (2014): 147-53.

26 D. S. Seldon & R. A. B. Leschen, 'Revision of the *Mecodema curvidens* species group (Coleoptera: Carabidae: Broscini)', *Zootaxa* 2829 (2011): 1-45.

27 A. J. Drummond et al., 'Evaluating a multigene environmental DNA approach for biodiversity assessment', *GigaScience* 4 (2015): 46.

## 6.2: Amphibians and reptiles

1 W. M. Hamilton, *Little Barrier Island (Hauturu)*, Bulletin of the New Zealand Department of Scientific and Industrial Research no. 137 (Wellington: Government Printer, 1961).

2 K. M. Ramstad et al., 'Species and cultural conservation in New Zealand: Maori traditional ecological knowledge of tuatara', *Conservation Biology* 21, no. 2 (2007): 455-64.

3 Reviewed by A. Cree, *Tuatara: Biology and conservation of a venerable survivor* (Christchurch: Canterbury University Press, 2014), 583.

4 O. von Wettstein, '*Sphenodon punctatus reischeki* nov. subsp', *Zoologisches Anzeiger* 143: 45-47.

5 J. M. Hay, S. D. Sarre, D. M. Lambert, F. W. Allendorf & C. H. Daugherty, 'Genetic diversity and taxonomy: A reassessment of species designation in tuatara (*Sphenodon*: Reptilia)', *Conservation Genetics* 11, no. 3 (2010): 1063-81.

6 J. A. Moore, N. J. Nelson, S. N. Keall & C. H. Daugherty, 'Implications of social dominance and multiple paternity for the genetic diversity of a captive-bred reptile population (tuatara)', *Conservation Genetics* 9, no. 5 (2008): 1243-51.

7 M. B. Thompson, G. C. Packard, M. J. Packard & B. Rose, 'Analysis of the nest environment of Tuatara, *Sphenodon punctatus*', *Journal of Zoology* 238 (1996): 239-51.

8 S. N. Keall, N. J. Nelson & C. H. Daugherty, 'Securing the future of threatened tuatara populations with artificial incubation', *Herpetological Conservation and Biology* 5, no. 3 (2010): 555-62.

9 N. J. Mitchell, N. J. Nelson, A. Cree, S. Pledger, S. N. Keall & C. H. Daugherty, 'Support for a rare pattern of temperature-dependent sex determination in archaic reptiles: Evidence from two species of tuatara (*Sphenodon*)', *Frontiers in Zoology* 3 (2006): 9. https://doi.org/10.1186/1742-9994-3-9.

10 J. A. Moore et al., 'Implications of social dominance'.

11 Ibid.

12 A. Cree, 'Low annual reproductive output in female reptiles from New Zealand', *New Zealand Journal of Zoology* 21 (1994): 351-72.

13 D. R. Towns, 'Interactions between geckos, honeydew scale insects and host plants revealed on islands in northern New Zealand, following eradication of introduced rats and rabbits', in *Turning the Tide: The eradication of invasive species*, eds C. R. Veitch & M. N. Clout (Gland, Switzerland & Cambridge, UK: IUCN SSC Invasive Species Specialist Group, 2002), 329-35.

## 6.3: Birds

1 W. L. Buller, *A History of the Birds of New Zealand*, 2nd ed., 2 vols (London: W. L. Buller, 1888).

2 A. Reischek, 'Notes on New Zealand Ornithology: Observations on *Pogonornis cincta* (Dubus): Stitch-bird (Tiora)', *Transactions of the New Zealand Institute* 18 (1886): 84-87.

3 Buller, *A History of the Birds of New Zealand*; Reischek, 'Notes on New Zealand Ornithology'; E. G. Turbott, 'Birds', in *Little Barrier Island (Hauturu)*, Bulletin of the New Zealand Department of Scientific and Industrial Research no. 137, ed. W. M. Hamilton (Wellington: Government Printer, 1961), 136-75.

4   D. Medway, 'The land bird fauna of Stephens Island, New Zealand in the early 1890s, and the cause of its demise', *Notornis* 51 (2004): 201–11.
5   W. R. B. Oliver, *New Zealand Birds*, 2nd ed. (Wellington: Reed, 1955); Medway, 'The land bird fauna of Stephens Island'.
6   Turbott, 'Birds'.
7   B. W. Hayward, 'Prehistoric man on the offshore islands of northern New Zealand and his impact on the biota', in *The Offshore Islands of Northern New Zealand*, eds A. E. Wright & R. E. Beever (New Zealand Department of Lands and Survey Information Series No. 16, 1986), 139–52.
8   C. Gaskin & M. J. Rayner, *Seabirds of the Hauraki Gulf: Natural history, research and conservation* (Auckland: Hauraki Gulf Forum, 2013).
9   J. L. Smith, C. P. H. Mulder & J. C. Ellis, 'Seabirds as ecosystem engineers: Nutrient inputs and physical disturbance', in *Seabird Islands: Ecology, invasion and restoration*, eds C. P. H. Mulder, W. B. Anderson, D. R. Towns & P. J. Bellingham (New York: Oxford University Press, 2011), 27–55.
10  M. J. Rayner et al., 'Contemporary and historic separation of transhemispheric migration between two genetically distinct seabird populations', *Nature Communications* 2 (2011): 332.
11  M. J. Rayner, C. J. F. Carragher & M. E. Hauber, 'Mitochondrial DNA analysis reveals genetic structure in two New Zealand Cook's petrel (*Pterodroma cookii*) populations', *Conservation Genetics* 11 (2010): 2073–77; M. J. Rayner et al., 'Human mediated extirpations across a species' historic range revealed by ancient DNA reinforces taxa boundaries and informs conservation management of an endemic New Zealand seabird' (in review *Ibis*).
12  Gaskin & Rayner, *Seabirds of the Hauraki Gulf*.
13  M. J. Rayner et al., 'Foraging ecology of the Cook's petrel *Pterodroma cookii* during the austral breeding season: A comparison of its two populations', *Marine Ecology Progress Series* 370 (2008): 271–84.
14  Rayner et al., 'Contemporary and historic separation of transhemispheric migration'.
15  M. J. Imber, J. A. West & W. J. Cooper, 'Cook's petrel (*Pterodroma cookii*): Historic distribution, breeding biology and effects of predators', *Notornis* 50 (2003): 221–30.
16  M. J. Rayner, M. N. Clout, R. K. Stamp, M. J. Imber, D. H. Brunton & M. E. Hauber, 'Predictive habitat modelling for the population census of a burrowing seabird: A study of the endangered Cook's petrel', *Biological Conservation* 138 (2007): 235–47.
17  Ibid.; M. J. Rayner, M. E. Hauber & M. N. Clout, 'Breeding habitat of the Cook's petrel (*Pterodroma cookii*) on Little Barrier Island (Hauturu): Implications for the conservation of a New Zealand endemic', *Emu — Austral Ornithology* 107 (2007): 59–68.
18  Turbott, 'Birds'.
19  M. J. Rayner, M. E. Hauber, M. J. Imber, R. K. Stamp & M. N. Clout, 'Spatial heterogeneity of mesopredator release within an oceanic island system', *Proceedings of the National Academy of Sciences of the United States of America* 104 (2007): 20862–65.
20  B. S. Greene, G. A. Taylor & R. Earl, 'Distribution, population status and trends of grey-faced petrel (*Pterodroma macroptera gouldi*) in the northern North Island, New Zealand', *Notornis* 62 (2015): 143–61.
21  Turbott, 'Birds'.
22  R. B. Sibson, 'A visit to Little Barrier', *New Zealand Bird Notes* 2 (1947): 134–44; E. G. Turbott, 'Birds of Little Barrier Island', *New Zealand Bird Notes* 2 (1947): 92–108; H. R. McKenzie, 'Little Barrier Island birds in winter', *New Zealand Bird Notes* 3 (1948): 4–9.
23  M. J. Rayner, B. J. Dunphy & T. J. Landers, 'Grey-faced petrel (*Pterodroma macroptera gouldi*) breeding on Little Barrier Island, New Zealand', *Notornis* 56 (2009): 222–23.
24  M. J. Imber, 'Breeding ecology and conservation of the black petrel (*Procellaria parkinsoni*)', *Notornis* 34 (1987): 19–39.
25  R. Freeman, T. Dennis, T. Landers, D. Thompson, E. Bell, M. Walker & T. Guilford, 'Black petrels (*Procellaria parkinsoni*) patrol the ocean shelf-break: GPS tracking of a vulnerable procellariiform seabird', *PLoS ONE* 5 (2010): e9236.
26  B. D. Heather & H. A. Robertson, *The Field Guide to the Birds of New Zealand* (Auckland: Viking, 1996).
27  Turbott, 'Birds'; Imber, 'Breeding ecology and conservation of the black petrel'.
28  M. J. Imber, I. McFadden, E. A. Bell & R. P. Scofield, 'Post-fledging migration, age of first return and recruitment, and results of inter-colony translocation of black petrels (*Procellaria parkinsoni*)', *Notornis* 50 (2003): 183–90; C. M. Miskelly & R. G. Powlesland, 'Conservation translocations of New Zealand birds, 1863–2012', *Notornis* 60 (2013): 3–28.
29  Y. Richard & E. R. Abraham, *Risk of Commercial Fisheries to New Zealand Seabird Populations (2006–2011)*, New Zealand Aquatic Environment and Biodiversity Report no. 109 (Wellington: Ministry for Primary Industries, 2013).

30 E. A. Bell, D. Burgin, J. Sim, K. Dunleavy, A. Fleishman & R. P. Scofield, 'Population trends, breeding distribution and habitat use of black petrels (*Procellaria parkinsoni*) — 2016/2017 operational report', in *New Zealand Aquatic Environment and Biodiversity Report No 198* (2018).
31 B. M. Stephenson et al., 'The New Zealand storm petrel (*Pealeornis maoriana* Mathews, 1932): First live capture and species assessment of an enigmatic seabird', *Notornis* 55 (2008): 191–206.
32 M. J. Rayner et al., 'Using miniaturized radiotelemetry to discover the breeding grounds of the endangered New Zealand Storm Petrel *Fregetta maoriana*', *Ibis* 157 (2015): 754–66.
33 B. C. Robertson, B. M. Stephenson & S. J. Goldstien, 'When rediscovery is not enough: Taxonomic uncertainty hinders conservation of a critically endangered bird', *Molecular Phylogenetics and Evolution* 61 (2011): 949–52.
34 M. J. Rayner et al., 'Brood patch and sex ratio observations indicate breeding provenance and timing in New Zealand storm petrel (*Fregetta maoriana*)', *Marine Ornithology* 41 (2013): 107–11.
35 C. P. Gaskin, N. Fitzgerald, E. Cameron & S. Heiss-Dunlop, 'Does the New Zealand storm-petrel (*Pealeornis maoriana*) breed in northern New Zealand?', *Notornis* 58 (2011): 104–12.
36 Stephenson et al., 'The New Zealand storm petrel'.
37 Rayner et al., 'Using miniaturized radiotelemetry'.
38 Turbott, 'Birds'.
39 Ibid.
40 Turbott, 'Birds of Little Barrier Island'; E. W. Dawson, 'Bird notes from Little Barrier', *Notornis* 4 (1950): 27–31.
41 Turbott, 'Birds'.
42 Ibid.
43 E. Dieffenbach, *Travels in New Zealand; with contributions to the geography, geology, botany, and natural history of that country* (London: John Murray, 1843).
44 F. W. Hutton, 'Notes on the birds of the Little Barrier Island', *Transactions of the New Zealand Institute* 1 (1869): 162; Turbott, 'Birds'.
45 Turbott, 'Birds'. Buller, *A History of the Birds of New Zealand*.
46 W. R. B. Oliver, 'The birds of Little Barrier Island', *Emu* 22 (1922): 45–51.
47 R. L. Palma, 'A new species of *Rallicola* (Insecta: Phthiraptera: Philopteridae) from North Island brown kiwi', *Journal of the Royal Society of New Zealand* 21 (1991): 313–22.
48 J. Herbert & C. H. Daugherty, 'Genetic variation, systematics and management of kiwi (*Apteryx* spp)' (Wellington: Department of Conservation, 1994).
49 R. Colbourne, 'Kiwi (*Apteryx* spp.) on offshore New Zealand islands: Populations, translocations and identification of potential release sites', DOC Research & Development Series 208 (Wellington: Department of Conservation, 2005).
50 Oliver, 'The birds of Little Barrier Island'; Turbott, 'Birds'; Miskelly & Powlesland, 'Conservation translocations of New Zealand birds'.
51 Turbott, 'Birds'.
52 A. Reischek, 'Description of the Little Barrier Island, the birds which inhabit it and the locality as a protection to them', *Transactions of the New Zealand Institute* 19 (1887): 181–84.
53 Turbott, 'Birds'.
54 Dawson, 'Bird notes from Little Barrier'.
55 Turbott, 'Birds'.
56 Ibid.
57 Hutton, 'Notes on the birds of the Little Barrier Island'; Reischek, 'Description of the Little Barrier Island'; Reischek, 'Notes on New Zealand Ornithology'.
58 Turbott, 'Birds'.
59 Ibid.
60 Ibid.
61 Hutton, 'Notes on the birds of the Little Barrier Island'.
62 Reischek, 'Description of the Little Barrier Island'.
63 C. M. Miskelly, 'The Little Barrier Island snipe', *Notornis* 35 (1988): 273–81.
64 S. A. B. Girardet, C. R. Veitch & J. L. Craig, 'Bird and rat numbers on Little Barrier Island, New Zealand, over the period of cat eradication 1976–80', *New Zealand Journal of Zoology* 28 (2001): 13–29.
65 R. G. Powlesland, D. V. Merton & J. F. Cockrem, 'A parrot apart: The natural history of the kakapo (*Strigops habroptilus*), and the context of its conservation management', *Notornis* 53 (2006): 3–26.
66 A. S. Wilkinson & A. K. Wilkinson, *Kapiti Bird Sanctuary: A natural history of the island* (Masterton: Masterton Printing Co., 1952).
67 D. K. Eason et al., 'Breeding biology of kakapo (*Strigops habroptilus*) on offshore island sanctuaries 1990–2002', *Notornis* 53 (2006): 27–36.
68 A. Ballance, *Kakapo: Rescued from the brink of extinction* (Nelson: Craig Potton Publishing, 2010).
69 T. C. Greene, 'Foraging ecology of the red-crowned parakeet (*Cyanoramphus novaezelandiae novaezelandiae*) and yellow-

crowned parakeet (*C. auriceps auriceps*) on Little Barrier Island, Hauraki Gulf, New Zealand', *New Zealand Journal of Ecology* 22 (1998): 161–71.
70   T. C. Greene, 'Breeding biology of red-crowned parakeets (*Cyanoramphus novaezelandiae novaezelandiae*) on Little Barrier Island, Hauraki Gulf, New Zealand', *Notornis* 50 (2003): 83–99.
71   C. R. Veitch & B. D. Bell, 'Eradication of introduced animals from the islands of New Zealand', in *Ecological Restoration of New Zealand Islands*, Conservation Sciences Publication No. 2, eds D. R. Towns, C. H. Daugherty & I. A. E. Atkinson (Wellington: Department of Conservation, 1990), 137–46.
72   R. H. Taylor, 'A reappraisal of the Orange-fronted Parakeet (*Cyanoramphus* sp.) — species or colour morph?', *Notornis* 45 (1998): 49–63.
73   I .G. McLean, 'Breeding behaviour of the long-tailed cuckoo on Little Barrier Island', *Notornis* 35 (1988): 89–98; Heather & Robertson, *The Field Guide to the Birds of New Zealand*.
74   R. J. Pierce, 'A relict population of Rifleman in Northland', *Notornis* 41 (1994): 234.
75   Girardet et al., 'Bird and rat numbers on Little Barrier Island'.
76   Reischek, 'Description of the Little Barrier Island'.
77   J. Innes, L. E. Molles & H. Speed, 'Translocations of North Island kokako, 1981–2011', *Notornis* 60 (2013): 107–14.
78   I. Flux, T. Thurley, K. McKenzie & J. McAulay, 'A kokako (*Callaeas wilsoni*) population estimate for Little Barrier Island (Hauturu): A trial of a novel sub-sampling design June 2013', unpublished report to Kokako Specialist Group Wellington: Department of Conservation, 2013).
79   Hutton, 'Notes on the birds of the Little Barrier Island'; Buller, *A History of the Birds of New Zealand*; Turbott, 'Birds'.
80   Turbott, 'Birds of Little Barrier Island'; Turbott, 'Birds'; C. R. Veitch, 'The eradication of feral cats (*Felis catus*) from Little Barrier Island, New Zealand', *New Zealand Journal of Zoology* 28 (2001): 1–12.
81   T. G. Lovegrove, 'A comparison of the effects of predation by Norway (*Rattus norvegicus*) and Polynesian rats (*R. exulans*) on the saddleback (*Philesturnus carunculatus*)', *Notornis* 43 (1996): 91–112.
82   Ibid.; C. M. King, *Immigrant Killers* (Auckland: Oxford University Press, 1984).
83   T. G. Lovegrove, 'Island releases of saddlebacks *Philesturnus carunculatus* in New Zealand', *Biological Conservation* 77 (1996): 151–57.
84   Toy et al., 'Changes in density of hihi (*Notiomystis cincta*), tieke (*Philesturnus rufusater*) and tui (*Prosthemadera novaeseelandiae*) on Little Barrier Island (Te Hauturu-o-Toi), Hauraki Gulf, Auckland, 2005–2013', *New Zealand Journal of Ecology* 42 (2018): 149–57.
85   Hutton, 'Notes on the birds of the Little Barrier Island'.
86   G. R. Angehr, 'A bird in the hand: Andreas Reischek and the stitchbird', *Notornis* 31 (1984): 300–11.
87   Reischek, 'Notes on New Zealand Ornithology'.
88   G. R. Angehr, *Ecology and Behaviour of the Stitchbird with Recommendations for Management and Future Research*, NRAC Post-Doctoral Fellow 1981–1984 report to New Zealand Wildlife Service, Department of Internal Affairs, Wellington, November 1984; G. R. Angehr, 'Ecology of honeyeaters on Little Barrier Island: A preliminary survey', in *The Offshore Islands of Northern New Zealand*, eds A. E. Wright and R. E. Beever (New Zealand Department of Lands and Survey Information Series No. 16, 1986), 1–11.
89   Angehr, *Ecology and Behaviour of the Stitchbird*; G. R. Angehr, 'Establishment of the Stitchbird on Hen Island', *Notornis* 31 (1984): 175–77; Miskelly & Powlesland, 'Conservation translocations of New Zealand birds'.
90   L. R. Doerr, K. M. Richardson, J. G. Ewen & D. P. Armstrong, 'Effects of supplementary feeding on reproductive success of hihi (stitchbird, *Notiomystis cincta*) at a mature forest reintroduction site', *New Zealand Journal of Ecology* 41 (2017): 34–40.
91   Toy et al., 'Changes in density of hihi'.
92   I. Castro, E. O. Minot, R. A. Fordham & T. R. Birkhead, 'Polygynandry, face-to-face copulation and sperm competition in the hihi *Notiomystis cincta* (Aves: Meliphagidae)', *Ibis* 138 (1996): 765–71; M. Low, 'Female resistance and male force: Context and patterns of copulation in the New Zealand stitchbird *Notiomystis cincta*', *Journal of Avian Biology* 36 (2005): 436–48.
93   H. Guthrie-Smith, *Birdlife on Island and Shore* (Edinburgh and London: W. Blackwood and Sons, 1925).
94   J. G. Ewen, I. Flux & P. G. P. Ericson, 'Systematic affinities of two enigmatic New Zealand passerines of high conservation priority, the hihi or stitchbird *Notiomystis cincta* and the kokako *Callaeas cinerea*', *Molecular Phylogenetics and Evolution* 40 (2006): 281–84.
95   A. C. Driskell, L. Christidis, B. J. Gill, W.

E. Boles, F. K. Barker & N. W. Longmore, 'A new endemic family of New Zealand passerine birds: Adding heat to a biodiversity hotspot', *Australian Journal of Zoology* 55 (2007): 73–78.
96 Turbott, 'Birds'.
97 E. G. Turbott, 'Notes on the occurrence of the bellbird in Northern New Zealand', *Notornis* 5 (1953): 175–78.
98 J. L. Craig, A. M. Stewart & M. E. Douglas, 'The foraging of New Zealand honeyeaters', *New Zealand Journal of Zoology* 8 (1981): 87–91.
99 Toy et al., 'Changes in density of hihi'.
100 Turbott, 'Birds'.
101 Miskelly & Powlesland, 'Conservation translocations of New Zealand birds'.
102 Turbott, 'Birds'.
103 Hutton, 'Notes on the birds of the Little Barrier Island'; Turbott, 'Birds'.
104 Oliver, *New Zealand Birds*.
105 Hutton, 'Notes on the birds of the Little Barrier Island'; Turbott, 'Birds'.
106 Reischek, 'Description of the Little Barrier Island'.
107 Turbott, 'Birds of Little Barrier Island'; Turbott, 'Birds'.
108 Turbott, 'Birds'.
109 J. M. Diamond & C. R. Veitch, 'Extinctions and introductions in the New Zealand avifauna: Cause and effect?', *Science* 211 (1981): 499–501.
110 C. M. Miskelly, 'Changes in the forest bird community of an urban sanctuary in response to pest mammal eradications and endemic bird reintroductions', *Notornis* 65 (2018): 132–51.
111 I. A. E. Atkinson, 'Ecological restoration on islands: Prerequisites for success', in *Ecological Restoration of New Zealand Islands*, Conservation Sciences Publication No. 2, eds D. R. Towns, C. H. Daugherty & I. A. E. Atkinson (Wellington: Department of Conservation, 1990), 73–90.

### 6.4: Bats

1 C. F. J. O'Donnell, K. M. Borkin, J. E. Christie, B. Lloyd, S. Parsons & R. A. Hitchmough, *Conservation Status of New Zealand Bats, 2017*, New Zealand Threat Classification Series 21 (Wellington: Department of Conservation, 2018), 1–8.
2 K. M. Borkin & S. Parsons, 'Sex-specific roost selection by bats in clearfell harvested plantation forest: Improved knowledge advises management', *Acta Chiropterologica* 13, no. 2 (2011): 373–83.
3 A. M. Arkins, A. P. Winnington, S. Anderson & M. N. Clout, 'Diet and nectarivorous foraging behaviour of the short-tailed bat (*Mystacina tuberculata*)', *Journal of Zoology* 247, no. 2 (1999): 183–87.
4 A. L. Gurau, 'The diet of the New Zealand long-tailed bat, *Chalinolobus tuberculatus*', MA thesis, Massey University, Palmerston North, 2014.
5 D. K. Riskin, S. Parsons, W. A. Schutt, G. G. Carter & J. W. Hermanson, 'Terrestrial locomotion of the New Zealand short-tailed bat *Mystacina tuberculata* and the common vampire bat *Desmodus rotundus*', *Journal of Experimental Biology* 209, pt. 9 (2006): 1725–36.
6 Z. Czenze & T. Thurley, 'Weather and demographics affect Dactylanthus flower visitation by New Zealand lesser short-tailed bats', *New Zealand Journal of Ecology* 42, no. 1 (2018): 81–84.
7 C. A. Toth, A. W. Santure, G. I. Holwell, D. E. Pattemore & S. Parsons, 'Courtship behaviour and display-site sharing appears conditional on body size in a lekking bat', *Animal Behaviour* 136 (2018): 13–19.
8 Arkins et al., 'Diet and nectarivorous foraging behaviour of the short-tailed bat'.
9 Z. J. Czenze et al., 'Spatiotemporal and demographic variation in the diet of New Zealand lesser short-tailed bats (*Mystacina tuberculata*)', *Ecology and Evolution* 8, no. 15 (2018): 7599–610.
10 D. E. Pattemore & D. S. Wilcove, 'Invasive rats and recent colonist birds partially compensate for the loss of endemic New Zealand pollinators', *Proceedings of the Royal Society B: Biological Sciences* 279, no. 1733 (2012): 1597–1605.
11 Arkins et al., 'Diet and nectarivorous foraging behaviour'.
12 Z. J. Czenze, R. M. Brigham, A. J. R. Hickey & S. Parsons, 'Winter climate affects torpor patterns and roost choice in New Zealand lesser short-tailed bats', *Journal of Zoology* 303, no. 3 (2017): 236–43.

### 6.5: Vegetation and vascular flora

1 T. Kirk, 'Catalogue of plants found on the south and south-east coasts of Little Barrier Island, December 1867', *Transactions of the New Zealand Institute* 1 (1869): 155–56.
2 W. M. Hamilton, *The Little Barrier Island, Hauturu*, Bulletin of the New Zealand Department of Scientific and Industrial Research no. 54 (Wellington: Government Printer, 1937).
3 W. M. Hamilton & I. A. Atkinson, 'Vegetation; List of plants observed on Little Barrier', in *Little Barrier Island (Hauturu)*, Bulletin of

4   R. E. Beever, A. E. Esler, M. E. Young & E. K. Cameron, 'Checklist of vascular plants recorded from Hauturu-o-Toi (Little Barrier Island)', *Auckland Botanical Society Bulletin* 30 (2012): 1–86.
5   D. J. Campbell, 'Seedling recovery after eradication of Pacific rats *Rattus exulans*, Little Barrier Island, New Zealand', in *Department of Conservation Internal Report to Warkworth Area Office* (Warkworth: Department of Conservation, 2009).
6   I. Atkinson & R. Stanley, 'Vegetation and threatened plants of Hauturu Little Barrier Island', *Auckland Botanical Society Journal* 73 (2018): 37–41.
7   For more extensive vegetation descriptions, see Hamilton & Atkinson, 'Vegetation; List of plants observed on Little Barrier'.
8   I. A. E. Atkinson, 'Successional processes induced by fires on the northern islands of New Zealand', *New Zealand Journal of Ecology* 28 (2004): 181–93.
9   Beever et al., 'Checklist of vascular plants recorded from Hauturu-o-Toi'.
10  P. J. de Lange, J. R. Rolfe, J. W. Barkla et al., 'Conservation status of New Zealand's indigenous vascular plants, 2017' (Wellington: Department of Conservation, 2018). https://www.doc.govt.nz/Documents/science-and-technical/nztcs22entire.pdf.
11  D. G. Drury, 'Annotated key to the New Zealand shrubby Senecioneae-Compositae and their wild and garden hybrids', *New Zealand Journal of Botany* 11 (1973): 731–84.
12  I. A. E. Atkinson, R. L. Bieleski & F. J. Newhook, '*Metrosideros parkinsonii* Buchan. on Little Barrier Island', *Transactions of the Royal Society of New Zealand, Botany* 1 (1962): 279–84.
13  M. Wilcox, 'Black beech (*Nothofagus solandri* var. *solandri*) outlier on Little Barrier Island and cultivated trees in Auckland', Auckland Botanical Society Bulletin 59 (Auckland: Auckland Botanical Society, 2004), 65–66.
14  D. Campbell & I. Atkinson, 'Depression of tree recruitment by the Pacific rat (*Rattus exulans* Peale) on New Zealand's northern offshore islands', *Biological Conservation* 107 (September 2002): 19–35.
15  Hamilton & Atkinson, 'Vegetation; List of plants observed on Little Barrier'.
16  Ibid.
17  R. E. Beever, 'Large-leaved plants of northern offshore islands, New Zealand', in *The Offshore Islands of Northern New Zealand*, eds A. E. Wright and R. E. Beever (New Zealand Department of Lands and Survey Information Series No. 16, 1986), 51–61.
18  See S. H. Anderson, D. Kelly, J. J. Ladley, S. Molloy & J. Terry, 'Cascading effects of bird functional extinction reduce pollination and plant density', *Science* 331 (2011): 1068–71.

### 6.6: Mosses

1   W. M. Hamilton, *The Little Barrier Island, Hauturu*, Bulletin of the New Zealand Department of Scientific and Industrial Research no. 54 (Wellington: Government Printer, 1937), 1–87; W. M. Hamilton, *Little Barrier Island (Hauturu)*, Bulletin of the New Zealand Department of Scientific and Industrial Research no. 137 (Wellington: Government Printer, 1961); J. E. Beever, 'Appendix II: Mosses', in *Checklist of the Vascular Plants Recorded from Hauturu-ō-Toi*, Auckland Botanical Society Bulletin no. 30, ed. M. E. Young (Auckland: Auckland Botanical Society, 2012), 101–110; J. E. Braggins, 'Appendix I: Liverworts', in *Checklist of the Vascular Plants Recorded from Hauturu-ō-Toi*, Auckland Botanical Society Bulletin no. 30, ed. M. E. Young (Auckland: Auckland Botanical Society, 2012), 89–100.
2   A. J. Fife, 'Mitteniaceae', in *Flora of New Zealand — Mosses*, by P. B. Heenan, I. Breitwieser & A. D. Wilton (Lincoln: Manaaki Whenua Press, 2015). http://dx.doi.org/10.7931/B1PP4N
3   E. S. Gibb, A. D. Wilton, I. Schönberger, A. J. Fife, D. S. Glenny, J. E. Beever et al., *Checklist of the New Zealand Flora — Hornworts, Liverworts and Mosses* (Lincoln: Manaaki Whenua Press, Landcare Research, 2018). http://dx.doi.org/10.7931/P14K9Z
4   Beever, 'Appendix II'; J. E. Beever & I. G. Stone, 'Studies of *Fissidens* (Bryophyta: Musci) in New Zealand: *F. taxifolius* Hedw. and *F. integerrimus* Mitt.', *New Zealand Journal of Botany* 30 (1992): 237–46.
5   J. R. Rolfe, A. J. Fife, J. E. Beever, P. J. Brownsey & R. A. Hitchmough, *Conservation Status of New Zealand Mosses, 2014*, New Zealand Threat Classification Series 13 (Wellington: Department of Conservation, 2016).
6   A. J. Townsend, P. J. de Lange, C. A. J. Duffy, C. M. Miskelly, J. Molloy & D. A. Norton, *New Zealand Threat Classification System Manual* (Wellington: Department of Conservation, 2008).

7  Rolfe et al., *Conservation Status of New Zealand Mosses*.
8  Ibid.
9  *Checklist of the Vascular Plants Recorded from Hauturu-ō-Toi*, Auckland Botanical Society Bulletin no. 30 (Auckland: Auckland Botanical Society, 2012).

## 6.7: Liverworts and hornworts

1  J. E. Braggins, 'Liverworts and hornworts', in *Natural History of Rangitoto Island, Hauraki Gulf, Auckland, New Zealand*, Auckland Botanical Society Bulletin 27, ed. M. D. Wilcox (Auckland: Auckland Botanical Society, 2007), 127–34.
2  J. E. Braggins, 'Liverworts and Hornworts of Hauturu, Appendix I', in *Checklist of the Vascular Plants Recorded From Hauturu-ō-Toi*, Auckland Botanical Society Bulletin no. 30, ed. M. E. Young (Auckland: Auckland Botanical Society, 2012) 89–100.
3  J. J. Engel & D. Glenny, 'A flora of the liverworts and hornworts of New Zealand, vol. 1', in *Monographs in Systematic Botany*, vol. 110 (St Louis, Missouri: Missouri Botanical Garden Press, 2008), 1–897.

## 6.8: Lichens

1  According to D. J. Galloway, *Flora of New Zealand — Lichens*, revised 2nd edn (Manaaki Whenua Press, Landcare Research, 2007).

## 6.9: Fungi

1  F. J. Newhook, 'Climate and soil type in relation to *Phytophthora* attack on pine trees', *Proceedings of the New Zealand Ecological Society* 7 (1960): 14–15; F. D. Podger & F. J. Newhook, '*Phytophthora cinnamomi* in indigenous plant communities in New Zealand', *New Zealand Journal of Botany* 9, no. 4 (1971): 625–38. DOI: 10.1080/0028825X.1971.10430225.
2  F. J. Newhook, '*Phytophthora cinnamomi* in New Zealand', in *Root Diseases and Soil-Borne Pathogens*, T. A. Toussoun, R. V. Bega & P. E. Nelson (Berkeley: University of California Press, 1970), 173–76.
3  T. Beauchamp & N. Waipara, '*Phytophthora* in Forests and Natural Ecosystems, Esquel', 7th Meeting of the IUFRO Working Party, 07-02-09 Proceedings, Argentina, 10–14 November 2014.
4  J. K. Perrott & D. P. Armstrong, '*Aspergillus fumigatus* densities in relation to forest succession and edge effects: Implications for wildlife health in modified environments', *EcoHealth* 8 (2011): 290–300. https://doi.org/10.1007/s10393-011-0716-8.
5  M. A. Aiken, 'Fungi of Little Barrier Island', *Tane* 6 (1952): 82; M. A. Aiken & O. M. Shreeves, 'Fungi of Little Barrier Island', *Tane* 6 (1954): 82–83.
6  P. R. Johnston & P. F. Cannon, 'New *Phyllachora* species from *Myrsine* and *Rostkovia* from New Zealand', *New Zealand Journal of Botany* 42, no. 5 (2004): 921–33. DOI: 10.1080/0028825X.2004.9512938.
7  J. M. Dingley, 'The Hypocreales of New Zealand. II. The genus *Nectria*', *Transactions and Proceedings of the Royal Society of New Zealand* 79, no. 2 (1951): 177–202.
8  J. M. Dingley, 'The Hypocreales of New Zealand. V. The Genera *Cordyceps* and *Torrubiella*', *Transactions and Proceedings of the Royal Society of New Zealand* 81, no. 3 (1953): 329–43.
9  A. J. Drummond et al., 'Evaluating a multigene environmental DNA approach for biodiversity assessment', *GigaScience* 4 (2015): 46. DOI: 10.1186/s13742-015-0086-1.
10  A. Dopheide, 'DNA Meta-Barcoding and Molecular Ecology in a Forested Island Ecosystem', PhD thesis, University of Auckland, 2016.

## 6.10: Stream vertebrates

1  M. McGlynn, B. McDowall, C. Roberts, I. Petrov & S. Marsh, '*Vertebrate stream survey of Hauturu*', unpublished data (Wellington: Department of Conservation and National Institute for Water and Atmospherics, 2000).
2  Lyn Wade, 'A survey of aquatic invertebrates and fish in a selection of intermittent streams on Hauturu-o-Toi/ Little Barrier Island', unpublished data, 2014 (accessible at Unitec and Northtec libraries).
3  J. M. Waters & G. P. Wallis, 'Cladogenesis and loss of the marine life-history phase in freshwater galaxiid fishes (Osmeriformes: Galaxiidae)', *Evolution Journal* 55, no. 3 (March 2001): 587–97.
4  Department of Conservation, 'Conservation Status of New Zealand freshwater fish, 2013', *New Zealand Threat Classification Series* 7 (Wellington: Department of Conservation, 2014).

## 6.11: Seaweeds

1  V. Dellow, 'Seaweed vegetation of Little Barrier Island', *Auckland Botanical Society Quarterly Newsletter* 7, no. 2 (1950): 4–6; V. Dellow, 'Marine algal ecology of the Hauraki Gulf, New Zealand', *Transactions of the Royal Society of New Zealand* 83 (1955): 1–91; C. B. Trevarthen, 'Features of the marine ecology of Little Barrier, Mayor and Hen Islands', *Tane* 6 (1953–1954): 34–60; C. B. Trevarthen, 'Shore ecology', *in Little Barrier Island*, Bulletin of the New Zealand Department of Scientific and

Industrial Research no. 137, ed. W. M. Hamilton (Wellington: Government Printer, 1961), 122–31; P. L. Bergquist, 'The marine algal ecology of some islands of the Hauraki Gulf', *Proceedings of the New Zealand Ecological Society* 7 (1959): 43–45; P. L. Bergquist, 'Notes on the marine algal ecology of some exposed rocky shores of Northland, New Zealand', *Botanica Marina* 1, nos. 3–4 (1960): 86–94; F. I. Dromgoole, 'The ecology of the sublittoral boulder beaches at the Little Barrier Island', *Tane* 10 (1964): 70–78.
2  J. Morton, *Seashore Ecology of New Zealand and the Pacific* (Auckland: David Bateman Ltd, 2004).
3  A. D. Cotton, 'Marine algae from the north of New Zealand and the Kermadecs', *Kew Bulletin* 61 (1912): 256–64.
4  Dellow, 'Seaweed vegetation of Little Barrier Island'.
5  M. D. Wilcox, 'Seaweeds of Auckland', Auckland Botanical Society Bulletin no. 33 (Auckland: Auckland Botanical Society, 2018).
6  Dellow, 'Marine algal ecology of the Hauraki Gulf'.
7  Ibid.
8  Trevarthen, 'Features of the marine ecology of Little Barrier, Mayor and Hen Islands'.
9  Dromgoole, 'The ecology of the sublittoral boulder beaches'.

### Chapter 8: The future

1  Department of Conservation, 'Te Hauturu-o-Toi/Little Barrier Island Nature Reserve 2017 Management Plan' (Wellington: Department of Conservation, 2017). Available at: https://www.doc.govt.nz/about-us/our-policies-and-plans/statutory-plans/statutory-plan-publications/conservation-management-plans/te-hauturu-o-toi-little-barrier-island-conservation-management-plan/

# Bibliography

Aiken, M.A. 'Fungi of Little Barrier Island'. *Tane* 6 (1952): 82.

Aiken, M.A. & O. M. Shreeves. 'Fungi of Little Barrier Island'. *Tane* 6 (1954): 82–83.

Allwood, J., D. M. Gleeson, G. Mayer, S. Daniels, J. R. Beggs & T. R. Buckley. 'Support for vicariant origins of the New Zealand Onychophora'. *Journal of Biogeography* 37 (2010): 669–81.

Amiot, C. & W. Ji. 'New host record for *Ornithomya variegata* (Diptera: Hippoboscidae) in New Zealand with a review of previous records in Australasia'. *Notornis* 62 (2015): 47–50.

Anderson, S. H., D. Kelly, J. J. Ladley, S. Molloy & J. Terry. 'Cascading effects of bird functional extinction reduce pollination and plant density'. *Science* 331 (2011): 1068–71.

Angehr, G. R. 'A bird in the hand: Andreas Reischek and the stitchbird'. *Notornis* 31 (1984): 300–11.

Angehr, G. R. 'Establishment of the Stitchbird on Hen Island'. *Notornis* 31 (1984): 175–77.

Angehr, G. R. *Ecology and Behaviour of the Stitchbird with Recommendations for Management and Future Research*. NRAC Post-Doctoral Fellow 1981–1984 report to New Zealand Wildlife Service, Department of Internal affairs, Wellington, November 1984.

Angehr, G. R. 'Ecology of honeyeaters on Little Barrier Island: A preliminary survey'. In *The Offshore Islands of Northern New Zealand*, edited by A. E. Wright and R. E. Beever, 1–11. New Zealand Department of Lands and Survey Information Series 16, 1986.

Arkins, A. M., A. P. Winnington, S. Anderson & M. N. Clout. 'Diet and nectarivorous foraging behaviour of the short-tailed bat (*Mystacina tuberculata*)'. *Journal of Zoology* 247, no. 2 (1999): 183–87.

Atkinson, I. A. E. 'Spread of the ship rat (*Rattus r. rattus*) in New Zealand'. *Journal of the Royal Society of New Zealand* 3 (1973): 457–72.

Atkinson, I. A. E. 'Ecological restoration on islands: Prerequisites for success'. In *Ecological Restoration of New Zealand Islands*, Conservation Sciences Publication No. 2, edited by D. R. Towns, C. H. Daugherty & I. A. E. Atkinson, 73–90. Wellington: Department of Conservation, 1990.

Atkinson, I. A. E. 'Successional processes induced by fires on the northern islands of New Zealand'. *New Zealand Journal of Ecology* 28 (2004): 181–93.

Atkinson, I. A. E. & R. Stanley. 'Vegetation and threatened plants of Hauturu Little Barrier Island'. *Auckland Botanical Society Journal* 73 (2018): 37–41.

Atkinson, I. A. E., R. L. Bieleski & F. J. Newhook. '*Metrosideros parkinsonii* Buchan. on Little Barrier Island'. *Transactions of the Royal Society of New Zealand*, Botany 1 (1962): 279–84.

Ballance, A. *Kakapo: Rescued from the brink of extinction*. Nelson: Craig Potton Publishing, 2010.

Beauchamp, T. & N. Waipara. '*Phytophthora* in Forests and Natural Ecosystems'. 7th Meeting of the IUFRO Working Party 07-02-09 Proceedings. Esquel, Argentina, 10–14 November 2014.

Beever, J. E. 'Appendix II: Mosses'. In *Checklist of the Vascular Plants Recorded from Hauturu-ō-Toi (Little Barrier Island)*, Auckland Botanical Society Bulletin no. 30, edited by R. E. Beever, A. E. Esler, M. E. Young & E. K. Cameron, 101–10. Auckland: Auckland Botanical Society, 2012.

Beever, J. E. & I. G. Stone. 'Studies of *Fissidens* (Bryophyta: Musci) in New Zealand: *F. taxifolius* Hedw. and *F. integerrimus* Mitt.' *New Zealand Journal of Botany* 30 (1992): 237–46.

Beever, R.E. 'Large-leaved plants of northern offshore islands, New Zealand'. In *The Offshore Islands of Northern New Zealand*, edited by A. E. Wright and R. E. Beever, 51–61. New Zealand Department of Lands and Survey Information Series 16, 1986.

Beever, R. E., A. E. Esler, M. E. Young & E. K. Cameron (eds). *Checklist of the Vascular Plants Recorded from Hauturu-ō-Toi (Little Barrier Island)*, Auckland Botanical Society Bulletin no. 30. Auckland: Auckland Botanical Society, 2012.

Bell, B. D. & D. H. Brathwaite. 'The birds of Great Barrier and Arid Islands'. *Notornis* 10 (1964): 363–83.

Bell, E. A., D. Burgin, J. Sim, K. Dunleavy, A. Fleishman & R. P. Scofield. 'Population trends, breeding distribution and habitat use of black petrels (*Procellaria parkinsoni*) — 2016/2017 operational report'. In *New Zealand Aquatic Environment and Biodiversity Report No 198*, 2018.

Bergquist, P. L. 'The marine algal ecology of some islands of the Hauraki Gulf'. *Proceedings of the New Zealand Ecological Society* 7 (1959): 43–45.

Bergquist, P. L. 'Notes on the marine algal ecology of some exposed rocky shores of Northland, New Zealand'. *Botanica Marina* 1, nos 3–4 (1960): 86–94.

Bishop, L. J. 'Partial confirmation of two of Reischek's Little Barrier records'. *Notornis* 10 (1963): 306.

Borkin, K. M. & S. Parsons. 'Sex-specific roost selection by bats in clearfell harvested plantation forest: Improved knowledge advises management'. *Acta Chiropterologica* 13, no. 2 (2011): 373–83.

Braggins, J. E. 'Liverworts and hornworts'. In *Natural History of Rangitoto Island, Hauraki Gulf, Auckland, New Zealand*, Auckland Botanical Society Bulletin no. 27, edited by M. D. Wilcox, 127–34. Auckland: Auckland Botanical Society, 2007.

Braggins, J. E. 'Appendix I: Liverworts and Hornworts of Hauturu'. In *Checklist of the Vascular Plants Recorded from Hauturu-ō-Toi (Little Barrier Island)*, Auckland Botanical Society Bulletin no. 30, edited by R. E. Beever, A. E. Esler, M. E. Young & E. K. Cameron, 89–100. Auckland: Auckland Botanical Society, 2012.

Brown, N. 'Little Barrier Island Annual Reptile Monitoring Programme Jan/Feb 2013 Report'. Warkworth: Department of Conservation, 2013.

Brunton, D. H., B. A. Evans & W. Ji. 'Assessing natural dispersal of New Zealand bellbirds using song type and song playbacks'. *New Zealand Journal of Ecology* 32, no. 2 (2008): 147–54.

Buckley, T. R. & S. Bartlam. 'Revising the Threat Classification status of Data Deficient earthworms from the Auckland and Northland regions'. Landcare Research Contract Report: LC0910/140, Lincoln, 2010.

Buller, W. L. *A History of the Birds of New Zealand* (2nd ed.), 2 vols. London: W. L. Buller, 1888.

Campbell, D. & I. Atkinson. 'Depression of tree recruitment by the Pacific rat (*Rattus exulans* Peale) on New Zealand's northern offshore islands'. *Biological Conservation* 107 (September 2002): 19–35.

Campbell, D. J. 'Seedling recovery after eradication of Pacific rats *Rattus exulans*, Little Barrier Island, New Zealand'. In *Department of Conservation Internal Report to Warkworth Area Office*. Warkworth: Department of Conservation, 2009.

Campbell, D. J., H. Moller, G. W. Ramsay & J. C. Watt. 'Observations on foods of kiore (*Rattus exulans*) found in husking stations on northern offshore islands of New Zealand'. *New Zealand Journal of Ecology* 7 (1984): 131–38.

Castro, I., O. E. Minot, E. A. Fordham & T. R. Birkhead. 'Polygynandry, face-to-face copulation and sperm competition in the hihi *Notiomystis cincta* (Aves: Meliphagidae)'. *Ibis* 138 (1996): 765–71.

Colbourne, R. 'Kiwi (*Apteryx* spp.) on offshore New Zealand islands. Populations, translocations and identification of potential release sites'. In *Department of Conservation Research & Development Series 208*. Wellington: Department of Conservation, 2005.

Cotton, A. D. 'Marine algae from the north of New Zealand and the Kermadecs'. *Kew Bulletin* 61 (1912): 256–64.

Craig, D. A., R. E. G. Craig & T. K. Crosby. 'Simuliidae (Insecta: Diptera)'. *Fauna of New Zealand* 68 (2012).

Craig, J. L., A. M. Stewart & M. E. Douglas. 'The foraging of New Zealand honeyeaters'. *New Zealand Journal of Zoology* 8 (1981): 87–91.

Cree, A. 'Low annual reproductive output in female reptiles from New Zealand'. *New Zealand Journal of Zoology* 21 (1994): 351–72.

Cree, A. *Tuatara: Biology and conservation of a venerable survivor*. Christchurch: Canterbury University Press, 2014.

Cromarty, P. L. et al., 'Eradication planning for invasive alien animal species on islands — the approach developed by the New Zealand Department of Conservation'. In *Turning the Tide: The eradication of invasive species*, edited by C. R. Veitch & M. N. Clout, 85–91. IUCN SSC Invasive Species Specialist Group, 2002.

Czenze, Z. & T. Thurley. 'Weather and demographics affect Dactylanthus flower visitation by New Zealand lesser short-tailed bats'. *New Zealand Journal of Ecology* 42, no. 1 (2018): 81–84.

Czenze, Z. J., J. L. Tucker, E. L. Clare, J. E. Littlefair, D. Hemprich-Bennett, H. F. M. Oliveira et al. 'Spatiotemporal and demographic variation in the diet of New Zealand lesser short-tailed bats (*Mystacina tuberculata*)'. *Ecology and Evolution* 8, no. 15 (2018): 7599–610.

Czenze, Z. J., R. M. Brigham, A. J. R. Hickey & S. Parsons. 'Winter climate affects torpor patterns and roost choice in New Zealand lesser short-tailed bats'. *Journal of Zoology* 303, no. 3 (2017): 236–43.

Daugherty, C. et al. 'Conservation of tuatara on Hauturu (Little Barrier Island)'. *New Zealand Journal of Zoology* 28 (2001): 362–63.

Dawson, E. W. 'Bird notes from Little Barrier'. *Notornis* 4 (1950): 27–31.

de Lange, P. J., J. R. Rolfe, J. W. Barkla et al. *Conservation status of New Zealand's indigenous vascular plants, 2017*. Wellington: Department of Conservation, 2017. https://www.doc.govt.nz/Documents/science-and-technical/nztcs22entire.pdf (accessed August 2018).

Delago, J. A. & R. L. Palmer. 'A revision of the *Podaena* Ordesh (Insecta: Coleoptera: Hydraenidae)'. *Zootaxa* 2678 (2010): 1–47.

Dellow, V. 'Seaweed vegetation of Little Barrier Island'. *Auckland Botanical Society Quarterly Newsletter* 7, no. 2 (1950): 4–6.

Dellow, V. 'Marine algal ecology of the Hauraki Gulf, New Zealand'. *Transactions of the Royal Society of New Zealand* 83 (1955): 1–91.

Delmiglio, C. & M. N. Pearson. 'Effects and incidence of Cucumber mosaic virus, Watermelon mosaic virus and Zucchini yellow mosaic virus in New Zealand's only native cucurbit, *Sicyos australis*'. *Australasian Plant Pathology* 35 (2006): 1–7.

Department of Conservation. 'Conservation Status of New Zealand freshwater fish, 2013'. *New Zealand Threat Classification Series 7*. Wellington: Department of Conservation, 2014.

Diamond, J. M. & C. R. Veitch. 'Extinctions and introductions in the New Zealand avifauna: Cause and effect?' *Science* 211 (1981): 499–501.

Dick, C. W. & B. D. Patterson. 'Bat flies: Obligate ectoparasites of bats'. In *Micromammals and Macroparasites*, edited by S. Morand, B. R. Krasnov & R. Poulin. Tokyo: Springer, 2006.

Dieffenbach, E. *Travels in New Zealand; with contributions to the geography, geology, botany, and natural history of that country*. London: John Murray, 1843.

Dingley, J. M. 'The Hypocreales of New Zealand. II. The genus *Nectria*.' *Transactions and Proceedings of the Royal Society of New Zealand* 79, no. 2 (1951): 177–202.

Dingley, J. M. 'The Hypocreales of New Zealand. V. The Genera *Cordyceps* and *Torrubiella*.' *Transactions and Proceedings of the Royal Society of New Zealand* 81, no. 3 (1953): 329–43.

Doerr, L. R., K. M. Richardson, J. G. Ewen & D. P. Armstrong. 'Effects of supplementary feeding on reproductive success of hihi (stitchbird, *Notiomystis cincta*) at a mature forest reintroduction site'. *New Zealand Journal of Ecology* 41 (2017): 34–40.

Dopheide, A. 'DNA Meta-Barcoding and Molecular Ecology in a Forested Island Ecosystem'. PhD thesis, University of Auckland, 2016.

Driskell, A. C., L. Christidis, B. J. Gill, W. E. Boles, F. K. Barker & N. W. Longmore. 'A new endemic family of New Zealand passerine birds: Adding heat to a biodiversity hotspot'. *Australian Journal of Zoology* 55 (2007): 73–78.

Dromgoole, F. I. 'The ecology of the sublittoral boulder beaches at the Little Barrier Island'. *Tane* 10 (1964): 70–78.

Drummond, A. J., R. D. Newcomb, T. R. Buckley, X. Dong, A. Dopheide et al. 'Evaluating a multigene environmental DNA approach for biodiversity assessment'. *GigaScience* 4 (2015): 46.

Drury, D. G. 'Annotated key to the New Zealand shrubby Senecioneae-Compositae and their wild and garden hybrids'. *New Zealand Journal of Botany* 11 (1973): 731–84.

Eason, D. K., G. P. Elliott, D. V. Merton, P. W. Jansen, G. A. Harper & R. J. Moorhouse. 'Breeding biology of kakapo (*Strigops habroptilus*) on offshore island sanctuaries 1990–2002'. *Notornis* 53 (2006): 27–36.

Engel, J. J. & D. Glenny. 'A flora of the liverworts and hornworts of New Zealand, vol. 1'. In *Monographs in Systematic Botany*, vol. 110, 1–897. St Louis, Missouri: Missouri Botanical Garden Press, 2008.

Ewen, J. G., I. Flux & P. G. P. Ericson. 'Systematic affinities of two enigmatic New Zealand passerines of high conservation priority, the hihi or stitchbird *Notiomystis cincta* and the kokako *Callaeas cinerea*'. *Molecular Phylogenetics and Evolution* 40 (2006): 281–84.

Fife, A. J. 'Mitteniaceae'. In *Flora of New Zealand — Mosses*, by P. B. Heenan, I. Breitwieser & A. D. Wilton. Lincoln: Manaaki Whenua Press, Landcare Research, 2015. http://dx.doi.org/10.7931/B1PP4N

Flux, I., T. Thurley, K. McKenzie & J. McAulay. 'A kokako (*Callaeas wilsoni*) population estimate for Little Barrier Island (Hauturu): A trial of a novel sub-sampling design June 2013'. Unpublished report to Kokako Specialist Group. Wellington: Department of Conservation, 2013.

Freeman, R., T. Dennis, T. Landers, D. Thompson, E. Bell, M. Walker & T. Guilford. 'Black petrels (*Procellaria parkinsoni*) patrol the ocean shelf-break: GPS tracking of a vulnerable procellariiform seabird'. *PLoS ONE* 5 (2010): e9236.

Fukami, T. et al. 'Above and below ground impacts of introduced predators in seabird dominated island ecosystems'. *Ecology Letters* 9 (2006): 1299–1307.

Galloway, D. J. *Flora of New Zealand — Lichens*, revised 2nd edn. Lincoln: Manaaki Whenua Press, Landcare Research, 2007.

Gaskin, C. & M. J. Rayner. *Seabirds of the Hauraki Gulf: Natural history, research and conservation*. Auckland: Hauraki Gulf Forum, 2013.

Gaskin, C. P., N. Fitzgerald, E. Cameron & S. Heiss-Dunlop. 'Does the New Zealand storm-petrel (*Pealeornis maoriana*) breed in northern New Zealand?' *Notornis* 58 (2011): 104–12.

Gaze, P. *Tuatara Recovery Plan, 2001–2011*. Wellington: Biodiversity Recovery Unit, Department of Conservation, 2001.

Gibb, E. S., A. D. Wilton, I. Schönberger, A. J. Fife, D. S. Glenny, J. E. Beever et al. *Checklist of the New Zealand Flora — Hornworts, Liverworts and Mosses*. Lincoln: Manaaki Whenua Press, Landcare Research, 2018. http://dx.doi.org/10.7931/P14K9Z.

Gibbs, G. W. 'Micropteridgidae (Lepidoptera) of the Southwestern Pacific: A revision with the establishment of five new genera from Australia, New Caledonia and New Zealand'. *Zootaxa* 2520 (2010): 1–48.

Gibbs, G. W. *New Zealand Butterflies: Identification and natural history*. Auckland: Collins, 1980.

Gill, B. J. (convener). *Checklist of the Birds of New Zealand, Norfolk and Macquarie Islands, and the Ross Dependency, Antarctica*. Ornithological Society of New Zealand Inc., 4th edn. Wellington: Te Papa Press, 2010.

Gillies, C. 'Advances in New Zealand mammalogy 1990–2000: House cat'. *Journal of the Royal Society of New Zealand* 31 (2001): 205–18.

Girardet, S. A. B., C. R. Veitch & J. L. Craig. 'Bird and rat numbers on Little Barrier Island, New Zealand, over the period of cat eradication 1976–80'. *New Zealand Journal of Zoology* 28 (2001): 13–29.

Gravatt, D. J. 'Further estimates of the breeding density of land birds on Little Barrier Island'. *Tane* 16 (1970): 105–13.

Green, C. J., G. W. Gibbs & P. A. Barratt. 'Wetapunga (*Deinacrida heterocantha*) population changes following Pacific rat (*Rattus exulans*) eradication on Little Barrier Island'. In *Island Invasives: Eradication and management*, edited by C. R. Veitch, M. N. Clout & D. R. Towns, 305–08. Gland, Switzerland: ICUN, 2011.

Greene, B. S., G. A. Taylor & R. Earl. 'Distribution, population status and trends of grey-faced petrel (*Pterodroma macroptera gouldi*) in the northern North Island, New Zealand'. *Notornis* 62 (2015): 143–61.

Greene, T. C. 'Foraging ecology of the red-crowned parakeet (*Cyanoramphus novaezelandiae novaezelandiae*) and yellow-crowned parakeet (*C. auriceps auriceps*) on Little Barrier Island, Hauraki Gulf, New Zealand'. *New Zealand Journal of Ecology* 22 (1998): 161–71.

Greene, T. C. 'Breeding biology of red-crowned parakeets (*Cyanoramphus novaezelandiae novaezelandiae*) on Little Barrier Island, Hauraki Gulf, New Zealand'. *Notornis* 50 (2003): 83–99.

Griffiths, R. 'Little Barrier Island Kiore Eradication — Post Operational Report'. Warkworth: Department of Conservation, 2004.

Griffiths, R. et al. 'Hauturu Kiore Eradication — Post Eradication Target Species Monitoring and

Assessment of Operational Success'. Warkworth: Department of Conservation, 2006.

Griffiths, R., et al. 'Costs and benefits for biodiversity following rat and cat eradication on Te Hauturu-o-Toi/Little Barrier Island'. In *Island Invasives: Scaling up to meet the challenge*, edited by C. R. Veitch, M. N. Clout, A. R. Martin, J. C. Russell & C. J. West, 558–67. Gland, Switzerland: IUCN, 2019.

Gurau, A. L. 'The diet of the New Zealand long-tailed bat, *Chalinolobus tuberculatus*'. MA thesis, Massey University, Palmerston North, 2014.

Guthrie-Smith, H. *Birdlife on Island and Shore*. Edinburgh and London: W. Blackwood and Sons, 1925.

Hamilton, W. M. *The Little Barrier Island, Hauturu*. Bulletin of New Zealand Department of Scientific and Industrial Research no. 54. Wellington: Government Printer, 1937.

Hamilton, W. M. *Little Barrier Island (Hauturu)*. Bulletin of New Zealand Department of Scientific and Industrial Research no. 137. Wellington: Government Printer, 1961.

Hamilton, W. M. & I. A. Atkinson. 'Vegetation; List of plants observed on Little Barrier'. In *Little Barrier Island (Hauturu)*, Bulletin of New Zealand Department of Scientific and Industrial Research no. 137, compiled by W. M. Hamilton, 87–121. Wellington: Government Printer, 1961.

Hartnett, D. E., F. H. MacDonald, N. A. Martin, G. P. Walker & D. F. Ward. 'A survey of the adventive parasitoid *Meteorus pulchricornis* (Hymenoptera: Braconidae) and larval parasitoids of native Lepidoptera, New Zealand'. *New Zealand Journal of Zoology* (March 2018): 326–40. DOI:10.1080/03014223.2018.1426021

Hay, J. M., S. D. Sarre, D. M. Lambert, F. W. Allendorf & C. H. Daugherty. 'Genetic diversity and taxonomy: A reassessment of species designation in tuatara (*Sphenodon*: Reptilia)'. *Conservation Genetics* 11, no. 3 (2010): 1063–81.

Hay, J. R., M. E. Douglas & P. J. Bellingham. 'The North Island kokako (*Callaeas cinerea wilsoni*) on northern Great Barrier Island'. *Journal of the Royal Society of New Zealand* 15 (1985): 291–93.

Hayward, B. W. 'Prehistoric man on the offshore islands of northern New Zealand and his impact on the biota'. In *The Offshore Islands of Northern New Zealand*, edited by A. E. Wright and R. E. Beever, 139–152. New Zealand Department of Lands and Survey Information Series 16, 1986.

Hayward, B. W., A. E. Wright & G. C. Hayward. 'Lichens of Little Barrier Island (Hauturu), northern New Zealand'. *Records of the Auckland Institute and Museum* 28 (1991): 185–99.

Heather, B. D. & H. A. Robertson. *The Field Guide to the Birds of New Zealand*. Auckland: Viking, 1996.

Herbert, J. & C. H. Daugherty. 'Genetic variation, systematics and management of kiwi (*Apteryx* spp)'. Wellington: Department of Conservation, 1994.

Hoare, J. M. 'Hauturu (Little Barrier Island) Reptile Survey — 23 February to 3 March 2009'. Warkworth: Department of Conservation, 2009.

Holloway, B. A. 'A new bat-fly family from New Zealand (Diptera: Mystacinobiidae)'. *New Zealand Journal of Zoology* 3, no. 4 (1976): 279–301.

Hopgood, A. M. & R. H. Barron. 'Notes on the geology of Little Barrier Island'. *Tane* 6 (1954): 7–19.

Hutton, F. W. 'Notes on the birds of the Little Barrier Island'. *Transactions of the New Zealand Institute* 1 (1869): 162.

Hutton, F. W. 'Comments on a snipe from Little Barrier Island'. *Transactions of the New Zealand Institute* 3 (1871): 86.

Imber, M. J. 'Breeding ecology and conservation of the black petrel (*Procellaria parkinsoni*)'. *Notornis* 34 (1987): 19–39.

Imber, M. J., I. McFadden, E. A. Bell & R. P. Scofield. 'Post-fledging migration, age of first return and recruitment, and results of inter-colony translocation of black petrels (*Procellaria parkinsoni*)'. *Notornis* 50 (2003a): 183–90.

Imber, M. J., J. A. West & W. J. Cooper. 'Cook's petrel (*Pterodroma cookii*): Historic distribution, breeding biology and effects of predators'. *Notornis* 50 (2003b): 221–30.

Innes, J. et al. 'Successful recovery of North Island kokako *Callaeas cinerea wilsoni* populations, by adaptive management'. *Biological Conservation* 87 (1999): 201–14.

Innes, J., L. E. Molles & H. Speed. 'Translocations of North Island kokako, 1981–2011'. *Notornis* 60 (2013): 107–14.

Johnson, R. *Report on the Crown Acquisition of Hauturu (Little Barrier Island)*. Report commissioned by the Waitangi Tribunal, WAI 567 A1, 1999.

Johnston P. R. 'Rhytismataceae in New Zealand 2. The genus *Lophodermium* on indigenous plants'. *New Zealand Journal of Botany* 27, no. 2 (1989): 243–74. DOI: 10.1080/0028825X.1989.10410377.

Johnston, P. R. & P. F. Cannon. 'New *Phyllachora* species from *Myrsine* and *Rostkovia* from New Zealand'. *New Zealand Journal of Botany* 42, no. 5 (2004): 921–33. DOI: 10.1080/0028825X.2004.9512938.

Keall, S. N., N. J. Nelson & C. H. Daugherty. 'Securing the future of threatened tuatara populations with artificial incubation'. *Herpetological Conservation and Biology* 5, no. 3 (2010): 555–62.

Kear, D. 'Geology'. In *Little Barrier Island (Hauturu)*, Bulletin of New Zealand Department of Scientific and Industrial Research no. 137, compiled by W. M. Hamilton. Wellington: Government Printer, 1961.

Kikkawa, J. 'The breeding density of land birds on Little Barrier Island'. *Physiology and Ecology* 12 (1964): 127–38.

King, C. M. *Immigrant Killers*. Auckland: Oxford University Press, 1984.

Kirk, T. 'Catalogue of plants found on the south and south-east coasts of Little Barrier Island, December 1867'. *Transactions of the New Zealand Institute* 1 (1869): 155–56.

Lindsay, J. & P. Moore. 'Geological features of Little Barrier Island, Hauraki Gulf'. *Tane* 35 (1995): 25–38.

Lindsay, J. M., T. J. Worthington, I. E. M. Smith & P. M. Black. 'Geology, petrology, and petrogenesis of Little Barrier Island, Hauraki Gulf, New Zealand'. *New Zealand Journal of Geology and Geophysics* 42, no. 2 (1999): 155–68.

Lovegrove, T. G. 'A comparison of the effects of predation by Norway (*Rattus norvegicus*) and Polynesian rats (*R. exulans*) on the saddleback (*Philesturnus carunculatus*)'. *Notornis* 43 (1996a): 91–112.

Lovegrove, T. G. 'Island releases of saddlebacks *Philesturnus carunculatus* in New Zealand'. *Biological Conservation* 77 (1996b): 151–57.

Low, M. 'Female resistance and male force: Context and patterns of copulation in the New Zealand stitchbird *Notiomystis cincta*'. *Journal of Avian Biology* 36 (2005): 436–48.

Lücking, R., B. P. Hodkinson & S. D. Leavitt. 'The 2016 classification of lichenized fungi in the Ascomycota and Basidiomycota — Approaching one thousand genera'. *The Bryologist* 119, no. 4 (2016): 361–416.

McCallum, J. 'The penetration of exotic passerines into modified forests on Little Barrier Island, northern New Zealand'. *Tane* 28 (1982): 37–51.

McGlynn, M., B. McDowall, C. Roberts, I. Petrov & S. Marsh. 'Vertebrate Stream Survey of Hauturu'. Wellington: Department of Conservation & National Institute for Water and Atmospherics, 2000.

McKenzie, H. R. 'Little Barrier Island birds in winter'. *New Zealand Bird Notes* 3 (1948): 4–9.

McLean, I. G. 'Breeding behaviour of the long-tailed cuckoo on Little Barrier Island'. *Notornis* 35 (1988): 89–98.

Medway, D. 'The land bird fauna of Stephens Island, New Zealand in the early 1890s, and the cause of its demise'. *Notornis* 51 (2004): 201–11.

Milligan, E. N. & J. J. Sumich. 'Mollusc species of Little Barrier Island'. In 'Studies of Three Offshore Islands', edited by T. C. Chambers & R. L. Bieleski. *Tane* 6 (1953–1954): 119–26.

Miskelly, C. 'The Little Barrier Island snipe'. *Notornis* 35 (1988): 273–81.

Miskelly, C. M. 'Changes in the forest bird community of an urban sanctuary in response to pest mammal eradications and endemic bird reintroductions'. *Notornis* 65 (2018): 132–51.

Miskelly, C. M. & R. G. Powlesland. 'Conservation translocations of New Zealand birds, 1863–2012'. *Notornis* 60 (2013): 3–28.

Mitchell, N. J., N. J. Nelson, A. Cree, S. Pledger, S. N. Keall & C. H. Daugherty. 'Support for a rare pattern of temperature-dependent sex determination in archaic reptiles: Evidence from two species of tuatara (*Sphenodon*)'. *Frontiers in Zoology* 3 (2006): 9. https://doi.org/10.1186/1742-9994-3-9.

Monge-Nájera, J. & B. Morera-Brenes. 'Velvet worms (Onychophora) in folklore and art: Geographic pattern, types of cultural reference and public perception'. *British Journal of Education, Society & Behavioural Science* 10, no. 3 (2015): 1–9.

Moore, J. A., N. J. Nelson, S. N. Keall & C. H. Daugherty. 'Implications of social dominance and multiple paternity for the genetic diversity of a captive-bred reptile population (tuatara)'. *Conservation Genetics* 9, no. 5 (2008): 1243–51.

Morton, J. *Seashore Ecology of New Zealand and the Pacific*. Auckland: David Bateman, 2004.

Newhook, F. J. 'Climate and soil type in relation to *Phytophthora* attack on pine trees'. *Proceedings of the New Zealand Ecological Society* 7 (1960): 14–15.

Newhook, F. J. '*Phytophthora cinnamomi* in New Zealand'. In *Root Diseases and Soil-Borne Pathogens*, edited by T. A. Toussoun, R. V. Bega & P. E. Nelson, 173–76. Berkeley: University of California Press, 1970.

O'Donnell, C. F. J., K. M. Borkin, J. E. Christie, B. Lloyd, S. Parsons & R. A. Hitchmough. *Conservation Status of New Zealand Bats, 2017*, New Zealand Threat Classification Series 21. Wellington: Department of Conservation, 2018.

Oliver, W. R. B. 'The birds of Little Barrier Island'. *Emu* 22 (1922): 45–51.

Oliver, W. R. B. *New Zealand Birds*, 2nd edn. Wellington: Reed, 1955.

Ortiz-Catedral, L. 'Homing of a red-crowned parakeet (*Cyanoramphus novaezelandiae*) from Motuihe Island to Little Barrier Island, New Zealand'. *Notornis* 57 (2010): 48–49.

Painting, C. J. & G. I. Holwell. 'Observations on the ecology and behaviour of the New Zealand giraffe weevil (*Lasiorhynchus barbicornis*)'. *New Zealand Journal of Zoology* 41, no. 2 (2014): 147–53.

Palma, R. L. 'A new species of *Rallicola* (Insecta: Phthiraptera: Philopteridae) from North Island brown kiwi'. *Journal of the Royal Society of New Zealand* 21 (1991): 313–22.

Pattemore, D. E. & D. S. Wilcove. 'Invasive rats and recent colonist birds partially compensate for the loss of endemic New Zealand pollinators'. *Proceedings of the Royal Society B: Biological Sciences* 279 (2011): 1597–605.

Perrott, J. K. & D. P. Armstrong. '*Aspergillus fumigatus* densities in relation to forest succession and edge effects: Implications for wildlife health in modified environments'. *EcoHealth* 8 (2011): 290–300. https://doi.org/10.1007/s10393-011-0716-8.

Pierce, R. J. 'A relict population of Rifleman in Northland'. *Notornis* 41 (1994): 234.

Podger, F. D. & F. J. Newhook. '*Phytophthora cinnamomi* in indigenous plant communities in New Zealand'. *New Zealand Journal of Botany* 9, no. 4 (1971): 625–38. DOI: 10.1080/0028825X.1971.10430225.

Powlesland, R. G., D. V. Merton & J. F. Cockrem. 'A parrot apart: The natural history of the kakapo (*Strigops habroptilus*), and the context of its conservation management'. *Notornis* 53 (2006): 3–26.

Ramstad, K. M., N. J. Nelson, G. Paine, D. Beech, A. Paul, P. Paul, F. W. Allendorf & C. H. Daugherty. 'Species and cultural conservation in New Zealand: Maori traditional ecological knowledge of tuatara'. *Conservation Biology* 21, no. 2 (2007): 455–64.

Rayner, M. J., A. van Loenen, L. D. Shepherd, I. Cubrinovska, R. P. Scofield, A. J. D. Tennyson, M. Bunce & T. E. Steeves. 'Human mediated extirpations across a species' historic range revealed by ancient DNA reinforces taxa boundaries and informs conservation management of an endemic New Zealand seabird'. *Ibis*, in review.

Rayner, M. J., B. J. Dunphy & T. J. Landers. 'Grey-faced petrel (*Pterodroma macroptera gouldi*) breeding on Little Barrier Island, New Zealand'. *Notornis* 56 (2009): 222–23.

Rayner, M. J., C. F. J. Carragher & M. E. Hauber. 'Mitochondrial DNA analysis reveals genetic structure in two New Zealand Cook's petrel (*Pterodroma cookii*) populations'. *Conservation Genetics* 11 (2010): 2073–77.

Rayner, M. J., C. P. Gaskin, B. M. Stephenson, N. B. Fitzgerald, T. J. Landers, B. C. Robertson, P. R. Scofield, S. M. H. Ismar & M. J. Imber. 'Brood patch and sex ratio observations indicate breeding provenance and timing in New Zealand storm petrel (*Fregetta maoriana*)'. *Marine Ornithology* 41 (2013): 107–11.

Rayner, M. J., C. P. Gaskin, N. B. Fitzgerald, K. A. Baird, M. M. Berg, D. Boyle, L. Joyce et al. 'Using miniaturized radiotelemetry to discover the breeding grounds of the endangered New Zealand Storm Petrel *Fregetta maoriana*'. *Ibis* 157 (2015): 754–66.

Rayner, M. J., M. E. Hauber, M. J. Imber, R. K. Stamp & M. N. Clout. 'Spatial heterogeneity of mesopredator release within an oceanic island system'. *Proceedings of the National Academy of Sciences of the United States of America* 104 (2007): 20862–65. DOI:10.1073/pnas.0707414105.

Rayner, M. J., M. E. Hauber & M. N. Clout. 'Breeding habitat of the Cook's petrel (*Pterodroma cookii*) on Little Barrier Island (Hauturu): Implications for the conservation of a New Zealand endemic'. *Emu — Austral Ornithology* 107 (2007): 59–68.

Rayner, M. J., M. E. Hauber, M. N. Clout, D. S. Seldon, S. Van Dijken, S. Bury & R. A. Phillips. 'Foraging ecology of the Cook's petrel *Pterodroma cookii* during the austral breeding season: A comparison of its two populations'. *Marine Ecology Progress Series* 370 (2008): 271–84.

Rayner, M. J., M. E. Hauber, T. E. Steeves, H. A. Lawrence, D. R. Thompson, P. M. Sagar, S. J. Bury et al. 'Contemporary and historic separation of transhemispheric migration between two genetically distinct seabird populations'. *Nature Communications* 2 (2011): 332.

Rayner, M. J., M. N. Clout, R. K. Stamp, M. J. Imber, D. H. Brunton & M. E. Hauber. 'Predictive habitat modelling for the population census of a burrowing seabird: A study of the endangered Cook's petrel'. *Biological Conservation* 138 (2007): 235–47.

Reed, S. M. 'Oriental cuckoo on Little Barrier Island'. *Notornis* 19 (1972): 88.

Reischek, A. 'Notes on New Zealand ornithology: Observations on *Pogonornis cincta* (Dubus): Stitch-bird (Tiora)'. *Transactions of the New

Zealand Institute 18 (1886): 84–87.

Reischek, A. 'Observations on *Sphenodon punctatum*, fringe-backed lizard (Tuatara)'. *Transactions of the New Zealand Institute* 18 (1886): 108–10.

Reischek, A. 'Description of the Little Barrier Island, the birds which inhabit it and the locality as a protection to them'. *Transactions of the New Zealand Institute* 19 (1887): 181–84.

Reischek, A. *Yesterdays in Maoriland*, translated and edited by H. E. L. Priday. London: Jonathan Cape, 1930.

Richard, Y. & E. R. Abraham. *Risk of Commercial Fisheries to New Zealand Seabird Populations (2006–2011)*, New Zealand Aquatic Environment and Biodiversity Report no. 109. Wellington: Ministry for Primary Industries, 2013.

Riskin, D. K., S. Parsons, W. A. Schutt, G. G. Carter & J. W. Hermanson. 'Terrestrial locomotion of the New Zealand short-tailed bat *Mystacina tuberculata* and the common vampire bat *Desmodus rotundus*'. *Journal of Experimental Biology* 209, pt 9 (2006): 1725–36.

Robertson, B. C., B. M. Stephenson & S. J. Goldstien. 'When rediscovery is not enough: Taxonomic uncertainty hinders conservation of a critically endangered bird'. *Molecular Phylogenetics and Evolution* 61 (2011): 949–52.

Rolfe, J. R., A. J. Fife, J. E. Beever, J. Patrick, P. J. Brownsey & R. A. Hitchmough. *Conservation Status of New Zealand Mosses, 2014*, New Zealand Threat Classification Series 13. Wellington: Department of Conservation, 2016.

Seldon, D. S. & R. A. B. Leschen. 'Revision of the *Mecodema curvidens* species group (Coleoptera: Carabidae: Broscini)'. *Zootaxa* 2829 (2011): 1–45.

Sibson, R. 'A visit to Little Barrier Island'. *New Zealand Birds Notes* 2 (1947): 134–44.

Sibson, R. B. 'A visit to Little Barrier'. *Notornis* 3 (1949): 151–55.

Sinclair, B. J. 'Ceratomirinae (Diptera: Empidoidea: Brachystomatidae)'. *Fauna of New Zealand* 74 (2017).

Smith, J. L., C. P. H. Mulder & J. C. Ellis. 'Seabirds as ecosystem engineers: Nutrient inputs and physical disturbance'. In *Seabird Islands: Ecology, invasion and restoration*, edited by C. P. H. Mulder, W. B. Anderson, D. R. Towns & P. J. Bellingham, 27–55. New York: Oxford University Press, 2011.

Smuts-Kennedy, C. & K. A. Parker. 'Reconstructing biodiversity on Maungatautari'. *Notornis* 60 (2013): 107–14.

Smuts-Kennedy, C. & T. G. Lovegrove. 'A barn owl (*Tyto alba*) on Little Barrier Island'. *Notornis* 43 (1996): 49–50.

Stanley, R., P. J. de Lange & E. K. Cameron. 'Auckland Regional Threatened & Uncommon vascular plant list'. *Auckland Botanical Society Journal* 60 (2005): 152–57.

Stephenson, B. M., C. P. Gaskin, R. Griffiths, H. Jamieson, K. A. Baird, R. L. Palma & M. J. Imber. 'The New Zealand storm petrel (*Pealeornis maoriana* Mathews, 1932): First live capture and species assessment of an enigmatic seabird'. *Notornis* 55 (2008): 191–206.

Stevens, M. I., D. J. Winter, J. McCartney & P. Greenslade. 'New Zealand's giant Collembolla: New information on distribution and morphology for *Holacanthella* Börner, 1906 (Neanuridae: Uchidanurinae)'. *New Zealand Journal of Zoology* 34, no. 1 (2007): 63–78.

Taylor, R. H. 'A reappraisal of the Orange-fronted Parakeet (*Cyanoramphus* sp.) — species or colour morph?' *Notornis* 45 (1998): 49–63.

Taylor, S., I. Castro & R. Griffiths. 'Hihi/stitchbird (*Notiomystis cincta*) recovery plan 2004–09'. In *Threatened Species Recovery Plan 54*, Department of Conservation, 31. Wellington: Department of Conservation, 2005.

Thompson, M. B., G. C. Packard, M. J. Packard & B. Rose. 'Analysis of the nest environment of Tuatara, *Sphenodon punctatus*'. *Journal of Zoology, London* 238 (1996): 239–51.

Toth, C. A., A. W. Santure, G. I. Holwell, D. E. Pattemore & S. Parsons. 'Courtship behaviour and display-site sharing appears conditional on body size in a lekking bat'. *Animal Behaviour* 136 (2018): 13–19.

Towns, D. & K. Broome. 'From small Maria to massive Campbell: Forty years of rat eradications from New Zealand islands'. *New Zealand Journal of Zoology* 30 (2003): 377–98.

Towns, D. R. 'Interactions between geckos, honeydew scale insects and host plants revealed on islands in northern New Zealand, following eradication of introduced rats and rabbits'. In *Turning the Tide: The eradication of invasive species*, edited by C. R. Veitch & M. N. Clout, 329–35. Cambridge, UK: IUCN SSC Invasive Species Specialist Group, 2002.

Townsend, A. J., P. J. de Lange, C. A. J. Duffy, C. M. Miskelly, J. Molloy & D. A. Norton. *New Zealand Threat Classification System Manual*. Wellington: Department of Conservation, 2008.

Toy, R., T. C. Greene, B. S. Greene, A. Warren & R. Griffiths. 'Changes in density of hihi

(*Notiomystis cincta*), tieke (*Philesturnus rufusater*) and tui (*Prosthemadera novaeseelandiae*) on Little Barrier Island (Te Hauturu-o-Toi), Hauraki Gulf, Auckland, 2005–2013'. *New Zealand Journal of Ecology* 42 (2018): 149–57.

Trevarthen, C. B. 'Features of the marine ecology of Little Barrier, Mayor and Hen Islands'. *Tane* 6 (1953–1954): 34–60.

Trevarthen, C. B. 'Shore ecology'. In *Little Barrier Island (Hauturu)*, Bulletin of New Zealand Department of Scientific and Industrial Research no. 137, compiled by W. M. Hamilton, 122–31. Wellington: Government Printer, 1961.

Turbott, E. G. 'Birds of Little Barrier Island'. *New Zealand Bird Notes* 2 (1947): 92–108.

Turbott, E. G. 'Notes on the occurrence of the bellbird in Northern New Zealand'. *Notornis* 5 (1953): 175–78.

Turbott, E. G. 'Birds'. In *Little Barrier Island (Hauturu)*, Bulletin of New Zealand Department of Scientific and Industrial Research no. 137, compiled by W. M. Hamilton, 136–75. Wellington: Government Printer, 1961.

Veitch, C. R. 'The eradication of feral cats (*Felis catus*) from Little Barrier Island, New Zealand'. *New Zealand Journal of Zoology* 28 (2001): 1–12.

Veitch, C. R. & B. D. Bell. 'Eradication of introduced animals from the islands of New Zealand'. In *Ecological Restoration of New Zealand Islands*, Conservation Sciences Publication no. 2, edited by D. R. Towns, C. H. Daugherty & I. A. E. Atkinson, 137–46. Wellington: Department of Conservation, 1990.

von Wettstein, O. '*Sphenodon punctatus reischeki* nov. subsp.' *Zoologisches Anzeiger* 143 (1943): 45–47.

Wade, L. *A Survey of Aquatic Invertebrates and Fish in a Selection of Intermittent Streams on Hauturu/Little Barrier Island*. Whangarei: NorthTec, Applied and Environmental Sciences, 2014.

Wade, L. *Hauturu (Little Barrier Island) Kiwi Survey July 2009*. Warkworth: Department of Conservation, 2009.

Ward D. F. & A. Ramón-Laca. 'Molecular identification of the prey range of the invasive Asian paper wasp'. *Ecology and Evolution* 3, no. 13 (2013): 4408–14.

Waters, J. M. & G. P. Wallis. 'Cladogenesis and loss of the marine life-history phase in freshwater galaxiid fishes (Osmeriformes: Galaxiidae)'. *Evolution* 55, no. 3 (2001): 587–97.

Watts, C. & D. Thornburrow. 'Habitat use, behaviour and movement patterns of a threatened New Zealand giant weta, *Deinacrida heteracantha* (Anostostomatidae: Orthoptera)'. *Journal of Orthoptera Research* 20, no. 1 (2011): 127–35.

Watts, C., I. Stringer, G. Sherley, G. Gibbs & C. Green. 'History of weta (Orthoptera: Anostostomatidae) translocation in New Zealand: Lessons learned, islands as sanctuaries and the future'. *Journal of Insect Conservation* 12 (2008): 359–70. DOI 10.1007/s10841-008-9154-5.

Wilcox, M. 'Black beech (*Nothofagus solandri* var. *solandri*) outlier on Little Barrier Island and cultivated trees in Auckland'. *Auckland Botanical Society Journal* 59 (2004): 65–66.

Wilcox, M. D. 'Seaweeds of Auckland', Auckland Botanical Society Bulletin no. 33. Auckland: Auckland Botanical Society, 2018.

Wilkinson, A. S. & A. K. Wilkinson. *Kapiti Bird Sanctuary: A natural history of the island*. Masterton: Masterton Printing Co., 1952.

Wilson, G. *Little Barrier Island Weed Management Report 2006*. Auckland: Department of Conservation, 2007.

Winterbourn, M. J. 'A survey of the stream fauna of Little Barrier Island'. *Tane* 10 (1964): 59–69.

Wise, K. A. J. 'Aquatic insects of Little Barrier Island'. *Records of the Auckland Institute and Museum* 4, no. 6 (1956): 321–27.

Worthy, T. H. & R. N. Holdaway. *The Lost World of the Moa: Prehistoric life of New Zealand*. Bloomington, Indiana: Indiana University Press, 2002.

Wright, S. D. 'Regeneration of three northern canopy species'. PhD thesis, University of Auckland, 1993.

# About the editors

**LYN WADE** has been a member of the Little Barrier Island (Hauturu) Supporters' Trust since its inception in 1997; she has been a trustee since 2005 and is the current chairperson. Lyn is a registered nurse and has a Diploma in Conservation Management from Northtec and a Bachelor of Applied Science, Biodiversity Management from Unitec. Her first visit to Hauturu was in 1956 alongside her father, Bill Hamilton, in the course of researching his DSIR Bulletin 137 'Little Barrier Island (Hauturu)'. Lyn has made multiple visits to the island in various capacities since then. In 2018 Lyn was awarded a QSM for her services to conservation, in particular for work she has done on Te Hauturu-o-Toi.

**DICK VEITCH** spent his working career with the New Zealand Wildlife Service, now part of the Department of Conservation. After completing a Wildlife Traineeship he worked in various positions around the country with the final goal of work on the northern offshore islands. His work as a threatened species officer took him to most parts of New Zealand, its outlying islands and quite a few other islands around the world. His first contact with Hauturu was with the team carrying out the capture and transfer of kiwi to Pōnui Island, where they now thrive. He later managed the cat eradication project, transfer of hihi to other islands, the return of tīeke to Hauturu and the transfer of kōkako from North Island forests to Hauturu. Dick is now retired, but is still actively involved with restoration projects on Hauturu.

# About the contributors

**JESSICA BEEVER**, née Spragg, has Pākehā and Māori (Ngāti Toa, Te Ati Awa and Ngāti Mutunga) ancestry. She has an MSc in Botany from the University of Auckland, and a PhD in Plant Science from the University of Leeds. Jessica's first visits to Hauturu were as a student, and resulted in her first scientific publication, with her future husband Ross Beever. Subsequently she specialised in mosses, and contributed to a 'Bulletin on Plants recorded from Te Hauturu-o-Toi' (Auckland Botanical Society), based on herbarium data and collections and observations she made on later visits. She is research associate at Manaaki Whenua — Landcare Research, currently working on native and introduced mosses for its *Flora of New Zealand: Mosses*.

**JOHN E. BRAGGINS'** first interest was ferns, until he discovered liverworts while a BSc student. Fascinated by their form and colour, his interest in them grew and they have been his main area of research for the past decades. John has been able to work in the field in most of the country with local and visiting experts, and has expanded his work on these remarkable plants. Living in New Zealand with its rich liverwort flora (some 10 per cent of all species) means that many species are present in almost all of New Zealand's forests.

**PETER BUCHANAN** is a fungal scientist (mycologist) working for Manaaki Whenua — Landcare Research in Auckland. He is science team leader for 30 science staff studying Aotearoa New Zealand's fungal, bacterial, plant and invertebrate biodiversity. Staff also manage five national collections and databases of these organisms, in which many specimens and records from Hauturu are held. Peter has visited Hauturu as a mycologist in the 1980s, and more recently as a member of the Hauturu Supporters' Trust. He is active in engaging the public in biodiversity awareness; he assisted in establishing BioBlitz events in 2004 and annual NZ Fungal Forays in 1986. Peter encourages school students to understand fungal ecology and traditional uses, while advocating for conservation of our threatened fungi.

**EWEN CAMERON** has been curator of botany at Auckland Museum for over 27 years. He has had a lifelong interest in natural history, especially botany. He has a particular passion for identifying and collecting wild plants (both indigenous and naturalised) of northern New Zealand, especially from the offshore islands. Over 17,000 of his collections are held in the museum's herbarium and he has documented the flora and vegetation of more than 60 islands and islets of the wider Hauraki Gulf. The Auckland Museum herbarium holds more than 1600 vascular plant specimens from Hauturu. Ewen first visited the island in 1981 and has revisited several times, including early in 2018 to search for new naturalised plants and to document the vegetation of Te Maraeroa.

**ROGER GRACE** is a semi-retired consultant marine biologist with extensive knowledge of the marine life of northeastern New Zealand. He has stayed on many northern islands with the Offshore Islands Research Group, and has studied and photographed fish and benthic life of the region for many years. He is a frequent speaker to clubs and societies, has written numerous magazine articles, and has a passion for marine conservation and promoting no-take marine areas or marine reserves. Since the mid-1970s he has monitored recovery of fish, crayfish and kelp forests at Tāwharanui Marine Reserve, and contrasted lack of recovery at Mimiwhangata Marine Park, where recreational fishing is allowed. Roger received the Queen's Service Medal for public service in 2005.

**RICHARD GRIFFITHS** gained his MSc at Lincoln University in 1996. He began working for the Department of Conservation in 1998 and over the next 13 years spearheaded a number of threatened species programmes and island restoration projects, including the operations to remove Pacific rats from Hauturu and eight pest species from Rangitoto and Motutapu. Richard led the Hihi Recovery Group between 2000 and 2007, overseeing the species' successful reintroduction to the mainland after a 120-year absence. He now works for the non-profit organisation Island Conservation where he leads a team of project managers and island restoration specialists to prevent extinctions through the removal of invasive vertebrates from islands across the world. He remains an honorary member of the Department of Conservation's Island Eradication Advisory Group.

**BRUCE W. HAYWARD** is a geologist and marine ecologist. He and his wife Glenys Hayward developed their interest in lichens in the 1970s and 1980s. Their lichen collection is deposited in the Auckland Museum Herbarium. Bruce spent two weeks on Hauturu with the Auckland University Field Club and the Offshore Islands Research Group in the 1980s. During these visits his fieldwork focused on the lichen biodiversity and ecological distribution and the pre-European archaeology. The updated results of this published lichen study are presented here.

**SUSAN KEALL** is senior technical officer in conservation ecology at Victoria University of Wellington. She participated in the first survey for tuatara on Hauturu in 1991, which captured four adults. Since 1994, when the first viable eggs were laid by the captive adult tuatara on Hauturu, she has been involved in incubation of the eggs and care of the young hatchlings at Victoria University before their return to Hauturu. In 2006 after Hauturu was declared kiore free, Susan helped release the first juvenile tuatara from the captive rearing programme into the wild on the island. She also participated in an island-wide survey for tuatara in 2015.

**JAN LINDSAY** is a volcanologist in the School of Environment at the University of Auckland. Her MSc thesis from the University of Auckland was on the geology of Little Barrier Island/Hauturu, and she has a PhD in Geosciences from the University of Giessen, Germany. She has held positions at GNS Science in Taupō, the GeoResearch Centre in Potsdam, Germany, and the University of the West Indies in Trinidad. She is past president of the Geoscience Society of New Zealand. Jan's current research focuses on making society more resilient to volcanic hazards and she has worked on projects in the area of volcanic geology, hazard and risk in New Zealand, Chile, the Lesser Antilles and Saudi Arabia.

**TIM LOVEGROVE** first visited Hauturu with the King's College Bird Club in 1972, and this began his love affair with this magical place. As a member of Dick Veitch's many Wildlife Service trips to Hauturu during the 1970s and 80s, he helped catch hihi for transfers to Taranga, Cuvier and Kāpiti; he has counted the bird transects many times and has volunteered on the cat eradication project. In 1984–85 he monitored the first tīeke release on Hauturu, and in 1992–93 he was a field manager on the kākāpō programme. This work took him all over Hauturu, so he got to know it pretty well. In his current role with the Auckland Council he helped establish the Tāwharanui Open Sanctuary, where kākā and bellbirds have successfully self-colonised from Hauturu.

**PAULA MORRIS** (Ngāti Wai, Ngāti Manuhiri, Ngāti Whātua) is the author of the story collection *Forbidden Cities* (2008); the long-form essay *On Coming Home* (2015); and eight novels, including *Rangatira* (2011), winner of best work of fiction at both the 2012 New Zealand Post Book Awards and Ngā Kupu Ora Māori Book Awards. Her most recent book is *False River* (2017), a collection of stories and essays around the subject of secret histories. She teaches creative writing at the University of Auckland, sits on the Māori Literature Trust, Mātātuhi Foundation and New Zealand Book Awards Trust, and is the founder of the Academy of New Zealand Literature (www.anzliterature.com). She was awarded the 2018 Katherine Mansfield Menton Fellowship, and was appointed an MZNM in 2019.

**NICOLA NELSON** is associate professor in the School of Biological Sciences at Victoria University of Wellington. Her research interests are in the field of ecology and evolution in general, and specifically on the effects of temperature on reptile ecology, population ecology and conservation. She is a scientific advisor for the Hauturu Supporters' Trust. Nicky participated in one of the first surveys for tuatara in 1992, and has been involved in conservation management of tuatara on Hauturu ever since.

**KEVIN PARKER** first visited Hauturu as a kākāpō volunteer in 1998. It was fitting that his first visit was to assist in the management of a translocated species, as most of his subsequent visits have been translocation-focused, either researching translocated species on the island or catching birds for translocation to other sites. He is a member of the Hihi Recovery Group and the Kōkako Specialist Group. Hauturu is very special to both of these species, as home to the only remaining natural self-sustaining population of hihi and the largest and most secure population of kōkako. There is at least one species that Kevin hopes to one day translocate back to Hauturu: the snipe.

**STUART PARSONS** completed his PhD on New Zealand bats, with Hauturu as one of his principal study sites. He worked with several other researchers on the island who helped to unravel the mysteries of New Zealand bats, including Alina Arkins and Andrew Winnington. Since completing his PhD, Stuart has worked at a number of universities in New Zealand, Australia and the UK. During his time at the University of Auckland, several of his students studied bats on Hauturu, contributing further to our understanding of these amazing animals, and the importance of the island. Stuart is currently professor and head of the School of Earth, Environmental and Biological Sciences at Queensland University of Technology in Brisbane.

**MATT RAYNER** is curator of land vertebrates at the Auckland War Memorial Museum and a research associate at the School of Biological Sciences, University of Auckland. He is a specialist in ornithology, vertebrate zoology and conservation biology, and has a strong history of involvement with Hauturu — he has conducted field research on the island's birds since 2004. Matt was a trustee of the Little Barrier Island Supporters' Trust between 2009 and 2015 and remains a scientific advisor for research on the island.

**DAVID S. SELDON** is senior tutor at the School of Biological Sciences, University of Auckland where, as a research entomologist, he specialises in the taxonomy and biogeography of ground beetles (Carabidae) in New Zealand, New Caledonia, Australia and South America. He is a Fellow of the Linnean Society, a research associate at Manaaki Whenua — Landcare Research and Auckland War Memorial Museum, and a scientific advisor to the Little Barrier Island (Hauturu) Supporters' Trust. Dave has conducted numerous studies of the insect fauna on Hauturu, including describing an island-endemic ground beetle species.

**DAVID TOWNS'** interest in reptiles began with studies for an MSc on the ecology and reproductive biology of Suter's skink at Auckland University. After switching to freshwater ecology for his PhD, and about six years overseas, he was lured back to New Zealand to lead lizard conservation research for the NZ Wildlife Service. The advent of the Department of Conservation (DOC) in 1987 led to a 30-year career based in the Mercury Islands, investigating the causes of population declines of threatened lizards, the roles of rodents (particularly kiore) on island faunas, methods and the effects of rodent eradications and the role of reptile translocations in island ecosystem restoration. In 2002, he was required by DOC to testify as an expert witness at the resource consent hearing for eradication of kiore from Hauturu.

**RICHARD WALLE** is the current ranger on Te Hauturu-o-Toi. He has been privileged to live and work on the island for the past eight years with his wife Leigh Joyce and their children Mahina and Liam. During that time he has been in charge of the tuatara headstart programme. Leigh and Richard both work on the job at times and complement each other's skills: Leigh has a PhD on kākāpō and has worked extensively in bio-diversity, while Richard, a carpenter by trade, has worked in conser-vation over the past 30 years. They have a strong connection to the South Island West Coast where they have a place at Ōkārito and Fox Glacier.

**MIKE WILCOX** is an honorary research associate at the Auckland Museum specialising in marine algae, and author of *Seaweeds of Auckland*, published in 2018 by the Auckland Botanical Society. He had a career (1959–2004) in forestry as a forester, scientist, research director and international forestry consultant. Included in his 262 publications on forestry and botanical subjects are books on the natural history of Rangitoto Island (2007) and the urban forests of Auckland (2012), and several articles about the seaweeds of the Hauraki Gulf islands. His interest in marine biology was sparked by John Morton (seashore) and Valentine Chapman (algae) at the University of Auckland in the early 1960s.

**MAUREEN YOUNG** is a retired primary school teacher who has had a life-long interest in New Zealand's native flora. She has been a member of the Auckland Botanical Society for 34 years, many of those as vice-president. While she was sailing as a youngster in Kawau Bay, Hauturu was a tantalising silhouette on the horizon, and in later years she developed a desire to explore the island and record the plants there. Permission was obtained to make a herbarium of island plants to be kept on the island for the benefit of scientists. Some plants, not seen there for a hundred years, were rediscovered. At the same time, along with Ewen Cameron, she published a checklist of the island flora updated from a DSIR typescript.

# Acknowledgements

**LYN WADE & DICK VEITCH:** Our thanks to the Little Barrier Island (Hauturu) Supporters' Trust, who instigated and funded this book. As with any project, there are so many people to acknowledge; one always hopes to have remembered to mention everyone.

Firstly thanks to Ruud Kleinpaste, a past trustee of the Little Barrier Island Supporters' Trust and now its Patron, for writing the foreword for this book.

We have been blessed with a great team of scientists whose enthusiasm for our island sanctuary shows in their writing. Their names are attached to their chapters, and it has been a privilege to work with them. Thanks to Mook Hohneck, Chair for Ngāti Manuhiri Settlement Trust, and Nicola MacDonald of Ngāti Rehua for providing a mihi, and to Paula Morris for her history of the long Māori association with Te Hauturu-o-Toi.

We have had a number of both professional and amateur photographers who have been willing to share their awesome photos of this unique island. Special thanks to Alison Wesley, Simon Fordham, Neil Fitzgerald, Warren Judd, Liz Whitwell, Andris Apse, Dylan van Winkel, Olivier Ball, Lesley Baigent, Colin O'Donnell, Andrew Digby, Richard Griffiths, David Mudge, David Stone, Dhahara Ratanunga, Karen Baird, Lyn Wade, and Dick Veitch, who also obtained access to many historic photos of life on Hauturu.

Thanks to generous funding from Foundation North and Chisholm Whitney Family Charitable Trust towards costs of creating this book, both have given major assistance to other projects for Hauturu over the years, and also to the Stout Trust. Special thanks to the families of Don Binney, Karen and Rex Mirams and Rod Bieleski for memorial donations that were given to the Supporters' Trust that have helped fund this book project.

Thanks to our publishers, Massey University Press: Anna Bowbyes, Emily Goldthorpe, Tracey Borgfeldt and in particular Nicola Legat, a previous trustee, for their expertise and assistance. We are sure they have gone the extra mile.

Thanks to Manaaki Whenua — Landcare Research for permission to use material from DSIR Bulletin 137, and to the Binney family for permission to use an image of Don Binney's painting *Kaka*.

Of course, there are all the support crews in the background: partners, co-workers, boat operators, the Little Barrier Island (Hauturu) Supporters' Trust, past and present island rangers and Department of Conservation staff, all who gave help and moral support but are too numerous to mention.

**DAVE S. SELDON:** I thank Robert Hoare for preparing an excellent document on the Lepidoptera of Hauturu, Thomas Buckley and Scott Bartlam (on worms), and Rich Leschen (Manaaki Whenua) for much appreciated articles.

I thank the following staff from Auckland Museum: Ruby Moore for making available the museum records in an easily accessible digital format, John Early for suggestions on possible sources of taxa, and Matt Rayner for editing the first draft. Thanks to Zoe Hilton (Cawthron Institute) for the invertebrate and other photographs. I appreciate the comments made and additional information on wētā and spiders from Chris Green (Department of Conservation).

**EWEN CAMERON & MAUREEN YOUNG:** EKC thanks the herbarium staff of the following herbaria for electronic specimen data, images of specific specimens and answering frequent requests: Kate Boardman and Ines Schönberger of the Allan Herbarium Manaaki Whenua (CHR); Jennifer Tate of the Massey University Dame Ella Campbell Herbarium (MPN), Leon Perrie of the Museum of New Zealand Te Papa Tongarewa herbarium (WELT); Rhys Gardner and Peter de Lange for assistance with identifications; ranger Richard Walle for local information; Dhahara Ranatunga for field assistance and support; Jessica Beever, John Braggins, Richard Walle and Shane Wright for supplying images; Joshua Salter for commenting on a draft of this chapter; and Dick Veitch for the vegetation map.

**JESSICA BEEVER:** I wish to thank numerous field companions for assistance and good company on Hauturu, including Graham Beever, Rosemary Beever, Ross Beever, Val and Rod Bieleski, John Braggins, Patrick Brownsey, Ewen Cameron, Helen Cogle, John Eiseman, Ross Ferguson, Bruce Hayward, Glenys Hayward, Anne Grace, Phil Moore, Paul Newman, Barbara Polly, Christopher Spragg, Dick Veitch, Lyn Wade and Anthony Wright. In particular I am grateful to John Braggins for much useful discussion on the bryophytes of Hauturu. Vascular plant identifications have been provided by Ross Beever, John Braggins and Patrick Brownsey. Special thanks are due to many herbarium staff, especially at AK (Auckland Museum), CHR (Manaaki Whenua, Lincoln), PDD (Manaaki Whenua, Tāmaki) and WELT (Museum of New Zealand Te Papa Tongarewa) for searching for specimens, arranging loans, giving me electronic access to their Hauturu specimen databases, and for accessioning my own bryophyte collections. Dhahara Ranatunga made important observations on *Fissidens taxifolius* on Hauturu. Colleagues at Manaaki Whenua, Tāmaki, have provided a supportive environment for research. I wish to thank Allan Fife (Manaaki Whenua, Lincoln) for generously sharing his wide bryological knowledge and for providing me with free access to his draft manuscript for the 'Flora of New Zealand — Mosses'. This research would not have been possible without the logistical support and provision of permits by the Department of Conservation and its predecessors. It is a pleasure to acknowledge the tangata whenua, Ngāti Manuhiri and Ngāti Wai, for their tautoko of my Hauturu studies. A Mobil Environmental Award supported the expedition in 1990. Additional funding for this research has come from the Foundation for Research, Science and Technology (FoRST) and its successors the Ministry for Science and Innovation, and the Ministry of Business, Innovation and Employment.

**JOHN BRAGGINS:** Thanks go to M. A. M. Renner for new records and advice with names, M. von Konrat for advice on *Frullania* names, E. K. Cameron for collections and support, and D. Glenny for identifications. In particular, I acknowledge Jessica Beever, my colleague on the main collecting trips, for her support, collections and, particularly, her critique of the manuscript. Thanks to the Auckland Museum for housing and curating the collections.

**BRUCE HAYWARD:** I thank Anthony Wright and Glenys Hayward for assistance with the lichens that we collected and identified together in 1981 and 1990. Many specialist lichenologists have assisted in the identification of Hauturu lichens for the original study and subsequently, particularly David Galloway (Lobariaceae, Pannariaceae, Physciaceae), Doug Verdon (Collemetaceae), Mason Hale (Parmeliaceae), Thorsten Lumbsch (saxicolous crustose lichens), Mats Wedin (Sphaerophoraceae), Sam Hammer (Cladoniaceae), Alan Archer (Cladoniaceae, Pertusariaceae), Harrie Sipman (*Megalospora*), Arve Elvebakk (Pannariaceae) and Daniel Bennett (*Chrysothrix*). Ewen Cameron and Dhahara Ranatunga assisted with extracting a list of Little Barrier lichens held in the Auckland Museum herbarium.

# Image credits

Thank you to all those who contributed photographs to this book: Andris Apse (AA), Abseil Access team (AbAc), Alex Dobbins (AxD), Andrew Digby (AD), Alexander Turnbull Library (ATL), Alison Wesley (AW), Bruce Hayward (BH), Betty Wisnesky (BW), Colin Miskelly (CM), Colin O'Donnell (CO), Dick Veitch (CRV), Chris Smuts-Kennedy (CSK), David Mudge (DM), Don MacKay (DMc), Department of Conservation (DOC), Dhahara Ranatunga (DR), Dave S. Seldon (DS), Darryl Torckler (DT), Dylan van Winkel (DvW), Ewen Cameron (EC), Edin Whitehead (EW), Frances Shakespear (FS), Irene Petrove (IP), John E. Braggins (JB), Jan Lindsay (JL), Ken Bigwood (KB), Kendall Clements (KC), Kate McAlpine (KM), Kurt Salmond (KS), Lesley Baigent (LB), L. C. W. Jensen (LJ), Lyn Wade (LW), Liz Whitwell (LWh), May Parkin (MP), Martin Sanders (MS), Mike Wilcox (MW), Neil Fitzgerald (NF), Noel Macdonald (NM), Olivier Ball (OB), Oscar Thomas (OT), Pete Barrow (PB), Phil Bendle (PBe), Rodger Blanshard (RB), Richard Griffiths (RG), Richard Walle (RW), Scott Bartlam (SB), Simon Fordham (SF), Stuart Parsons (SP), Terry Greene (TG), Tim Lovegrove (TL), Warren Judd (WJ), W. M. Hamilton (WMH).

# Index

Please note that complete lists of species can be found in Chapter 9. Only species mentioned in the main text are indexed. Numbers in **bold** indicate illustrations.

Abseil Access 83
Adams, James 186
akakura (*Metrosideros fulgens*) **203**
akeake (*Dodonaea viscosa*) **198**
akepiro (*Olearia furfuracea*) 172, **204**
algae 105, 244–48, 253–54, **256**, **263** (fig), 276; brown 248; as food source 118, 130, 258; green 248; red 245–46; species table 343–44; species:
  *Abroteia suborbiculare* 246
  *Adamsiella chauvinii* 246
  *Aphanocladia delicatula* 246
  *Apophlaea sinclairii* 245
  *Bachelotia antillarum* 248
  *Bangia atropurpurea* 246
  *Bostrychia intricata* 246
  *Bryopsis plumosa* 248
  *Carpophyllum flexuosum* **263** (fig)
  *Carpophyllum maschalocarpum* (flapjack) 246, 248, 254, **263** (fig)
  *Carpophyllum plumosum* 246, 248, **263** (fig)
  *Catenella: fusiformis* 246
  *Catenella nipae* 246
  *Catenellopsis oligarthra* 245
  *Caulacanthus ustulatus* 245, 246
  *Caulerpa geminata* 248
  *Centroceras clavulatum* 245
  *Ceramium uncinatum* 246
  *Champia novae-zelandiae* 246
  *Chondria lanceolata* 246
  *Cladhymenia coronata* 246
  *Codium convulatum* 248
  *Colpomenia peregrina* 246
  *Corallina ferreyrae* (corallines) 246
  *Cystophora retroflexa* 246, 248
  *Cystophora torulosa* 246
  *Dasya subtilis* 246
  *Dasyclonium ovalifolium* 246
  *Derbesia novae-zelandiae* 248
  *Dictyota kunthii* 246, 248
  *Dipterosiphonia dendritica* 246
  *Ecklonia radiate* (kelp) **244**, 246, 248, **252**, 254, **256**, **263**
  *Feldmannia mitchelliae* 248
  *Gayliella flaccida* 246
  *Gelidium caulacantheum* 245
  *Gigartina minuta* 245
  *Halopteris virgata* 246, 248
  *Hapalospongidion saxigenum* 246
  *Haraldiophyllum crispatum* 246
  *Hildenbrandia* 245
  *Hormosira banksii* 246
  *Hymenena variolosa* 246
  *Isactis plana* 248
  *Jania sagittata* 246
  *Laurencia distichophylla* 246
  *Liagora harveyana* 245
  'Lithothamnia' (non-geniculate coralline 'pink-paint') 246
  *Lomentaria caespitosa* 246
  *Lychaete herpestica* 248
  *Melanthalia abscissa* 246
  *Melosira moniliformis* 248
  *Microdictyon mutabile* 248
  *Nancythalia humilis* 246
  *Nothogenia pulvinata* 245
  *Perithalia capillaris* 245
  *Petalonia binghamiae* 246
  *Phacelocarpus labillardierei* 246
  *Pleurostichidium falkenbergii* 246
  *Plocamium angustum* 246
  *Polysiphonia scopulorum* 246
  *Pterocladia lucida* 246
  *Pterocladiella capillacea* 246
  *Pyropia plicata* 246
  *Ralfsia verrucosa* 246, **247**
  *Rhizopogonia asperata* 246
  *Rhodochorton purpureum* 246
  *Sargassum sinclairii* 246
  *Scytothamnus australis* 246
  *Spatoglossum chapmanii* 246
  *Spatoglossum* sp. 246, 248, **249**
  *Splachnidium rugosum* 246
  *Spyridia filamentosa* 246
  *Symphyocladia marchantioides* 246
  *Ulva australis* (Pacific sea lettuce) **247**, 248, **255**
  *Ulva procera* 248
  *Vertebrata aterrima* 246
  *Vidalia colensoi* 246
  *Xiphophora chondrophylla* 245, 246, **263** (fig)
  *Zonaria aureomarginata* 246
  *Zonaria turneriana* 246, 248
amphibians 125–26
Anderson, Michael 149
*Androstoma empetrifolia* 185
Angehr, George 152
angiosperms (*table*) **187**
ants 63, 110, 119; Argentine ant 63; formicid (large ant) 110
Aotea, Mount 184–85, 221
Aotea/Great Barrier Island 17, 18, 115, 146, 149, 154; and black petrels 85, 139; and chevron skink 129; geographical limits for vascular plants 184–85; and plant species 179, 183, 229; translocations 56, 81, 139, 151
*Aporostylis bifolia* 185
Arachnida (spiders and others) 116; as food source 162 *see also* spiders
*Archeria racemosa* 178, 179, **181**, 184, 185, **195**
arthropods, New Zealand Collection 105
asparagus, climbing (*Asparagus scandens*) 75, **79**, 80–81, 89, 180, 272
*Astelia* species: *banksia* (coastal astelia) 175, **208**, 219; *hastata* 178, **209**; *microsperma* **209**; *solandri* 178; *trinerva* **209**
Atkinson, Ian 158, 172, 186; and John Campbell 183; and R. Stanley 167–69, **187** (table); and W. M. Hamilton 167, 184
Attenborough, David **57**
Auckland Institute and Museum 38
Auckland Museum Records 230
Auckland Museum/War Memorial Museum 20, 49; collections 35, 39, **104**, **107**, 119, 145; herbarium 105, 216, 226
Auckland University Field Club 49, 230
Australasian Virtual Herbarium 106
Awaroa Point **101**, 219
Awaroa Stream 36, 95, 114, 219, 226, **280–81**
āwheto (*Ophiocordyceps robertsii*) 235

Baker, Shirley 234
Bald Rock *see* Hauruia, Mount
bangalow palm (*Archontophoenix cunninghamiana*) 180
Barrow, Ben 60
Barrow, Hamish 60
Barrow, Peter 60, **65**, 66
Barrow, Tim 60
bats: 118, 161–64, 186; species table 295; greater short-tailed (*Mystacina robusta*) 161; lesser short-tailed (*M. tuberculata*) 118, **160**, 161–64, **163**; long-tailed (*Chalinolobus tuberculatus*) 162, **163**
beech (*Nothofagus* sp.) 132, **222**; black beech (*Fuscospora solandri*) 176, 182–8, 183, 184; hard beech/tawhairaunui (*F. truncata*) 176, **187** (table), **199**, 233, 235
bees 28, 35, 119, **120**; honeybee (*Apis mellifera*) 119; *Lasioglossum* 119; New Zealand native (*Leioproctus* sp.) 119, **120**
beetles (Coleoptera): 109, 119, **120**, 121–23, 162; aquatic 114–15; darkling (*Tenebrionid* sp) **120**; ground (Carabidae) 122; huhu (*Prionoplus reticularis*) **120**, 121; New Zealand striped longhorn

New Zealand striped longhorn (*Coptomma lineatum*) 120; *Mecodema*: *aoteanoho* 122; *M. haunoho* 122; minute moss (Hydraenidae) 114; moss (*Podaena hauturu* and *P. latipalpus*) 115; riffle (Elmidae) 114; rove (Staphylinidae) 122
Beever, Jessica 216, 226
Beever, Ross 184, 186
bellbird/korimako (*Anthornis melanura*) 153, **155**; diet, predators 149, 152, 156; feeding of 40, 45, 56, 60; population of 135, 136, 153–54, 157, 269; at Tāwharanui 87, 270
Bellingham, R M 226
Bergquist, P. L. 245
Bermuda buttercup (*Oxalis pes-caprae*) 171
berry saltbush (*Chenopodium triandrum*) 175, **197**
Bigwood, Ken 49
bindweed: *Calystegia sepium* subsp. *roseata* 171; hybrid (*C. sepium* subsp. *roseata* x *C. tuguriorum*) 169, 171, 176; shore (*C. soldanella*) 175, **196**
biodiversity, threats to management of 76–89
biosecurity 63, 82–84, 180, 270, 272
bird catcher/parapara (*Pisonia brunoniania*) 72, 88, 175, 183, **205**
birds: feeding of 45, 49, 56–58, 60, 146, 153; introduced species 156–57; mobility of 270; seabirds 136–141; species table 293–95; tamed 53–54, 56, **57**, **58**, 59, 78 *see also* species, translocations
blackbird (*Turdus merula*) 76, 87, 156, 157; shot as pests 43
Blanshard, Ani 50, **52**, 53–54
Blanshard, David 50, **52**, 53–54
Blanshard, Gina 50, **52**, 53–54, 143
Blanshard, Lisa 50, **52**, 53–54
Blanshard, Rodger 50, 53–54, 66
Blanshard, Susi 50, **52**, 53–54
Blundell, Oscar 44
boat services 38, 40, 44, 49, 53, 64
boatbuilding 40, 50
boats 45, 50, **52**, **61**, 63, 64, **67**; owned by residents 26, **37**, 39, 40, **41**, 50
*Bolivar* (whaleboat) **37**, 39
Bolt, Alice 38
Boscawen, Hugh 36
Braggins, John 216, 226
Brazilian fireweed (*Erechtites valerianifolia*) 180
brooms (*Carmichelia* sp.) **196**, 235; giant-flowered broom (*C. williamsii*) 88, 182, **196**; native broom (*C. australis*) **196**
Brown, Nichollette 63, **65**, 66
brown teal/pāteke (*Anas chlorotis*) 78, 87, **142**, 144
Browne, Simon Welton 38

Brownsey, Patrick J. 186, 216
bryophytes 172, 178, 179, 215, 216, 221
Buddle, G. A. 141
bugs 118–19
buildings **32–33**; biosecurity shed **83**, 84; boatsheds **37**, 45, 50, 54, 61; houses **37**, 38–39, **41**, 45–46, **47**, 54, **55**; staff quarters 50, **52**, 53, 56, 76; utility sheds 43, 49, 54, **55**, 60
Bull, Peter 141
Buller, Walter 35, 135, 143
bush lawyer (*Rubus cissoides*) **206**
bush lily/kakaha: *Astelia fragrans* 178, **209**; *Astelia hastata* 178, **209**
butterflies **86**, 88, 109, 110, 121, 162; coastal copper (*Lycaena salustius*) **108**; forest ringlet (*Dodonidia helmsii*) **86**, 88, 121

cabbage tree/tī kōuka (*Cordyline australis*) 179, 185; bush/tī ngahere (*C. banksii*) **210**; dwarf/tī rauriki (*C. pumilio*) **210**
caddisflies 88, 114; Chathamiidae sp. 88; *Oxyethira albiceps* 88
Californian thistle (*Cirsium arvense*) 169
Cameron, Ewen 186, 226
Campbell, John 167, 183
*Cardamine chlorina* 184
carpet grass (*Axonopus fissifolius*) 171
*Carpha alpina* 185
Castle Rock 184
cats (*Felis catus*) 18, 75; eradication of 45, 54, 56, 76–77, 85; feline enteritis 54, 56, 76; as predators 125, 137, 139, 145, 148, 151, 152, 154
cattle *see* livestock
*Celmisia*: *adamsii* var. *adamsii* 184; *adamsii* var. *rugulosa* 184; *incana* 185; *major* 184
centipedes 111–12; giant (*Cormocephalus rubriceps*) 112, **113**; stone (Geophilomorpha) 111–12
chaffinch (*Fringilla coelebs*) 156, 157
Chambers, T. C. 216
*Checklist of the New Zealand Flora* 220
Cheeseman, Thomas 104, 143, 186
*Chenopodium trigonum* 175
chiton, giant (*Eudoxochiton nobilis*) 254
clam, purple (*Venericardia purpurata*) 257
Cleaver, Lilian 44
Cleaver, William 44, 66, 76
clematis: *Clematis cunninghamii* (yellow) **197**; *C. paniculata* **197**
cloud forest *see* moss forest
coastal forest and scrub 175
cockroaches (Blattodea) 118
cocksfoot (*Dactylis glomerata*) 169
columbine 45

comb star (*Astropecten polyacanthus*) 257
communications 45–46, **47**, 53, 54, 56, 59; internet 60, 64
coneheads *see* Protura
conifers **187** (table)
Conservation Management Plan 2017 273
convolvulus (*Calystegia sepium* ssp. *roseata*) 45, **196**; climbing (*C. tuguriorum*) **196**
Cook's turban (*Cookia sulcata*) 254, 258
*Coprosma*: *colensoi* 185; *dodonaeifolia* 184; *foetidissima* 185; *macrocaroa* x *C. propinqua* **197**; *neglecta* 182, 184; *repens* 169, 183, 185, **197**; *rhamnoides* 172, 175, **197**; *see also* kanono, karamū, māmāngi
corals **252**; *Alcyonium aurantiacum* **256**; black coral 261; *Culicea rubeola* **256**
Cotton, A. D. 245
crayfish (*Jasus edwardsii*) 254, 262
Crayfish Rock 245, 248
Cree, Alison 130
creeping mallow (*Modiola caroliniana*) 171
cryptic soil dwellers 111–12
cucumber mosaic virus 182; cucumber, native/māwhai (*Sicyos mawhai*) 182, **207**
cuckoo: long-tailed/koekoeā (*Eudynamys taitensis*) 148–49, **150**, 154, 270; shining/pīpīwharauroa (*Chrysococcyx lucidus*) 148, **150**, 270
Cuvier Island *see* Repanga/Cuvier Island

*Dactylanthus taylorii* 82, 89, **160**, 162, 164, 182, 186, **198**; and bats 162, 164, 186
dandelion (*Taraxacum officinale*) 171
*Danhatchia australis* 183
Database of Island Invasive Species Eradications 269
Dawson, E. W. 141, 144
de Graaf, Andre 66
deaths on Hauturu 44, 54
Dellow, Vivienne 245, 246
Department of Agriculture 36, 38
Department of Conservation (DOC) 77, 123, 139, 140, 167, 220–21; and Hauraki Gulf Maritime Park 59; and Ngāti Manuhiri 20; and Seachange initiative 264
Department of Internal Affairs 49
Department of Lands and Survey 36, 38, 44, 49, 84
Department of Scientific and Industrial Research (DSIR) 234
Diamond, J. M., and C. R. Veitch 157
*Dichelachne crinita* 175
Dieffenbach, Ernst 143

390  **Hauturu**

Dingley, Joan M. 216, 234, 235
DNA sampling 105, 123, 139, 140, 237, 272
Dobbins, Alex 56, **58**, 66
Dobbins, Mark (Phred) 56, **58**
Dobbins, Mike 56, **58**
Dobbins, Toni 56
Dodson, Helen 60
dog cockle (*Tacetona laticostata*) 257
dogs (kurī) 71, 72, 143; trained wildlife 146
dolphin: bottlenose (*Tursiops truncatus*) 258; common (*Delphinus delphis*) 258
Dominion Museum 245
Dopheide, Andrew 237
*Dracophyllum*: *latifolium* **198**; *patens* 185; *traversii* (mountain neinei) 179, 185, **199**
179, 185, **199**
Drew, John 54
Dromgoole, F. I. 245, 248
Drummond, A. J. 123
DSIR Bulletins 44, 53
dunnock (*Prunella modularis*) 156, 157

eagle ray (*Myliobatus tenuicaudatus*) 258
earthstars (*Geastrum* sp.) 234, **239**
earthworms: *Anisochaeta gigantea* (giant) 111, **113**; *A. shakespeari* 111; *Aporrectodea caliginosa* 111; *Eisenia fetida* 111
East Cape 76, 221, 226
eastern rosella (*Platycercus eximius*) 53
eels 243; longfin (*Anguilla dieffenbachii*) 243
*Epacris sinclairii* 185
epiphytes 178, 179, 216, 218, 224, 246; *Dicnemon calycinum* 218; *Dicranoloma menziesii* 218; *Holomitrium perichaetiale* 218, 219
Esler, Alan 186

fairy prion/tītī wainui (*Pachyptila turtur*) 141, 157
falcon/kārearea, New Zealand (*Falco novaeseelandiae*) 35, 53, 144, 158
Falla, Robert A. 44, 143
fantail, North Island/pīwakawaka (*Rhipidura fuliginosa*) 136, 154, **155**, 157
feather star (*Comanthus novaezelandiae*) 257
fennel (*Foeniculum vulgare*) 80
fernbird, North Island/mātātā (*Bowdleria punctata*) 158
ferns and lycophytes 121, 178, 179, **187** (*table*), **188–93**, 218; species table 296–300; *see also* tree ferns; species:
  *Adiantum aethiopicumi* (true maidenhair) **188**
  *Adiantum cunninghamii* (common maidenhair) **188**;

*Adiantum diaphanum* (small maidenhair) **188**
*Adiantum hispidulum* (rosy maidenhair) **188**
*Arthropteris tenella* (jointed) **188**
*Asplenium bulbiferum* (hen and chickens) **188**
*Asplenium flaccidum* (hanging spleenwort) **188**
*Asplenium haurakiense* (coastal spleenwort) 175, **188**
*Asplenium lamprophyllum* **188**
*Asplenium oblongifolium* (shining spleenwort) 175, 176, **189**
*Asplenium polyodon* (sickle spleenwort) **189**
*Blechnum blechnoides* **189**
*Blechnum chambersii* **189**
*Blechnum colensoi* 183, 184, 185
*Blechnum discolor* (crown fern/piupiu) **189**
*Blechnum filiforme* (thread fern) **189**
*Blechnum fraseri* **189**
*Blechnum nigrum* **189**, 218
*Blechnum novae-zelandiae* (kiokio) **189**
*Blechnum procerum* **190**
*Doodia australis* (rasp) 169, **190**
*Gleichenia microphylla* (carrier tangle/waewae kākā) **190**
*Hymenophyllum dilatatum* **190**
*Hymenophyllum lyallii* 185
*Hymenophyllum nephrophyllum* (kidney/rauranga) **190**
*Hymenophyllum villosum* 185
*Hypolepis ambigua* 169, **190**
*Lastreopsis glabella* **190**
*Lastreopsis hispada* (hairy legs) **190**
*Lindsaea linearis* **191**
*Lindsaea trichomanoides* **191**
*Lindsaea viridis* 184, **191**
*Microsorum pustulatum* (hounds tongue) 175, **191**, 219
*Microsorum scandens* (fragrant/mokimoki) **191**
*Notogrammis heterophylla* **191**
*Paesia scaberula* **192**
*Pellaea rotundifolia* **192**
*Phlegmariurus varius* (tassel/iwituna) **192**
*Pneumatopteris pennigera* (gully/pākau) **192**
*Polystichum wawranum* (shield fern) **192**
*Pteridum esculentum* (bracken/rārahu) 169, **192**
*Pteris comans* **192**
*Pteris tremula* (turawera) **192**
*Pyrrosia eleagnifolia* (leather) 175, **192**
*Rumohra adiantiformis* **193**
*Schizaea bifida* (bifurcated comb) **193**
*Schizaea dichotoma* (fan) **193**
*Sticherus cunninghamii* (umbrella/

tapuwae kōtuku) **193**
*Tmesipteris elongata* (chain) **193**
*Tmesipteris* spp. 218
fescue, tall (*Lolium arundinaceum*) 169
firewood 46, **48**, 49; commercial harvesting of 36, 72
Firman, Grace 43
Firman, Thomas Pierson 43, 66
Fisheries NZ 262
fishes, marine:
  banded wrasse (*Notolabrus fucicola*) 258
  blue maomao (*Scorpis violaceous*) 258
  butterfish/greenbone (*Odax pullus*) 258
  butterfly perch (*Caesioperca lepidoptera*) 258
  demoiselle (*Chromis dispilus*) 258
  goatfish (*Upeneichthys lineatus*) 258
  hāpuku (*Polyprion oxygeneios*) 262
  hiwihiwi/kelpfish (*Macronemus marmoratus*) 258
  kingfish (*Seriola grandis*) **256**; pelagic (*S. lalandi* 258
  leatherjacket (*Meuschenia scaber*) **252**, **256**, 258
  mado (*Atypicthys strigatus*) **256**
  marblefish (*Aplodactylus arctidens*) 258
  paketi/spotty (*Notolabrus celidotus*) 258
  parore (*Girella tricuspidata*) 258
  pichards (*Sardinops neopilchardus*) **259**, 260
  pink maomao (*Caprodon longimanus*) 258
  porae (*Nemadactylus douglasii*) 258
  red moki (*Cheilodactylus spectabilis*) **255**, 258
  red pigfish (*Bodianus unimaculatus*) 258
  sandager's wrasse (*Coris sandeyeri*) 258
  silver drummer (*Kyphosus sydneyanus*) 258
  snapper (*Chrysophrys auratus*) 254, 258, 261–62
  spotty/paketi (*Notolabrus celidotus*) 258
  sweep (*Scorpis lineolata*) 258
  trevally (*Pseudocaranx georgianus*) 258, 262
  yellowtail (*Seriola lalandi*) 258
  yellowtail jack mackerel (*Trachurus novaezelandiae*) 258
fishing: impacts, management of 254, 258, 260–64, **265**
five finger/whauwhaupaku (*Pseudopanax arboreus*) 172, 178, **206**
flax (*Phormium* sp.) 122: harakeke (*P. tenax*) **212**, 219; mountain/wharariki (*P. cookianum*) 179, **212**

fleabane (*Erigeron sumatrensis*) 180
fleas (Siphonaptera) 118
flies (Diptera) 115, 118, 162;
  *Austrosimulium australense* 115;
  avian parasitic fly (*Ornithomya variegatus*) 118; black flies (Simuliidae) 115; *Ceratomerus* spp. 115; dance flies (Brachystomatidae) 115; Dixidae (aquatic gnats) 115; Dolichopodidae (*Parentia*) 118; larvae 110; New Zealand bat fly (*Mystacinobia zelandica*) 118, 164
flora: absentee species 183–84; botanists/collectors **186** (table); lost species 179–80; threatened species 182–83; vascular species **182** (fig), **187** (table); vascular species table 296–317
Forest Research Institute 234
forest successions 172–79, **177** (fig)
Francis, Marie 60, 66
frogs 125, 126; Archey's (*Leiopelma archeyi*) 126; Hochstetter's (*L. hochstetteri*) 125
fungi 233–37; species table 330–42; species: *Armillaria limonea* **238**; *Armillaria novaezelandiae* (harore) 235
  *Aseroe rubra* (puapuatai/ stinkhorn fungi) 235, **238**
  *Aspergillus fumigatus* 234
  *Clavulinopsis* sp. **239**
  *Clavulinopsis sulcata* (coral fungus) 233, **238**
  *Clonostachys ralfsii* 235
  *Cordyceps hauturu* 235, **236** (fig)
  *Cordyceps kirkii* 235, **238**
  *Cortinarius gemmeus* 237
  *Cortinarius* sp. 233, **238**
  *Cyptotrama* sp. 233, **238**
  *Dichomitus newhookii* 237
  *Entoloma blandiodorum* 237
  *Entoloma consanguineum* 237
  *Entoloma hochstetteri* (werewere -kōkako) **232**, 235, **239**
  *Favolaschia calocer* (orange pore) **239**
  *Fomes hemitephrus* 234, **239**
  *Ganoderma* sp. 234, **239**
  *Geastrum* sp. 234, **239**
  *Hebeloma* sp. 233, **238**
  *Hygrocybe* sp. 233, **239**
  *Hymenochaete magnahypha* 237
  *Hypholoma australianum* 233, **240**
  *Ileodictyon cibearium* (basket, matakupenga) 235, **240**
  *Laetiporus portentosus* (pūtawa) 235, **240**
  *Lophodermium hauturuanum* 235, **236** (fig)
  *Morganella compacta* (puffballs, pukurau) 235, **240**
  *Nectria hauturu* 235
  *Nidula niveotomentosa* (bird's nest fungi) 234, **240**
  *Ophiocordyceps robertsii* 235
  *Ophiocordyceps sinensis* 235
  *Phyllachora hauturu* ssp. 234, 235
  *Phytophthora agathidicida* 63, 71, 234, 272
  *Phytophthora cinnamomi* 234
  *Pluteus velutinornatus* 233, **241**
  *Podoserpula pusio* **241**
  *Podoserpula petalodes* 234
  *Pycnoporus coccineus* 234, **241**
  *Rossbeevera pachycarpa* 233, **241**
  slime mould plasmodium 234, **241**
  *Stereum veriscolor* 234, **241**
  *Sticta filix* 230, **231**
  *Sticta latifrons* 230
  *Sticta livida* 229
  *Stirtoniella kelica* 230
  *Tubaria perplexa* 237
fur seal (*Arctocephalus forsteri*) 258

galaxids 243 see kōkopu, banded
gardens, domestic 38, 39–40, 180
Gardiner, Hugh 60, 66
geckos 80, 87, 129–31, 132; common/ raukawa (*Woodworthia maculata*) 87, **127**, 129, 130; Duvaucel's (*Hoplodactylus duvaucelii*) 87, **127**, 130; elegant (*Naultinus elegans*) **124**, **133**; forest (*Mokopirirakau granulatus*) 87, **127**, 130, 132; Pacific (*Dactylocnemis pacificus*) 87, **127**, 130, 132
geology 92–101; geological map **92**
*Geranium*: *dissectum* 171; *homeanum* 169, 171; *molle* 171
giant weta see wētāpunga
Gibbs, G. W. 88
Gilmour, John 234
glasswort (*Salicornia quinqueflora*) 175
Glithrow, Mark 60, 66
glowworm (*Arachnocampa luminosa*) 119; parasite *Betyla fulva* 119
gnats see flies
goats 148
goldfinch, European (*Carduelis carduelis*) 156
*Gonocarpus incanus* 172, **200**
gorse (*Ulex europaeus*) 44, 80
Gossage, Aroha 24, 31
Gossage, Star 31
Grace, Roger 263
Grave Stream (Turners Creek) **27**, 43, 169
Gravely tractor **48**, 49, **51**
Great Barrier Island see Aotea/Great Barrier Island
Greene, Terry 141, 144, 146, 148
greenfinch, European (*Carduelis chloris*) 156
grey warbler/riroriro (*Gerygone igata*) 87, 136, 148, 153, **155**, 157
Griffiths, Richard 130, 140
gull: red-billed (*Larus novaehollandiae*) 260, 262; southern black-backed/karoro (*L. dominicanus*) 141, **142**
Guthrie-Smith, Herbert 44, 152, 154

hakea, prickly (*Hakea sericea*) 180
hakeke (*Auricularia cornea*) 235, **238**
*Halocarpus*: *biformis* 185; *kirkii* 185
Hamilton, W. M. (Bill) 13, **58**, **61**; DSIR Bulletins 20, 44, 53, 167; and Hamilton Track 59; and mosses 216; vascular plant recording **186** (fig)
Hamilton, W. M., and Ian Atkinson 36, 75, 184, **186** (fig); Hauturu vegetation zones 167–79, **168** (fig), **187** (table)
Hamilton Track 59, **61**, 76, 121
hangehange (*Geniostoma ligustrifolium*) 172, 176, **200**
hanging valleys 93, 99
Haowhenua: Breccia 94–95, 96, **97**, 99, **101**
Haowhenua Dacite 93–94, 95, **97**, 99, **101**
Haowhenua Point 95, **97**
Haowhenua Stream 139, 218
harakeke see flax
Hardgrave, Frank 45
Hardgrave, Len 45–46, **47**, 66, 80, 84
Hardgrave, Martha 45
Hardgrave, Rene 45, 46, **47**
Hardgrave, William H. 45–46, **47**, 66, 75, 80
harrier see swamp harrier
Hauraki Gulf, ecology of 260–62, 264; high peaks flora 184–85
Hauraki Gulf Maritime Park (HGMP) 54, 56, 59, 76–77
Hauruia, Mount (Bald Rock) **90–91**, 99, **101**, 218, 226
Hauruia Stream 93
Hauturu, Mount 93, 184–85, 226, **266–67**
Hauturu vegetation zones **187** (table)
Hayward, Bruce 183, 230
Hayward, Glenys 230
*Hebe*: *macrocarpa* var. *latisepala* **200**; *pubescens* subsp. *sejuncta* 179, **181**, 184, **200**
heketara (*Olearia rani*) **204**; and tōwai-tawaroa forest 178
helicopters, use of 77, 78–80, **79**, 80–81; emergency landing area 171, 221; island survey 29
Hen and Chicken Islands 81, 116, 130, 146, 148, 153, 220, 273; translocations 56, 151
Henry, Richard 81
herbaria 105–06, 216, 226
Herbert, J., and C. H. Daugherty 143
Herekohu, Mount ('The Thumb') 8
*Hine Moana* (boat) **67**
Hingaia Reef 245
Hingaia Rockslide 96, **98**, 99, 176, **187** (fig), 220
Hobson, Mount/Hirakimata 229
Hoare, Robert 121
honeydew scale: *Coelostomidia wairoensis* 130, 132; *C. zealandica* 132

Horak, Egon  235, 237
Horneman, F.  39
hornworts  226; *Dendroceros granulatus*  226; *Phaeoceros carolinianus*  226; species table  324
horopito (*Psedowintera axillaris*)  178, **206**
horse mussels (*Atrina zelandica*)  261
Horseshoe Bay  99
houpara (*Pseudopanax lessonii*)  176, 183, 185
huia (*Heteralocha acutirostris*)  72, 81, 135
Hunter-Blair, Robert  43, 66, 82, 84
Hut Bay Creek Stream  114
Hutton, F. W.  35, 143, 144, 149, 151, 154, 156, 157; and snipe  145
hutu (*Ascarina lucida*)  185, **195**
Hynes, C. (Phyllis)  186

ice plant, New Zealand (*Disphyma australe*)  175, **198**
Imber, Mike  56, **57**, 137, 139
Indian doab (*Cynodon dactylon*)  169
inkweed (*Phytolacca octandra*)  180
invertebrates: aquatic  114–15, forest  115–23, 154
iris: *Libertia*: *grandiflora* (native iris) **211**; *micrantha*  185, **212**
Island Conservation  269
IUCN Red List of Threatened Species  106

Japanese honeysuckle  272
jasmine, New Zealand (*Parsonsia capsularis*)  **204**
Johnston, Peter  **236** (fig), 237; and P. F. Cannon  234–35
Joyce, Leigh  **65**

kahakaha (*Astelia* sp.)  164
kahawai (*Arripis trutta*)  258, 262
kahikatea (*Dacrycarpus dacrydioides*)  60, **62**, 88, 183, 185, **194**
*Kaka* (painting, Don Binney)  **4**
kākā, North Island (*Nestor meridionalis*)  35, 136, 146, **159**, 270; feeding of  49, 56, **58**, 59, 60, 153; predators of  144; 'Rat Bag'  59, 78
kākāpō (*Strigops habroptilis*)  **57**, **65**, 77, 146, **147**; Richard Henry  **57**; translocations  56, 63, 81, 135
kākāriki (*Cyanoramphus* spp.)  136, 144, 149, 156, 157: orange-fronted (*C. malherbi*)  148; red-crowned (*C. novaezelandiae*)  **6–7**, 136, **147**, 148; yellow-crowned (*C. auriceps*) **147**, 148
kanono (*Coprosma autumnalis*)  **197**
kānuka (*Kunzea robusta*)  169, **173**; in forest succession  85, 176, **177** (fig), 185; as habitat  153, 154, 157, 182, **201**, 218, 230, 233; kānuka–mānuka forest  72, 172, **187**; and lizards

**129** (table), 130, 132; sand (*K. amathicola*)  180; at Te Maraeroa  36, 46, 54, 72
Kāpiti Island  72, 77, 135, 148, 221; translocations  146, 152
kāpuka (*Griselinia littoralis*)  179, 185
karaka (*Corynocarpus laevigatus*)  88, 146, 169, 175, 183, **198**
karamū (*Coprosma robusta*)  175, **198**; coastal (*C. macrocarpa*)  88, 169, 175, 176, 183; shining (*C. lucida*)  **197**
kareao *see* supplejack
kārearea *see* falcon, New Zealand
karo (*Pittosporum crassifolium*)  122, 132, 175, 183, 185, **205**
katydids  110
kauri: *Agathis australis*  **12**, **68–69**, **194**, 272, 278; dieback (*Phytophthora agathidicida*)  63, 71, 234, 272; and epiphytes  132, 218, 230; in forests  157, 172, **177** (fig), 178, 179; and fungi  234; gum  44, 72, 172; in Hauturu vegetation zones  **187** (table); kauri–hard beech forests  176, 182, 218; logging of  36, 40, 72–73, 76, 85, 149
Kauri Ridge  121, 226
kawakawa (*Piper excelsum*)  146, 169, 175, **205**
*Kelleria dieffenbachii*  185
Kendrick, Bryce  235
kererū *see* pigeon, New Zealand
kiekie (*Freycinetia banksii*)  176, 178, **211**
kikuyu (*Cenchrus clandestinus*)  221
kina barrens  254, **255**, 257, 262, **263** (fig), 276. *see also* sea urchins/kina
kingfisher, New Zealand/kōtare (*Todiramphus sanctus*)  149, **150**
kiore *see* rats
Kiriraukawa, Mount  226
Kiriraukawa Gorge  99, **101**
Kiriraukawa Stream  93, 99
Kirk, Thomas  35, 143, 167, 186, 226
Kirk's daisy/kohurangi (*Brachyglottis kirkii*)  166, 178, 182–83, **195**, **196**
kiwi  35, 39, 72, 87: albino  143; great spotted (*Apteryx haastii*)  53, 143; North Island brown (*Apteryx mantelli*)  141–43, **142**; South Island brown  143; translocations of  54, 82, 143; white  44, 53, 143
kiwifruit (*Actinidia chinensis* var. *deliciosa*)  180
kohekohe (*Dysoxylum spectabile*)  88, 146, 169, 172, 175, 183, **199**; in forest succession  **177** (fig)
kōkako, North Island (*Callaeas wilsoni*)  18, 87, 149–51, **150**, 153, 157, **271**; translocations of  56, 81, 85, 135; and werewere-kōkako  235
kōkopu, banded (*Galaxias fasciatus*)  243
korokia (*Corokia buddleioides*)  **198**
koromiko: *Hebe macrocarpa* var. *latisepala*  **200**; *H. p.* subsp.

*sejuncta*)  179, **181**, 184, **200**; veronica  84
kōtukutuku (*Fuchsia excorticata*)  178, **199**
kōwhai (*Sophora* sp.)  84
krill (*Nictiphanes* sp.)  **259**, 260, 262
*Kunzea amathicola* (sand kānuka)  180; *ericoides*  88; hybrids  179; *sinclairii*  185 *see also* kānuka

Lady Alice Island  56, 151
*Lady Anne* (boat)  45, 50
lady's fingers (*Caladenia chlorostyla*)  **209**
Lamb Bay Creek  175
Landcare Research (CHR) herbarium  105
Landcare Research (PDD) *see* Manaaki Whenua
lawn daisy (*Bellis perennis*)  171
Layard, E. L.  143
lekking  162
*Lepidium*: *didymium* (twin cress)  171; *oleraceum*  **104**
*Lepidothamnus intermedius*  185
*Leptinella tenella*  **201**
*Leptopteris superba*  182
*Leptospermum*  **187** (table)
*Libocedrus bidwillii*  185
lice (Phthiraptera)  118; *Rallicola* (*Aptericola*) *rodericki*  143
lichens  178, **228**, 229–30, **231**; species table  325–29; species: *Arthonia cinnabarina*  229
*Arthothelium fusconigrum*  229
*Buellia* sp.  230
*Bundophoron australe*  230, **231**
*Bundophoron pantagonicum*  230
*Calicium robustellum*  229
*Caloplaca* sp.  230
*Cladia aggregata*  230
*Cladia sullivanii*  230
*Cladonia confusa*  230
*Cladonia nudicaulis*  230
*Coccocarpia*  230
*Coccotrema cucurbitula*  230, **231**
*Coenogonium zonatum*  229
*Heterodermia*  230
*Heterodermia isidiophora*  229
*Hypotrachyna* spp.  230; *Hypotrachyna immaculata*  229
*Lecanora*  230
*Lefidium tenerum*  230
*Leiorreuma exaltata*  230, **231**
*Lichina confinis*  230
*Lopadium monosporum*  229
*Megalaria grossa*  230
*Megalospora*  230
*Megalospora bartletti*  229
*Melaspilea subeffigurans*  229
*Menegazzia*  230
*Menegazzia neozelandica*  **231**
*Metus conglomeratus*  230
*Miltidea ceroplasta*  230
*Nephroma plumbeum*  230
*Opegrapha intertexta*  230
*Pannaria*  230

*Pannaria elixii* 230
*Parmelinopsis* 230
*Parmotrema* 230
*Peltigera nana* 230
*Pertusaria* 230
*Phaeographis inusta* 229
*Physcia caesia* 230
*Pseudocyphellaria carpoloma* 230
*Pseudocyphellaria coronata* 230
*Pseudocyphellaria dissimilis* **228**, 230
*Pseudocyphellaria faveolata* 230
*Pseudocyphellaria glabra* 230
*Pseudocyphellaria montagnei* 230
*Pseudocyphellaria multifida* 230
*Pseudocyphellaria rubella* 230
*Ramalina australiensis* 230
*Ramalina celastri* 230
*Siphula decumbens* 230
*Siphula gracilis* 230
*Stereocaulon ramulosum* 230
*Teloschistes* 230
*Usnea inermis* 230
*Usnea molliuscula* 229
*Verrucaria microporoides* 230
*Xanthoparmelia* 230
*Xanthoparmelia australasica* **231**
*Xanthoparmelia mougeotina* 230
*Xanthoria* 230
limpets: *Cellana stellifera* 254; fragile (*Atalacmea fragilis*) 253; radiate (*Cellana radians* 253
Lindsay, J., and P. Moore 95
Lindsay, J. et al 93
Lion Rock 95, **97**, 141
'Little Barrier Cavalry Corps' 40, **41**
'Little Barrier Island, Hauturu': *DSIR Bulletin* 54 44; *DSIR Bulletin* 137 13, 53
Little Barrier Island (Hauturu) Supporters' Trust 59, 60, **62**, 81
Little Barrier Purchase Act 1894 28, 38
liverworts 215, 223–26; in Quintinia-tāwari-southern rātā forest 179; species table 321–24; species: *Anastrophyllopsi* sp. 223
*Anastrophyllopsi subcomplicata* 224
*Asterella* sp. 226
*Bazzania adnexa* 223, 224
*Bazzania hochstetteri* 224, **225**
*Bazzania tayloriana* 224
*Chandonanthus squarrosus* **222**
*Chiloscyphus lentus* 224
*Chiloscyphus semiteres* **225**
*Cuspidatula monodon* 224
*Echinolejeunea papillata* **225**
*Frullania rostellata* 224
*Frullania* sp. 224
*Funaria hygrometrica* 220
*Goebeliella* 224
*Gottschea* 223
*Heteroscyphus* 223
*Isotachis* 223
*Kurzia fragifolia* (*Telaranea fragifolia*) 223

*Leidolaena clavigera* **225**
*Leiomitra lanata* **225**
*Lembidium longifolium* 223
*Lepicolea* 223
*Lepidolaena clavigera* 224
*Lepidolaena* liverworts 223
*Lepidolaena taylorii* 224
*Lepidozia* 223, 224
*Lopholejeunea* 224
*Lunularia cruciata* (weed liverwort) 226
*Metalejeunea cucullata* 224
*Plagiochila* liverworts 223
*Radula strangulata* 224
*Schistochila* 223
*Schistochila glaucescens* **225**
*Schistochila nobilis* **225**
*Spruceanthus* 224
*Syzgiella colorata* 224
*Temnoma* 223
*Tetracymbaliella* 223
*Thysananthus anguiformis* 224, **225**
*Thysananthus* spp. 224
*Tricholepidozia* 224
livestock 36, 38, 40, **41**, 75, 156; removal of 60, 76, 85, 169
lizards 59–60, 129–31; distribution of 129 (table) see also geckos; skinks
*Lobelia: anceps* (shore lobelia) **202**; *physaloides* 184
logging see kauri
loquat (*Eriobotrya japonica*) 180
Lots Wife rock stack **98**, 141
*Lotus pedunculatus* 169, 171
Lower Valley Track **173**
Lush, P. Alison 186, 245
*Luzuriaga parviflora* 185
lycophytes 121, **187** (table); species table 296 see also mosses

*Macropiper excelsum* 88
Madell, Dot 53
māhoe (*Melicytus ramiflorus*) 169, 176, 178, **202**; coastal (*M. novae-zelandiae*) 176, 183, 185, **202**; narrow-leaved (*M. lanceolatus*) 183, 184, 185
maire (*Nestegis* spp.) 71, 88, 182, 183; coastal (*Nestegis apetala*) 71, 183; in forest succession **177** (fig); white (*N. lanceolata*) 88, 183
mairehau (*Leionema nudum*) **201**
maire-taike/mida (*Mida salicifolia*) 178, 182, **203**
Maki (Ngāti Manuhiri tipuna) 25–26
māmāngi (*Coprosma arborea*) 172, **197**; in forest succession **177** (fig)
mammals, marine 258, 260 see also fur seals, whales
Manaaki Whenua — Landcare Research 105, 119; fungarium PDD 233, 234, 237; herbarium CHR 105, 216
mangeao (*Listea calicaris*) **202**
*Manoao colensoi* 185
manta rays 276

mānuka (*Leptospermum scoparium*) **201**; as habitat 153, 154, 157, 218, 233; kānuka/mānuka forest 72, 172, **187** (table)
mānuka gumland scrub 172
Māori 25–31, 72, 112, 136, 172; and Hauturu fungi 234, 235 see also Ngāti Manuhiri
māpou (*Myrsine australis*) 169, 172, 175, 176, **204**, 234
maps **16**, **29**, **92**, **168**, **263**
marbleleaf/putaputawētā (*Carpodetus serratus*) 184
marine ecology zones **263** (fig)
Marine Spatial Plan **263**, 264
Marsh, Sid 60
Martinson, Paul 73
Massey University herbarium MPN 105, 216
Matheson, D. 44
*Matricaria discoidea* 172
Maud Island 146
mayflies (Ephemeroptera) 88, 114; *Ameletopsis perscitus* (yellow dun) 114; *Arachnocolus phillipsi* 88; *Ichthybotus hudsoni* 88; *Isothraulus abditus* 88; *Mauiulis luma* 88; *Neozephlebia scita* 88; *Zephlebia spectabilis* 88
McCallum, J 157
McCann, Charles 125
McCleod, A. J. 39
McGlynn, Mike 243
McInnes, Shane 60, **62**, 63, 66
McKenzie, Ross 49, 139, 153
McLean, Ian 149
Melville, Ronald et al 186
Mercury Bay weed (*Dichondra repens*) 171, **198**
Meteorological Service 45, 49, 59 see also weather reporting
*Metrosideros parkinsonii* 179, **181**, 183, 185 see also pōhutukawa, rātā
Mexican daisy (*Erigeron karvinskianus*) 75, 80
Mexican devil (*Ageratina adenophora*) 80–81, 180
mice (*Mus musculus*) 76, 85
microlaena (*Microlaena stipoides*) 169, 176
Milligan, E. N., and J. J. Sumich 115, 116
mingimingi (*Leucopogon fasiculatus*) 172, **201**
Ministry for Primary Industries (MPI) 139, 262
Ministry of Agriculture and Fisheries (MAF) 261–62
miro (*Prumnopitys ferruginea*) 146, 176, 178, **187** (table), **194**
Miskelly and Turbott 145
mist flower (*Ageratina riparia*) 80, 180
mistletoe (*Peraxilla tetrapetala*) 182, 184, 185, **204**; dwarf (*Korthalsella salicornioides*) **201**
Mitchell, Cathy 63

Mitchell, Peter **65**, 66
Moehau, Mount 184–85, 218
Mokohinau Islands 132, 179, 246, 253
monkey apple (*Syzygium smithii*) 180
Moon, Geoff 49
Moore, Lucy B. 184, 216, 219, 220
*Morelotia affinis* 172, **211**
morepork/ruru (*Ninox novaeseelandiae*) 76, 149, **150**; 'Wol' 53–54
Moreton Bay fig 272
morning star shell (*Tawera spissa*) 257
Morton, J. 245
moss forest 215, 223, **274**
mosses 214–21; DOC classified 220–21; epiphytic 179; habitats 216–20; species table 318–20; species: 
  *Achrophyllum dentatum* 218
  *Achrophyllum quadrifarium* 218
  *Austrohondaella limata* 216
  *Barbula unguiculata* 220
  *Bryum argenteum* 220
  *Bryum dichotomum* 219
  *Bryum duriusculum* 220
  *Bryum sauteri* 220
  *Calyptrochaeta apiculata* 219, 221
  *Calyptrochaeta cristata* 218
  *Camptochaete pulvinata* 219
  *Campylopodium medium* 219
  *Campylopus clavatus* 219
  *Campylopus introflexus* 219
  *Campylopus purpureocaulis* 216, 221
  *Campylopus pyriformis* 219, 220
  *Ceratodon purpureus* 220
  *Chenia leptophylla* 220
  *Cladomnion ericoides* 218
  *Ctenidium pubescens* 219
  *Cyathophorum bulbosum* 218
  Daltoniaceae 218
  *Dawsonia superba* **214**, 218, 219
  *Dendrohypopterygium filiculaeforme* 218
  *Dicnemon calycinum* 218
  *Dicranella vaginata* 219
  *Dicranoloma cylindropyxis* 216
  *Dicranoloma menziesii* 218
  *Dicranoloma robustum* 216, **217**
  *Didymodon australasiae* 219
  *Distichophyllum crispulum* 216, 218
  *Distichophyllum crispulum* var. *adnatum* 218
  *Distichophyllum microcarpum* 218
  *Distichophyllum pulchellum* 218
  *Distichophyllum rotundifolium* 218
  *Ditrichum difficile* 219
  *Echinodium umbrosum* 219
  *Eurhynchium praelongum* 220
  *Fissidens asplenioides* 219
  *Fissidens blechnoides* 219
  *Fissidens curvatus* 220
  *Fissidens pallidus* 219
  *Fissidens rigidulus* 219
  *Fissidens taxifolius* ('rogue Fissidens moss') **217**, 220, 221
  *Fissidens tenellus* 216, 219
  *Fissidens tenellus* var. *tenellus* 219
  Hookeriaceae 218
  *Hypnodendron arcuatum* 219
  *Hypnum chrysogaster* 218
  *Hypnum cupressiforme* 216, 219
  *Ischyrodon lepturus* 219, 220–21
  *Leptostomum macrocarpum* 219, 220
  *Leratia obtusifolia* 218
  *Lopidium concinnum* 218
  *Macromitrium brevicaule* 219, 220
  *Macromitrium gracile* 218, 219
  *Macromitrium ligulaefolium* 218
  *Macromitrium longipes* 218
  *Macromitrium prorepens* 219
  *Mittenia plumula* 218
  *Oligotrichum tenuirostre* 219
  *Orthorrhynchium elegans* 218
  *Papillaria flavolimbata* 218
  *Pleuridium subulatum* 220
  *Pogonatum sublatum* 219
  *Polytrichadelphus magellanicus* 219
  *Polytrichum commune* 218
  *Polytrichum juniperinum* 218
  *Ptychomitrium australe* 219, 220
  *Ptychomnion aciculare* ('pipe-cleaner moss') 216, **217**, 218
  *Racomitrium crispulum* 218
  *Racomitrium lanuginosum* 218
  *Rhaphidorrhynchium amoenum* 218
  *Rhizogonium pennatum* 216, 221
  *Rosulabryum capillare* 219
  *Rosulabryum subtomentosum* 216, **217**, 218, 219
  *Sematophyllum contiguum* 219
  *Sematophyllum homomallum* 219
  *Sematophyllum jolliffi* 219
  *Syntrichia laevipila* 219, 220
  *Syntrichia papillosa* 219, 220
  *Thuidium cymbifolium* 220, 221
  *Thuidium sparsum* 219
  *Tortella cirrhata* 220, 221
  *Tortella flavovirens* 219, 220
  *Tortella muralis* 220
  *Tortella truncata* 220
  *Trichomanes elongatum* 218
  *Trichomanes endlichrianum* 218
  *Weissia austrocrispa* 220
  *Weissia controversa* 216
  *Weissia controversa* var. *controversa* 220
  *Weymouthia cochlearifolia* 218
  *Wijkia extenuata* 218
  *Wilsoniella blindioides* 219
mothplant (*Araujia sericifera*) 75, 80, 272
moths (Lepidoptera) 109, 110, **120**, 121, 162; species: 
  *Circoxena ditrocha* 121
  *Hygraula nitens* 121
  *Izatha peroneanella* 121
  *Izatha* sp. (lichen tuft) 121
  *Pseudocoremia dugdalei* 121
  *Sabatina chelcophanes* 121
  *Tatasoma* sp. **120**
Mount Hauruia (Bald Rock) *see* Hauruia, Mount
Mount Ōrau *see* Ōrau, Mount
mountain five finger (*Pseudopanax colensoi*) 178; *P. colensoi* var. *colensoi* 184, 185
mown lawn 157, 171–72, 215
*Muehlenbeckia*: 145–46, **187** (table); *complexa* 85, **108**, 169, 171, 175, **203**
Mueller, Gerhard 28
Murison, John 54
Murman, Geordie 140
Museum of New Zealand Te Papa Tongarewa 35, 105, 216, 245
mushrooms 233–34, 235, 237 *see also* fungi
mustelids 144, 151 *see also* stoats
mycorrhizae 112
myna, common (*Acridotheres tristis*) 156
Myriapoda 162
myrtle rust (*Austropuccinia psidii*) 63, 234, 272

National Institute of Water and Atmospheric Research (NIWA) 243, 261
Native Land Court 28, 29
nature reserve status 20, 76, 85
Neilson, Keri 129
Nelson, Robert 43–44, 45, 66, 76, 80, 84, 143
Nelson family **47**
nerita, black (*Nerita atromentosa*) 253
*Nertera*: *depressa* **204**; *villosa* 185
New Zealand Arthropod Collection 105, 119
New Zealand Fungarium Collection 105
New Zealand Threat Classification System 106, 220–21
New Zealand Wildlife Service *see* Wildlife Service
Newhook, F. J. 216, 234
ngaio (*Myoporum laetum*) 132, 176, 183
Ngatamahine Point 36, 96, **97**, 245
Ngāti Manuhiri 20, 25–26, 60, 77, 126, 128; and Conservation Management Plan 273; Deed of Settlement 31; and ground beetle 122
Ngāti Rehua 77
Ngāti Wai 25–26, 77, 126
Ngāti Wai Trust Board 77
Ngāti Whātua 26
Ngorengore Point 95, **97**
*Nictiphanes* sp. **259**, 260, 262
nīkau (*Rhopalostylis sapida*) 88, 169,

172, 178, 183, **213**; Hauturu **173**, 184; and kereru 146
niniao (*Helichrysum lanceolatum*) **200**
North Auckland Seabird Flyway 137
*Notogrammitis*: *billardierei* 185; *pseudociliata* 185

Offshore Islands Research Group 230
*Olearia allomii* 185
Oliver, Walter R. B. 44, 186
Omaha 26, 28, 30, 38, 40
Omaha marae 30
*Oplismenus hirtellus* 175
Ōrau, Mount **19**, **266-67**
Ōrau Cove 96, 139, 220
Ōrau Gorge **79**, 80, 183
Ōrau Hut 180
orchids 178; species:
  autumn (*Earina autumnalis*) **211**
  *Drymoanthus adversus* **211**
  gnat (*Cyrtostylis oblonga*) **210**
  greenhood orchid (*Pterostylis graminea*) **212**
  helmet (*Corybas cheesemanii*) **210**
  horned (*Orthoceras novae-zeelandiae*) **212**
  onion (*Microtis unifolia*) **212**
  pixiecap (*Acianthus sinclairii*) **208**
  pygmy (*Bulbophyllum pygmaeum*) **210**
  small greenhood (*Pterostylis alobula*) **212**
  small greenhood (*Pterostylis trullifolia*) **212**
  spider (*Corybas acuminatus, C. rivularis, C. trilobus*) **210**
  spring (*Earina mucronata*) **211**
  sun (*Thelymitra longifolia*) **213**
  *Thelymitra longifolia* **212**, **213**
  winikā (*Dendrobium cunninghamii*) **211**
*Oreobolus pectinatus* 185
Orthoptera as food source 110, 162
*Ourisia macrophylla* spp. *macrophylla* 185
owl, laughing/whēkau (*Sceloglaux albifacies*) 35 *see also* morepork/ruru
*Oxalis*: *exilis* 171; *pes-caprae* (Bermuda buttercup) 171
oxen 36
oyster, Pacific (*Crassostrea gigas*) 254
oystercatcher, variable (*Haematopus unicolor*) 145
*Ozothamnus leptophyllus* 179, 180

Palma, Ricardo 143
Palmer, Lady Anne Sophia née Walpole 45
pampas grasses (*Cortaderia* spp.) 75, 80-81, 180, 272
papatāniwhaniwha (*Lageneophora pumila*) **201**
paradise shelduck/pūtangitangi (*Tadorna variegata*) 144

parapara *see* bird catcher/parapara
parasitoids 119
parataniwha (*Elatostema rugosium*) 184
Parihākoakoa Stream 180, 218, 226
Parihākoakoa Valley 140
Parkin, Charlie 46, **47**, **48**, 49-50, **51**, 60, 66
Parkin, May 46, 49, 50, **51**
Parkins Knoll 183
partridge, black (*Melanoperdix niger*) 82
paspalum (*Paspalum dilatatum*) 169, 171, **187** (table)
passerines, introduced 156-57
pasture, fernland/sedgeland 169-172
patē (*Schefflera digitata*) 178, **207**
pāteke *see* brown teal/pāteke
Pattemore, David 121
penguins 45, 141, **142**: little/kororā (*Eudyptula minor*) 141, **142**
*Pentachondra pumila* 185
peperomia (*Peperomia urvilleana*) 175, **204**, 219
peripatus/ngāokeoke (*Peripatoides sympatica*) 112, **113**
periwinkle (*Littorina unifasciata*) 253
petrels 44, 75, **138**:
  black/tāiko (*Procellaria parkinsoni*) 56, 136-37, **138**, 139, 270; predators 75, 85; translocations of 56, **57**
  black-winged (*Pterodroma nigripennis*) 157
  common diving/kuaka (*Pelecanoides urinatrix*) 75, 85, **138**, 141, 157
  Cook's petrel/tītī (*Pterodroma cookii*) 72, 85, 136-137, **138**, 270, **271**; breeding colony 179; predators 75, 77, 149
  grey-faced petrel/ōi (*Pterodroma gouldi*) 72, 75, 85, 136, 137-39, **138**, 275
  New Zealand storm (*Fregetta maoriana*) 85, **86**, **138**, 139-40, 273
  Pycroft's petrel 275
  white-faced storm (*Pelagodroma marina*) 75, 157, 275
Petries, Donald 38
Petrove, Irene 60, **62**, 66
Petrove, Natasha 60, **62**
Phasmatodea (stick insects) 118
Phillips, Henare 53-54
Phthiraptera (lice) 118
pigeon, New Zealand/kererū (*Hemiphaga novaeseelandiae*) 136, 144, 145-46, **147**, 270; 'Pidge' 56, **58**
pigeonwood/porokaiwhiri (*Hedycarya arborea*) 88, 183, **200**
*Pimelea*: *acra* 182, 184, 185; *urvilleana* (pinātoro) **205**
pine trees 45, 80
piopio (*Turnagra tanagra*) 72, 135
pipit, New Zealand (*Anthus novaeseelandiae*) 157, 158
*Pirate* (yacht) 40, **41**

pitfall traps 87, 123, 129, 130
*Pittosporum*: *cornifolium* (tāwhiri karo) **205**; *crassifolium* (karo) 122, 132, 175, 183, 185, **205**; *kirkii* 185, **205**; *umbellatum* (haekaro) 82, 88, 172, **177**, 183, **205**; in forest succession 177 (fig)
*Plagianthus divaricatus* (salt-marsh ribbonwood) 180
*Plagusua capensis* (red rock crab) 257
*Plantago lanceolata* 171
plants: introduced 45, 75, 84; translocations of 82
*Poa*: *anceps* 175; *colensoi* 185
*Podocarpus laetus* (Hall's tōtara) 179, **194**; *laetus* x *P. nivalis* 185
Pohl, Irene 263
pōhuehue (*Muehlenbeckia complexa*) 169, 171, 175
pōhutukawa (*Metrosideros excelsa*) **165**, **174**, **203**; in coastal forest 169, 175, 176, **187** (fig); as food source 130; as habitat 141, 219, 230; pollinated by bats 162, 164 *see also* Pua Mataahu pōhutukawa forest
Pōhutukawa Flat 36, **41**, 76, 96, **98**, 176, 226; bats trapped at 162; birds of 139, 140, 141, 151
Polly, B. 216
*Polygonum arenastrum* 172
*Polymastia granulosa* **263** (fig)
*Pomaderris*: *amoena* 172, **206**; *kumeraho* 183
Pōnui Island 54, 82
Poor Knights Islands 116, 153, 253; avifauna of 135-36, 145, **147**; marine reserves 278
poplar (*Populus nigra*) 45, 80
poroporo (*Solanum aviculare*) **207**
possum (*Trichosurus vulpecula*) 20
power systems 45-46, 49, **51**, 53, 54, **55**, 56, 60, **62**; solar power 63, 64
prairie grass (*Bromus catharticus*) 169
prickly mingimingi (*Leptecophylla juniperina*) 172
privet 272
Procellariiformes (tubenoses) 136-141 *see also* by species
Protura (coneheads) 112; *Eosentomon dawsonii* 112; *Gracilentulus gracilis* 112
*Prunella vulgaris* (selfheal) 171
*Pseudopanax discolor* 184, 185, **206** *see also* five finger, houpara, mountain five finger
*Pseudowintera*: *axillaris* 185; *colorata* 185
Pua Mataahu pōhutukawa forest **174**, 175-76, **187** (table)
Public Works Department 46
puka (*Griselinia lucida*) 178, **200**
pukatea (*Laurelia novae-zelandiae*) 178, **181**, 185, **201**
pūkeko (*Porphyrio melanotus*) 145,

pūriri (*Vitex lucens*) 169, 172, 175, **208**; kererū and 146; and rātā 176, 178

quail: Australian brown (*Coturnix ypsilophora*) 143–144; New Zealand/koreke (*Coturnix novaezelandiae*) 72, 144
quarantine controls *see* biosecurity
Queen rock *see* The Queen
quintinia/tāwheowheo (*Quintinia serrata*) 179, 185, **187** (table), **206**, 223; and tōwai-tawaroa forest 178
Quota Management System (QMS) 261

radio *see* communications; ZLD Auckland Radio
*Rahiri* (boat) 50
rail, banded/moho pererū (*Gallirallus philippensis*) 87, **142**, 144–45, **147**, 158
rails 135, **142**; extinct 135
*Rallicola* (*Aptericola*) *rodericki* (louse) 143
ramarama (*Lophomyrtus bullata*) **202**
rangers 49, 60; list of **66** (table)
rangiora (*Brachyglottis repanda*) 178, **196**
Rangitoto Island 220, 223, 224, 235, 272
*Ranunculus*: *amphitrichus* 185; *urvilleanus* 171, 182
rats: eradication programmes 45, 60, **65**, 77, **79**; impacts of 85, 87–88, 153, 154; logistics of 78–80, 146; Ngāti Manuhiri and 77, 128; species: black rat/ship's rat (*Rattus rattus*) 76, 85, 151, 153, 154, 272 kiore/Pacific rat (*R. exulans*) 18, 71–72, **73**, 75, 106, 109; as predators 125, 135, 137, 154; ruru and 149; as seed predator 167, 176, 183 Norway rat (*R. norvegicus*) 18, 85, 151
rātā (*Metrosideros* spp.) 178, 179, 185, **202**, **203**; carmine (*M. carminea*) **202**; in forest successions 176, **187** (table); northern tree (*M. robusta*) 178, **203**; northern rātā/pūriri-tawaroa forest 176, 178; small white/akatea (*M. perforata*) **203**; southern (*M. umbellata*) 178, 185, **203**; white/ akatea (*M. albiflora*) 179, 185, **202**; *see also* Quintinia-tāwari-southern rātā forest
rat's tail (*Sporobolus africanus*) 171
*Raukaua simplex* 183; var. *simplex* 184, 185
raukawa (*Raukaua edgerleyi*) **206**
raupō/bulrush (*Typha orientalis*) 179, 185
Rayner, Matt 137, 139

redpoll, common (*Carduelis flammea*) 156
red-crowned parakeet (Sammy) 49
redfin bully (*Gobiomorphus huttoni*) 243
Reischek, Andreas 140, 144, 149, 151–52, 156; 1887 bird count 157; collections 35, 143, 148; and tuatara 126; *Yesterdays in Maoriland* 26, 28
remote monitoring technologies 272
rengarenga lily (*Arthropodium cirratum*) 175, **208**
Renner, M. A. M. 226
Repanga/Cuvier Island 56, 81
reptiles *see* lizards; tuatara
Reserves Act 1977 77
Resolution Island 143, 146
rewarewa (*Knightia excelsa*) 88, 172, 183, **201**; in forest succession **177** (fig); pollinated by bats 164
ribbonwood, salt-marsh (*Plagianthus divaricatus*) 180
Richard Henry (kākāpō) 56, **57**
rifleman, North Island/titipounamu (*Acanthisitta chloris*) 87, 135, 149, **150**
rimu (*Dacrydium cupressinum*) 183–84, **194**
robin, North Island/toutouwai (*Petroica longipes*) 44, 87, 135, 153, 154–56, **155**, 157
Robinson, Charles 36, 38, 66
ruru *see* morepork/ruru
rushes: *Juncus*: *australis* 169; *novae-zelandiae* 185; *usitasus* 171
*Rytidosperma*: *setifolium* 185; spp. 172

saddleback, North Island/tīeke (*Philesturnus rufusater*) 18, **83**, **150**, 151, 152; populations of 87, 136, 153, 154, 157; translocations 56, 81, 85, 135
sawflies 119
scallops (*Pecten novaezelandiae*) 257, **263** (fig)
*Scandia rosifolia* 175
Scarlett, Will 59, 60, 66
*Schoenoplectus tabernaemontani* 185
school 50, **52**
Schuster, R M 226
Scofield, Paul 141
sea spurrey (*Spergularia tasmanica*) **207**
sea urchin/kina (*Evenchinus chloroticus*) 254, **255**, 262, **263** (fig) *see also* kina barrens
seabed zones **263** (fig)
seabirds 88, 136–41, 276 *see also* by species
Seachange initiative 264
seagrass (*Zostera* sp.) 261; *Z.* (*estuarine*) 180
seaweeds *see* algae
sedgeland 169, 171

sedges 156, 171, 172, 175, 176; species: *Carex coronopifolia* 180; *C. flagelliflora* 175; *C. horizontalis* 184, 185; *C. inversa* 171; *C. pumila* 180; *C. solandri* **209**; *C. virgata* 171; *Cyperus brevifolia* 171; *C. ustulatus* (giant umbrella sedge) 156, 171, **210**; *Ficinia nodosa* (coastal) 176; *Gahnia* sp. 121, 235; *G. pauciflora* 179, 185; *Lepidosperma laterale* 175; *Schoenus tendo* 172, 175
selfheal (*Prunella vulgaris*) 171
shag, pied/kāruhiruhi (*Phalacrocorax varius*) 141, **142**
Shag Track 80, 172, 221, 226
Shakespear, Blanche 38
Shakespear, Ethel 38
Shakespear, Frances 38, 39, 40–41, 186, 216
Shakespear, Helen 38
Shakespear, Ivy 38
Shakespear, Katherine 38
Shakespear, Robert H A 30, 38, 66, 75, 82, 143, 245
Shakespear, Robert H R (jnr) 38, 40, 43, 66
Shakespear, Ruby 38
Shakespear, Will 82
Shakespear Regional Park 158
shearwaters 75; flesh-footed 275; fluttering/pakahā (*Puffinus gavia*) 85, **138**, 141, 157, 275; little (*Puffinus assimilis*) 75, 141, 157, 275
sheep *see* livestock
shore bindweed (*Calystegia soldanella*) 175, **196**
shore celery (*Apium prostratum*) **195**
shore lobelia (*Lobelia anceps*) **202**
Sibson, Dick 49, 139, 153
silvereye (*Zosterops lateralis*) 87, 156, 157
Simpson Diversity Index 162
sixpenny scale (*Ctenochiton viridis*) 148
skinks 63, 80, **83**, 87, **131**; species: chevron (*Oligosoma homalonotum*) 59–60, **83**, 129, **131**, 132; 'Chevy' 87 copper (*O. aeneum*) 129, **131** Hauraki (*O. townsi*) 87, 129, 130, **131** McGregor's (*O. macgregori*) 132, 273 moko (*O. moco*) 87, 130, **131** ornate skink (*O. ornatum*) 129, **131** plague (*Lampropholis delicate*) 63 robust (*O. alani*) 132, 273 shore (*O. smithi*) 87, 129, 130, **131** Suter's (*O. suteri*) 130, **131**; striped (*O. striatum*) 60, 88, 129, **131**, 132
skylark, Eurasian (*Alauda arvensis*) 156, 157
slugs 71, 115, 116; leaf-veined (Athoracophoridae sp.) **117**; pāua (*Schizoglossa*) 116
Smith, Dave 54, 66

Smith, Edith M  38, 39, 186, 216, 245
Smith, Lorell  54
Smith, Lynette  54
Smith, Margaret  54
Smith, Pam  54
Smuts-Kennedy, Chris  59, 60, **62**, 66
Smuts-Kennedy, Robyn  59, **62**
snails  115, 116
sneezeweed (*Centipeda minima*)  172, **196**
snipe: Chatham Island  89; North Island (*Coenocorypha barrierensis*)  72, **73**, 107, 135, 145; surrogate species for  89, 158, 275
snowberry (*Gaultheria antipoda*)  **200**
*Solanum*: *americanum*  169, **207**; *opacum*  175  *see also* poroporo
song thrush (*Turdus philomelos*)  156
South Coast  99–100
sparrow, house (*Passer domesticus*)  156
spiders  116, 119; species: *Cambridgea* sp.  **117**; *Dolomedes minor*  119; *Dolomedes* sp.  **117**; *Hexathele hochstetteri* (tunnelweb)  **117**; *Migas insularis* (trapdoor)  119; nursery web spider (*Dolomedes minor*)  119; *Porrhothele* spp. (tunnelweb)  119; *Stanwellia hapua* (trapdoor)  119; *Uliodon* spp. (vagrant)  119
spinach, native (*Tetragonia implexicoma*)  175, **207**
sponges  **256**; **263** (*fig*); *Ancorina alata* **263**; *Callyspongia ramosa* **263**; *Raspalia* sp.  **263**; *Siphonochalina latituba* (tube)  **256**
spotless crake/pūweto (*Porzana tabuensis*)  72, 87, 145, **147**, 158
springtails (Collembola)  111–12; *Holacanthella duospinosa*  111–12
spurge, coastal/sand milkweed (*Euphorbia glauca*)  89, **174**, 175, **199**
Stamp, Rosalie  129
Stanley, R, and Ian Atkinson  167–69, **187** (*fig*)
starling, common (*Sturnus vulgaris*)  156
*Stellaria parviflora*  175
stick insects (Phasmatodea)  110, 118, **120**; *Clitarchus hookeri* **120**; *Spinotectarchus* sp.  120
stitchbird/hihi (*Notiomystis cincta*)  44, 45, **134**, 151–52, **155**; and *Aspergillus fumigatus* 234; and collectors  35, 38; survival of  85, 87, 135
stoats (*Mustela erminea*)  81, 154, 272
stratigraphy  93–94
summit ridge  **217**, **232**, **266–67**, **274**
summit scrub  **187** (*fig*)
Summit Track  45, 143, 226; renamed Hamilton Track  59

sundew (*Drosera auriculata*)  172, **199**
supplejack/kareao (*Ripogonum scandens*)  88, 178, 183, **212**
swamp harrier/kāhu (*Circus approximans*)  76, 144
Sykes, W (Bill)  186

tā moko (tattooing), pigment for  235
Taiawa, Wi  30
takahē (*Porphyrio hochstetteri*)  158
Tamihana, Kino  30
tānekaha (*Phyllocladus trichomanoides*)  88, 183, **194**
taraire (*Beilschmiedia tarairi*)  88, 146, **195**; in forest succession  169, 172, **177** (*fig*), 178
Taranga/Hen Island: avifauna of  135–36, 148, 149; and McGregor's skink  132, 273; translocations  56, 81, 151, 152
taupata (*Coprosma repens*)  169, 183, 185, **197**
taurepo (*Rhabdothamnus solandri*)  184, 186, **206**
tawa  146, 151, **187** (*table*) *see also* tawaroa
tawāpou (*Planchonella costata*)  175, 183, **206**
tāwari (*Ixerba brexioides*)  178, 179, 183, **200**
tawaroa (*Beilschmiedia tawaroa*)  176, 178, 184, **195**
Tawaroa Stream  176
*Tawera spissa* (morning star shell)  257
tawhairaunui (*Fuscospora* spp.)  233, 235
Tāwharanui Regional Park  87, 144, 145, 149, 153
tāwheowheo *see* quintinia/tāwheowheo
tawhero/tawa forest  **187** (*table*)
tāwhiri karo (*Pittosporum cornifolium*)  **205**
Taylor, Rowley  148
Te Ananuiarau Bay  96, **97**, 141, **255**
Te Hauturu-o-Toi/Little Barrier Island  **2–3**, 8, **14–15**, **22–23**, 24, **247**, **250–51**; Certificate of Title 1881  **29**; control, management of  28, 49, 59; expectations and issues  273–79; international significance of  269–70; named  17; as nature reserve  20, 76, 85; placenames of  **16**
Te Hue Point  59, 76, 96, **97**, 245
Te Hue Stream  183
Te Kawerau  26
Te Kiri  26
Te Kiri, Rahui  26–31, **27**; granddaughter 'Girlie'  60
Te Manu, Paratene  25–26, 28, 30
Te Maraeroa  **14**, **34**, **74**, 100, **265**; human modification of  72, 75–76; mosses at  219–20; regeneration of  185; vegetation of  35–36, 85, 169, 180

Te Roa, Ngapeka  **27**, 30
Te Tītoki Point  **24**, 100, **174**, 175, 180, 246, **277**; pōhutukawa forest  175
Te Waikohare Stream  38, 43, 50, 169; moss in  219
Te Wairere Stream  93, 95, 114
Tenetahi, Kiri  30, 38
Tenetahi, Wiremu  26, **27**, 29, 35; buildings of  **37**, 38–39; evicted from Hauturu  28, 30–31
tern, white-fronted/tara (*Sterna striata*)  141, **142**
ternlet, grey (*Procelsterna cerulea*)  157
'The Queen'  **98**, 99
'The Thumb' (Mount Herekohu)  8
*Thismia rodwayi*  182, 183
threatened species  106, 182–83
thrips (Thysanoptera)  118
thrush *see* song thrush
Thumb Track  76, 80, **214**, 220, 226
Thysanoptera (thrips)  118
Tirikakawa Stream  36, **102–03**, 141, **173**, 175, 219, 243; striped skink in  60
Tirikakawa Valley  162, 178
Tiritiri Matangi Island  82, 169, 234, 262, 264; hihi on  152, 234
titoki (*Alectryon excelsus*)  183
toatoa (*Phyllocladus toatoa*)  179, 185, **194**
toetoe (*Austroderia splendens*)  **209**
Toi Te Huatahi  17, 25
tomtit, North Island/miromiro (*Petroica macrocephala*)  87, 136, 154, **155**, 157
toro (*Myrsine salicina*)  178, 185, **204**
toropapa (*Alseuosmia macrophylla*)  82, 152, 172, 185, 186, **195**
toru (*Toronia toru*)  172, **207**
tōtara, Hall's (*Podocarpus laetus*)  179, **194**
Tourist Department  43, 44, 82, 84
tōwai (*Weinmannia silvicola*)  176, 185, **208**; tōwai-tawaroa forest  178, 223
Toy, R. et al  151, 152, 154
translocations  54, 63, 81–82, 158, 273, 275; of plants  82; of seabirds  139  *see also* by species
tree ferns: black/mamaku (*Cyathea medullaris*)  178, **190**; ponga (*C. dealbata*)  164, 172
tree-planting  60, **62**
trees  167–79, **171** (*fig*), **177** (*fig*) *see also* by species
Trevarthen, C. B.  245
*Tricomanes strictum*  185
*Trifolium*: *pratense*  172; *repens*  171
tuatara (*Sphenodon punctatus*)  18, **65**, **86**, 88, 126–29, **127**, 279; captive breeding programme  59, **61**, 64, 82, 126, 128; as predator  132; and predators  72, 75, 77, 125; tuatarium  **61**, **62**, 78
tubenoses (Procellariiformes)  136–41  *see also* by species

*Tucetona laticostata* (large dog cockle) 257
tūī (*Prosthemadera novaeseelandiae*) 153–54, **155**, 157, 270; diet of 45, 56, 60, 152; populations of 87, 136
Turbott, Graham 49, 135, 139, 140, 156; bird counts 141, 144, 153, 154, 156, 157; and C. Miskelly 145
tūrepo/large-leaved milktree (*Streblus banksii*) 71, 183
Turner, Mr 38
Turners Creek (Grave Stream) 43
tūrutu blueberry (*Dianella nigra*) **211**
Tutamoe Range 184, 218
tutu (*Coriaria arborea*) **198**
twin cress (*Lepidium didymium*) 171

underwater zones 263 (*fig*)

Valley Track 76, 221, 226
vascular plant species **182** (*fig*); recorders of **186** (*table*)
vegetation: overview 167–69; types **170** (*fig*); zones 169–79, **187** (*table*)
Veitch, Dick 157, 167, 168, 183
velvet worms (Onychophora) 112
veronica (koromiko) 84
*Veronica macrocarpa* (hebe) 162
*Veronica persicaria* 172; *serpyllifolia* 172
*Vicia sativa* 169
Victoria University of Wellington 82, 126
visitors 45, 49, 53–54, 64, 275 see also biosecurity
volcano 93–94

Wade, Lyn **65**, 114, 115, 243, 245
*Wahlenbergia vernicosa* 175
Waikohare Stream see Te Waikohare Stream
Waimaomao Bay 96, 140, 141
Waimaomao Rhyodacite 93–94, 96, **98**, 99
Waipawa Stream 36, 139, 143, 175, 226
Waipawa Track **268**
Walle, Liam 63, 183
Walle, Mahina 63
Walle, Richard 63, **65**, 66, **67**, 183
Warren, Mr 44, 46
wasps 50, 53, 59, 88, 109, 110, 119; species: Asian paper wasp (*Polistes chinensis*) 59, 110; Australian paper wasp (*P. humilis*) 110; black hunter wasp (*Priocnemis monachus*) 119; *Casinaria* sp. (Ichneumonidae) 119; *Certonotus fractinervis* 119; common wasp (*Vespula vulgaris*) 88, 109, 119; eradication of 153; German wasp (*V. germanica*) 53, 88, 110; golden hunter wasp (*Sphictostethus nitidus*) 119;
water supply systems 43, 46, 59, **61**
weather reporting 45, 46, **47**, 49, 59, 64

weeds 75–76; eradication programmes 60, 63, **79**, 80–81 see also biosecurity
weevils **107**, 119, **120**, 122; flax (*Anagotus fairburnii*) 122, 123; fungus (anthribids) 122; elephant (*Rhyncodes ursus*) 119; giraffe (*Lasiorhynchus barbicornis*) **107**, **120**, 121; *Stephanorhynchus crassus* 122
Weidman, Herbert George 44
weka (*Gallirallus* spp.) 82; (*Gallirallus australis*) 135
welcome swallow (*Hirundo neoxena*) 156
West Flat **42**
West Landing **42**, 246
wētā 109, 116, **117**, 118, 235; 109; Auckland cave (*Glymnoplectron acanthocera*) 116; Auckland tree (*Hemideina thoracica*) 109, 116, **117**; Cook Strait giant (*Deinacrida rugosa*) 235; ground (*Hemiandrus maculifrons*) 235; ground (*H. pallitarsus*) 109 see also wētāpunga
wētāpunga (*Deinacrida heteracantha*) 18, 80, **83**, 85, 116, **117**; and fungi 235; predators 88, 109, 149; translocations of 82
Whakarua Moutere World Heritage Site proposal 269–70, 276, 278
whales 260; Bryde's (*Balaenoptera edeni brydei*) 260; humpback (*Megaptera novaeangliae*) 260; orca (*Orcinus orca*) 260; pigmy blue (*Balaenoptera musculus brevicauda*) 260; southern right (*Eubalaena australis*) 260
wharangi (*Melicope ternata*) 175, **202**
Whēkau Stream 53, 94, 144
Whēkauwhēkau, Mount 226
Whenua Hou/Codfish Island 137, 146
Whitaker, Tony 125, 129, 130
white clover (*Trifolium repens*) 171
whitehead/pōpokotea (*Mohoua albicilla*) 44, 87, 135, 154, **155**, 157; and koekoeā 148–49; and ruru 149
Whitwell, Liz 63, 66
Wilcox, Mike 245
Wildlife branch, Department of Internal Affairs 49
Wildlife Service 54, 56, 76–77, 139, 152
wineberry/makomako (*Aristotelia serrata*) 178, **195**
Winterbourn, Mike 114, 115
wire vine/pōhuehue (*Muelenbeckia complexa*) 85, **108**, 169, 171, 175, **203**
Wise, K. A. J. 114, 115
Wisnesky, John 54, **55**, 66
Wisnesky family 54
Wood, Chippy 66
World Heritage Site proposal 269–70
Wormald, I. B. 216
worms 111–12, **113**; polychaete worms

257; tiger (*Eisenia fetida*) 111 see also earthworms
wren: bush (*Xenicus longipes*) 135, 158; flightless (*Pachyplichas jagmi*) 135; rock (*Xenicus gilviventris*) 158; Stephens Island wren (*Traversia lyalli*) 135
Wright, Anthony 186, 230
Wright, Henry 36, 66, 177

yellowhammer (*Emberiza citrinella*) 156
Yorkshire fog (*Holcus lanatus*) 169
Young, Maureen **173**, 186

Zealandia (Karori Sanctuary) 152
ZLD Auckland Radio 46, 47, 53, 59

**MASSEY UNIVERSITY PRESS**

First published in 2019 by Massey University Press
Reprinted 2021
Private Bag 102904, North Shore Mail Centre
Auckland 0745, New Zealand
www.masseypress.ac.nz

Published with the support of the
Little Barrier Island (Hauturu) Supporters' Trust

**HAUTURU**
LITTLE BARRIER ISLAND SUPPORTERS TRUST

Text © individual contributors, 2019
Images © as credited, 2019

Design by Kate Barraclough
Cover photograph: Te Titoki Point, Hauturu,
at sunset, by Andris Apse
This page: Tūī in harakeke, by Neil Fitzgerald

The moral right of the authors and photographers
has been asserted

All rights reserved. Except as provided by the Copyright
Act 1994, no part of this book may be reproduced,
stored in or introduced into a retrieval system or
transmitted in any form or by any means (electronic,
mechanical, photocopying, recording or otherwise)
without the prior written permission of both the
copyright owner(s) and the publisher.

A catalogue record for this book is available from
the National Library of New Zealand

Printed and bound in China by Everbest Investment Ltd

ISBN: 978-0-9951095-8-2